Cellulose (Nano)Composites

Cellulose (Nano)Composites

Editors

Denis Mihaela Panaitescu
Adriana Nicoleta Frone

MDPI • Basel • Beijing • Wuhan • Barcelona • Belgrade • Manchester • Tokyo • Cluj • Tianjin

Editors

Denis Mihaela Panaitescu
Polymers
National Institute for RD in
Chemistry and
Petrochemistry
Bucharest
Romania

Adriana Nicoleta Frone
Polymers
National Institute for RD in
Chemistry and
Petrochemistry
Bucharest
Romania

Editorial Office
MDPI
St. Alban-Anlage 66
4052 Basel, Switzerland

This is a reprint of articles from the Special Issue published online in the open access journal *Polymers* (ISSN 2073-4360) (available at: www.mdpi.com/journal/polymers/special_issues/Nanocellulose_Based_Materials).

For citation purposes, cite each article independently as indicated on the article page online and as indicated below:

LastName, A.A.; LastName, B.B.; LastName, C.C. Article Title. *Journal Name* **Year**, *Volume Number*, Page Range.

ISBN 978-3-0365-7949-8 (Hbk)
ISBN 978-3-0365-7948-1 (PDF)

© 2023 by the authors. Articles in this book are Open Access and distributed under the Creative Commons Attribution (CC BY) license, which allows users to download, copy and build upon published articles, as long as the author and publisher are properly credited, which ensures maximum dissemination and a wider impact of our publications.

The book as a whole is distributed by MDPI under the terms and conditions of the Creative Commons license CC BY-NC-ND.

Contents

About the Editors vii

Preface to "Cellulose (Nano)Composites" ix

Denis Mihaela Panaitescu and Adriana Nicoleta Frone
Cellulose (Nano)Composites
Reprinted from: *Polymers* 2023, *15*, 2512, doi:10.3390/polym15112512 1

Catalina Diana Usurelu, Stefania Badila, Adriana Nicoleta Frone and Denis Mihaela Panaitescu
Poly(3-hydroxybutyrate) Nanocomposites with Cellulose Nanocrystals
Reprinted from: *Polymers* 2022, *14*, 1974, doi:10.3390/polym14101974 3

Valentino Bervia Lunardi, Felycia Edi Soetaredjo, Jindrayani Nyoo Putro, Shella Permatasari Santoso, Maria Yuliana and Jaka Sunarso et al.
Nanocelluloses: Sources, Pretreatment, Isolations, Modification, and Its Application as the Drug Carriers
Reprinted from: *Polymers* 2021, *13*, 2052, doi:10.3390/polym13132052 33

Dingyuan Zheng, Yangyang Zhang, Yunfeng Guo and Jinquan Yue
Isolation and Characterization of Nanocellulose with a Novel Shape from Walnut (*Juglans Regia* L.) Shell Agricultural Waste
Reprinted from: *Polymers* 2019, *11*, 1130, doi:10.3390/polym11071130 81

Marina Reis de Andrade, Tatiana Barreto Rocha Nery, Taynã Isis de Santana e Santana, Ingrid Lessa Leal, Letícia Alencar Pereira Rodrigues and João Henrique de Oliveira Reis et al.
Effect of Cellulose Nanocrystals from Different Lignocellulosic Residues to Chitosan/Glycerol Films
Reprinted from: *Polymers* 2019, *11*, 658, doi:10.3390/polym11040658 95

Mohammad Hassan, Ragab E. Abou Zeid, Wafaa S. Abou-Elseoud, Enas Hassan, Linn Berglund and Kristiina Oksman
Effect of Unbleached Rice Straw Cellulose Nanofibers on the Properties of Polysulfone Membranes
Reprinted from: *Polymers* 2019, *11*, 938, doi:10.3390/polym11060938 111

Denis Mihaela Panaitescu, Sorin Vizireanu, Sergiu Alexandru Stoian, Cristian-Andi Nicolae, Augusta Raluca Gabor and Celina Maria Damian et al.
Poly(3-hydroxybutyrate) Modified by Plasma and TEMPO-Oxidized Celluloses
Reprinted from: *Polymers* 2020, *12*, 1510, doi:10.3390/polym12071510 125

Marius Stelian Popa, Adriana Nicoleta Frone, Ionut Cristian Radu, Paul Octavian Stanescu, Roxana Trușcă and Valentin Rădițoiu et al.
Microfibrillated Cellulose Grafted with Metacrylic Acid as a Modifier in Poly(3-hydroxybutyrate)
Reprinted from: *Polymers* 2021, *13*, 3970, doi:10.3390/polym13223970 141

Giyoung Shin, Da-Woon Jeong, Hyeri Kim, Seul-A Park, Semin Kim and Ju Young Lee et al.
Biosynthesis of Polyhydroxybutyrate with Cellulose Nanocrystals Using *Cupriavidus necator*
Reprinted from: *Polymers* 2021, *13*, 2604, doi:10.3390/polym13162604 161

Yuchen Jiang, Guihua Li, Chenyu Yang, Fangong Kong and Zaiwu Yuan
Multiresponsive Cellulose Nanocrystal Cross-Linked Copolymer Hydrogels for the Controlled Release of Dyes and Drugs
Reprinted from: *Polymers* **2021**, *13*, 1219, doi:10.3390/polym13081219 . . . . **169**

Ke Li, Huiyu Yang, Lang Jiang, Xin Liu, Peng Lang and Bo Deng et al.
Glycerin/NaOH Aqueous Solution as a Green Solvent System for Dissolution of Cellulose
Reprinted from: *Polymers* **2020**, *12*, 1735, doi:10.3390/polym12081735 . . . . **187**

Adriana Nicoleta Frone, Cristian Andi Nicolae, Mihaela Carmen Eremia, Vlad Tofan, Marius Ghiurea and Ioana Chiulan et al.
Low Molecular Weight and Polymeric Modifiers as Toughening Agents in Poly(3-Hydroxybutyrate) Films
Reprinted from: *Polymers* **2020**, *12*, 2446, doi:10.3390/polym12112446 . . . . **199**

Weizhe Wang, Lijie Li, Shaohua Jin, Yalun Wang, Guanchao Lan and Yu Chen
Study on Cellulose Acetate Butyrate/Plasticizer Systems by Molecular Dynamics Simulation and Experimental Characterization
Reprinted from: *Polymers* **2020**, *12*, 1272, doi:10.3390/polym12061272 . . . . **219**

About the Editors

Denis Mihaela Panaitescu

Denis Mihaela Panaitescu received her Ph.D. in materials engineering in 2001 from the Politehnica University of Bucharest, Romania. She is currently a Scientific Researcher I at the National Institute of R&D in Chemistry and Petrochemistry—ICECHIM, Polymer Composites and Nanocomposites Team. Her main research activities focus on the development of new products based on biopolymers and polymer composites with nanocellulose for biomedical and packaging application, the improvement of polymer–fiber interface, and the development of new technological solutions for environment protection. She is an author of about 120 peer-reviewed articles and 15 patents, and she has coordinated or has been a key member of about 20 national and international projects.

Adriana Nicoleta Frone

Frone Adriana Nicoleta received her Ph.D. in chemical engineering in 2012 from the Politehnica University of Bucharest, Romania. She is currently a Scientific Researcher I at the National Institute of R&D in Chemistry and Petrochemistry—ICECHIM, Polymer Composites and Nanocomposites Team. Her scientific interest is in the field of polymeric biomaterials, their characterization and applications, as well as the isolation and functionalization of nanocellulose. She has published over 60 scientific papers in international peer-reviewed journals, 3 book chapters and 10 patents in the fields of chemical sciences, materials science and chemical engineering.

Preface to "Cellulose (Nano)Composites"

Our environment has been severely affected by the intensive production and use of plastics derived from fossil fuels and their uncontrolled end-of-life disposal. The return to using natural products is a characteristic of the most recent decades, and nanocellulose occupies a privileged position among these intensively studied products. Nanocellulose is obtained from cellulose, which is the most abundant natural polymer, by applying different chemical, mechanical, enzymatic and, most often, combined methods. A huge effort has been invested in the application of nanocellulose as a modifier or reinforcing agent in polymer nanocomposites. This Special Issue brings together twelve original articles and studies that contribute to our understanding of the fundamental and technological knowledge of cellulose–polymer nanocomposites. The isolation of nanocellulose from cheap sources and, especially, from agro-food industry waste is an important step to be implemented for cost reduction and environmental protection. An appropriate surface treatment of nanocellulose is a key element for achieving a good interfacial adhesion and superior properties in polymer nanocomposites. The use of more appropriate and green solvent systems for cellulose, the use of biobased plasticizers and toughening agents in nanocellulose nanocomposites, and the use of molecular dynamics simulations for the prediction of the compatibility of cellulose blends are valuable methods for expanding the application of nanocellulose.

Denis Mihaela Panaitescu and Adriana Nicoleta Frone
Editors

Editorial

Cellulose (Nano)Composites

Denis Mihaela Panaitescu * and Adriana Nicoleta Frone *

National Institute for Research and Development in Chemistry and Petrochemistry—ICECHIM, 202 Splaiul Independentei, 060021 Bucharest, Romania
* Correspondence: panaitescu@icechim.ro (D.M.P.); adriana.frone@icechim.ro (A.N.F.)

The environment has been severely affected by the intensive production and use of plastics derived from fossil fuels, and their uncontrolled end-of-life disposal. Recent decades have seen a return to using natural products, and nanocellulose (NC) occupies a privileged position among these intensively studied products [1,2]. NC is obtained from cellulose, the most abundant natural polymer, by applying different chemical, mechanical, enzymatic, and most often, combined methods. A huge effort has been invested in the application of NC as a modifier or reinforcing agent in polymer nanocomposites [1,2]. However, the beneficial effects of NC on the properties of polymer nanocomposites are hindered by the poor dispersion of hydrophilic NC in hydrophobic polymer matrices. The Special Issue 'Cellulose (Nano)Composites' brings together twelve original articles and studies that contribute to advances in our collective fundamental and technological knowledge of cellulose-polymer nanocomposites. Isolation of NC from cheap sources and especially from agricultural and forestry waste is an important step to be implemented for cost reduction and environmental protection. NC has been isolated from walnut shells through a mechano-chemical treatment, followed by 2,6,6-Tetramethylpiperidinyloxy (TEMPO) oxidation and ultrasound, showing properties comparable to or better than those of NC obtained by sulfuric acid hydrolysis [3]. Cellulose nanocrystals (CNC) have been extracted from different byproducts of the agro-food industries, such as corncobs, corn husks, wheat bran and coconut shells, and used as reinforcements for chitosan films [4]. Rice straw nanofibers isolated from unbleached pulps had a good effect on the porosity, hydrophilicity, and antifouling of polysulfone membranes [5].

An appropriate surface treatment of NC is a key element for achieving a good interfacial adhesion and superior thermal and mechanical properties in polymer nanocomposites [6,7]. Cellulose has been surface-modified using plasma treatment in liquid, and its effect on the properties of poly(3-hydroxybutyrate) (PHB) was compared to that of TEMPO-oxidized cellulose [6]. In another study, microfibrillated cellulose was first treated with methacryloxypropyltrimethoxysilane, and then graft-polymerized with methacrylic acid, after which it showed better dispersion in a PHB matrix and induced a greater improvement in PHB properties compared to unmodified microfibrillated cellulose [7]. Direct preparation of PHB/CNC nanocomposites using *Cupriavidus necator* fermented in well-dispersed CNC-supplemented culture media was also proposed [8]. CNC was also used as a crosslinker in a poly(acrylic acid-co-acrylamide) multi-responsive composite hydrogel for the controlled release of dyes and drugs [9]. Other methods of expanding the application of NC include more appropriate and green solvent systems for cellulose, the use of biobased plasticizers and toughening agents in NC nanocomposites, or the use of molecular dynamics simulations for the prediction of the compatibility and mechanical properties of cellulose blends [10–12].

Conflicts of Interest: The authors declare no conflict of interest.

References

1. Usurelu, C.D.; Badila, S.; Frone, A.N.; Panaitescu, D.M. Poly(3-hydroxybutyrate) Nanocomposites with Cellulose Nanocrystals. *Polymers* **2022**, *14*, 1974. [CrossRef] [PubMed]
2. Lunardi, V.B.; Soetaredjo, F.E.; Putro, J.N.; Santoso, S.P.; Yuliana, M.; Sunarso, J.; Ju, Y.-H.; Ismadji, S. Nanocelluloses: Sources, Pretreatment, Isolations, Modification, and Its Application as the Drug Carriers. *Polymers* **2021**, *13*, 2052. [CrossRef] [PubMed]
3. Zheng, D.; Zhang, Y.; Guo, Y.; Yue, J. Isolation and Characterization of Nanocellulose with a Novel Shape from Walnut (*Juglans regia* L.) Shell Agricultural Waste. *Polymers* **2019**, *11*, 1130. [CrossRef] [PubMed]
4. de Andrade, M.R.; Nery, T.B.R.; de Santana e Santana, T.I.; Leal, I.L.; Rodrigues, L.A.P.; de Oliveira Reis, J.H.; Druzian, J.I.; Machado, B.A.S. Effect of Cellulose Nanocrystals from Different Lignocellulosic Residues to Chitosan/Glycerol Films. *Polymers* **2019**, *11*, 658. [CrossRef] [PubMed]
5. Hassan, M.; Abou Zeid, R.E.; Abou-Elseoud, W.S.; Hassan, E.; Berglund, L.; Oksman, K. Effect of Unbleached Rice Straw Cellulose Nanofibers on the Properties of Polysulfone Membranes. *Polymers* **2019**, *11*, 938. [CrossRef] [PubMed]
6. Panaitescu, D.M.; Vizireanu, S.; Stoian, S.A.; Nicolae, C.-A.; Gabor, A.R.; Damian, C.M.; Trusca, R.; Carpen, L.G.; Dinescu, G. Poly(3-hydroxybutyrate) Modified by Plasma and TEMPO-Oxidized Celluloses. *Polymers* **2020**, *12*, 1510. [CrossRef] [PubMed]
7. Popa, M.S.; Frone, A.N.; Radu, I.C.; Stanescu, P.O.; Trusca, R.; Raditoiu, V.; Nicolae, C.A.; Gabor, A.R.; Panaitescu, D.M. Microfibrillated Cellulose Grafted with Metacrylic Acid as a Modifier in Poly(3-hydroxybutyrate). *Polymers* **2021**, *13*, 3970. [CrossRef] [PubMed]
8. Shin, G.; Jeong, D.-W.; Kim, H.; Park, S.-A.; Kim, S.; Lee, J.Y.; Hwang, S.Y.; Park, J.; Oh, D.X. Biosynthesis of Polyhydroxybutyrate with Cellulose Nanocrystals Using *Cupriavidus necator*. *Polymers* **2021**, *13*, 2604. [CrossRef] [PubMed]
9. Jiang, Y.; Li, G.; Yang, C.; Kong, F.; Yuan, Z. Multiresponsive Cellulose Nanocrystal Cross-Linked Copolymer Hydrogels for the Controlled Release of Dyes and Drugs. *Polymers* **2021**, *13*, 1219. [CrossRef] [PubMed]
10. Li, K.; Yang, H.; Jiang, L.; Liu, X.; Lang, P.; Deng, B.; Li, N.; Xu, W. Glycerin/NaOH Aqueous Solution as a Green Solvent System for Dissolution of Cellulose. *Polymers* **2020**, *12*, 1735. [CrossRef] [PubMed]
11. Frone, A.N.; Nicolae, C.A.; Eremia, M.C.; Tofan, V.; Ghiurea, M.; Chiulan, I.; Radu, E.; Damian, C.M.; Panaitescu, D.M. Low Molecular Weight and Polymeric Modifiers as Toughening Agents in Poly(3-Hydroxybutyrate) Films. *Polymers* **2020**, *12*, 2446. [CrossRef] [PubMed]
12. Wang, W.; Li, L.; Jin, S.; Wang, Y.; Lan, G.; Chen, Y. Study on Cellulose Acetate Butyrate/Plasticizer Systems by Molecular Dynamics Simulation and Experimental Characterization. *Polymers* **2020**, *12*, 1272. [CrossRef] [PubMed]

Disclaimer/Publisher's Note: The statements, opinions and data contained in all publications are solely those of the individual author(s) and contributor(s) and not of MDPI and/or the editor(s). MDPI and/or the editor(s) disclaim responsibility for any injury to people or property resulting from any ideas, methods, instructions or products referred to in the content.

Review

Poly(3-hydroxybutyrate) Nanocomposites with Cellulose Nanocrystals

Catalina Diana Usurelu, Stefania Badila, Adriana Nicoleta Frone * and Denis Mihaela Panaitescu *

National Institute for Research and Development in Chemistry and Petrochemistry—ICECHIM, 202 Splaiul Independentei, 060021 Bucharest, Romania; usurelu_catalina@yahoo.ro (C.D.U.); stefania.badila@yahoo.com (S.B.)
* Correspondence: adriana.frone@icechim.ro (A.N.F.); panaitescu@icechim.ro (D.M.P.)

Abstract: Poly(3-hydroxybutyrate) (PHB) is one of the most promising substitutes for the petroleum-based polymers used in the packaging and biomedical fields due to its biodegradability, biocompatibility, good stiffness, and strength, along with its good gas-barrier properties. One route to overcome some of the PHB's weaknesses, such as its slow crystallization, brittleness, modest thermal stability, and low melt strength is the addition of cellulose nanocrystals (CNCs) and the production of PHB/CNCs nanocomposites. Choosing the adequate processing technology for the fabrication of the PHB/CNCs nanocomposites and a suitable surface treatment for the CNCs are key factors in obtaining a good interfacial adhesion, superior thermal stability, and mechanical performances for the resulting nanocomposites. The information provided in this review related to the preparation routes, thermal, mechanical, and barrier properties of the PHB/CNCs nanocomposites may represent a starting point in finding new strategies to reduce the manufacturing costs or to design better technological solutions for the production of these materials at industrial scale. It is outlined in this review that the use of low-value biomass resources in the obtaining of both PHB and CNCs might be a safe track for a circular and bio-based economy. Undoubtedly, the PHB/CNCs nanocomposites will be an important part of a greener future in terms of successful replacement of the conventional plastic materials in many engineering and biomedical applications.

Keywords: nanocomposites; polyhydroxyalkanoates; cellulose nanocrystals

Citation: Usurelu, C.D.; Badila, S.; Frone, A.N.; Panaitescu, D.M. Poly(3-hydroxybutyrate) Nanocomposites with Cellulose Nanocrystals. *Polymers* **2022**, *14*, 1974. https://doi.org/10.3390/polym14101974

Academic Editor: Nathanael Guigo

Received: 13 April 2022
Accepted: 10 May 2022
Published: 12 May 2022

Publisher's Note: MDPI stays neutral with regard to jurisdictional claims in published maps and institutional affiliations.

Copyright: © 2022 by the authors. Licensee MDPI, Basel, Switzerland. This article is an open access article distributed under the terms and conditions of the Creative Commons Attribution (CC BY) license (https://creativecommons.org/licenses/by/4.0/).

1. Introduction

The unprecedented rhythm of using plastics derived from fossil fuels has led to unmanageable environmental problems. These issues are significantly more serious in the packaging industry, where the continuous increase in the out-of-home food consumption has led to an enormous growth in the production and use of packaging materials. Moreover, the COVID-19 pandemic has accelerated the disposal of fossil-fuel-derived plastics from food packaging and medical fields [1,2]. Being the largest and growing consumer of plastics, the packaging industry represents the primary supplier of waste plastics from the environment [1–3]. All this mismanaged plastic waste is becoming a source of soil and groundwater pollution that eventually enters our food chain as microplastics [4]. Once inside the human body, the microplastics can cause toxicity, oxidative stress, cytokine secretion, cell damage, and inflammatory and immune reactions [4]. As a result, continuous research efforts are carried out worldwide for the development of more sustainable and environmentally friendly packaging materials derived from renewable feedstocks, including agro, microbial sources, and biomasses [3]. In this context, biopolymers synthesized via bacterial fermentation, like the ones from the polyhydroxyalkanoates (PHAs) family that includes poly(3-hydroxybutyrate) (PHB) and poly(3-hydroxybutyrate-co-3-hydroxyvalerate) (PHBV) polymers, are considered promising candidates for replacing the fossil fuel-based polymeric materials and addressing the waste disposal issue [5,6]. These biopolymers exhibit several features similar to those of fossil fuel-based polymers and, additionally, can be easily degraded by the action of enzymes and living organisms, eliminating the need for disposal systems [7]. Most importantly, their degradation products possess no danger to human beings or the environment. Moreover, being biocompatible, PHB is also suitable for use in a variety of medical applications [7,8].

PHB is by far the most intensively studied biopolymer from PHAs' family and its global market is predicted to reach $221.14 million by 2027 [9]. PHB exhibits remarkable properties such as optical activity, stiffness, and high oxygen barrier properties [8]. However, despite these desirable properties its widespread industrial application, especially in the food packaging sector, has been limited due to its inherent physical aging, low ductility, limited processability, and low crystallization rates in the absence of nucleating agents [7,8]. Multiple strategies to overcome PHB's drawbacks and to enhance its performance for targeted applications were proposed over time, including blending with other bio-based polymers [10], graft or block copolymerization, the use of specific additives (plasticizers, stabilizers, chain extenders, nucleating agents etc.), and the addition of natural micro-/nano- reinforcing agents [7,8].

PHB-based polymer nanocomposites, which are obtained by the incorporation of nanosized particles into the polymer matrix, exhibited clear improvement of properties compared to neat PHB, especially with respect to mechanical and barrier performances [11]. Owing to their outstanding properties, such as renewability, biodegradability, good thermal resistance, low density, high specific strength and stiffness, nontoxicity and lack of corrosion, cellulosic nano-reinforcements are some of the most suitable fillers for PHB [12,13]. In addition, they may contribute to preserving the properties of food products due to their barrier properties and ability to carry antioxidant and antibacterial agents. Different cellulose nanofillers, nanocrystals or nanofibers, were used to overcome some of PHB's weaknesses, such as its slow crystallization, brittleness, modest thermal stability, and low melt strength [14–18]. However, the strong hydrophilicity of cellulose nanofillers prevents their good dispersion in the hydrophobic PHB matrix and reduces their reinforcing ability. Therefore, surface treatment by TEMPO oxidation or plasma treatment [14], grafting with silanes [18,19], and polymers compatible with PHB [18] were tested as methods to improve the adhesion between the cellulose nanofillers and the PHB or PHBV matrix.

One route to improve the properties of PHB is the addition of cellulose nanocrystals (CNCs) as nano-reinforcements and the fabrication of PHB/CNCs nanocomposites. In this contribution we provide a comprehensive review on the PHB/CNCs nanocomposites intended for packaging application. Particular attention was paid to choosing the adequate processing technology for producing the PHB/CNCs nanocomposites and an appropriate surface treatment for the CNCs as key elements for achieving a good interfacial adhesion, superior thermal stability, and mechanical performances for the resulting formulations. Although there are several general reviews on biocomposites with cellulose nano-reinforcements, no review dealing with the PHB/CNCs nanocomposites was published so far. The information provided in this review may be considered as a basis for finding new strategies to reduce the manufacturing costs or to design better technological solutions for the large-scale production of PHB/CNCs nanocomposites.

2. Poly(3-hydroxybutyrate)

Poly(3-hydroxybutyrate) (PHB) is the most well-known and widely used member of PHAs [20], a family of biodegradable polyesters derived from the microbial fermentation of different carbon sources [21]. Although chemical synthesis is also possible [22], the main method of producing PHB is its extraction from various bacterial strains which are capable of synthesizing and accumulating PHB intracellularly, as carbon and energy reserve, under nutrient limiting conditions [23]. Due to its biodegradability in both soil and marine environments [24], non-toxicity, biocompatibility, and a melting temperature, elastic modulus and tensile strength similar to that of isotactic polypropylene [25], PHB has drawn increasing attention as an environmentally friendly substitute for petroleum-based thermoplastics, such as polypropylene (PP) and polyethylene (PE) [26]. The world's rising concern on plastic pollution and oil resources depletion supported the study of PHB—based materials for potential applications in packaging (films, bags, bottles, containers etc.) [27], biomedicine (surgical sutures, drug delivery systems, surgical meshes, wound dressings, scaffolds for tissue engineering etc.) [28] and agriculture (carriers for the slow release of pesticides, herbicides, fertilizers etc.) [29]. However, the high degree of crystallinity of PHB [30] that imparts brittleness [14], the thermal degradation at temperatures just above its melting temperature [31] that narrows its processing window [26], the low elongation at break [32] and the high production costs [14], are serious disadvantages which have restrained the use of PHB on a large scale [33].

In recent years, efforts have been made to eliminate these disadvantages. For example, in order to reduce the production costs of PHB, the replacement of the noble carbon sources such as glucose, mannose, and lactose [34] with low-cost agro-industrial byproducts and residues has been proposed [35]. Waste glycerol from biodiesel fuel production [36], corn waste [37], wheat straw [38], rice straw [39], dairy waste [40], sugarcane vinasse, and molasses [41] are just a few of the industrial and agricultural by-products and residues that were successfully employed as carbon sources in the obtaining of PHB by microbial fermentation [42]. Regarding the improvement of the mechanical properties and thermal stability of PHB, several methods have been developed. One method consists in the incorporation of secondary flexible monomers such as 3-hydroxyvalerate, 4-hydroxybutyrate, or 3-hydroxyhexanoate [43] in the main chain of PHB that led to the formation of PHBV [44], poly (3-hydroxybutyrate-co-4-hydroxybutyrate) (P4HB) [45], or poly(3-hydroxybutyrate-co-3-hydroxyhexanoate) (PHBHH) [44] copolymers. Despite the fact that these PHB copolymers have shown higher flexibility, lower melting temperatures and wider processing windows as compared to the pristine PHB [46], they still have poor mechanical properties or are difficult to be obtained through efficient synthesis processes, which has limited their utilization for commercial use [47]. A second strategy involves the use of petroleum-based (dioctyl adipate, dioctyl phthalate, dibutyl phthalate, polyethylene glycol, polyadipates etc.) or bio-based (glycerol, glycerol triacetate, triethyl citrate, soybean oil, epoxidized soybean oil etc.) [48] plasticizers, which are known to increase the flexibility of PHB and to decrease its glass transition temperature (T_g). However, plasticizers may degrade at temperatures lower than PHB, accelerating its thermal degradation during melt processing, or migrate at the surface of the material, altering its mechanical properties [49]. A third method implies the melt blending of PHB with polymers such as medium chain length-PHAs [10], poly(caprolactone) (PCL), polyethylene glycol (PEG), poly(butylene adipate-co-terephthalate) (PBAT), and poly(butylene succinate) (PBS) etc. [10,50,51]. In this case, it has been shown that the obtained blends have improved flexibility, toughness, and processability as compared to the pure PHB. However, in many situations the improvements were not at the anticipated level due to the poor miscibility between PHB and the second polymer in the melt state [52]. Another method involves the addition to PHB of various nanofillers such as titanium dioxide (TiO_2) [53], zinc oxide (ZnO) [54], carbon nanotubes (CNTs) [55], clays [56], and nanocellulose [16,17]. Using these nanofillers makes possible the obtaining of PHB-based nanocomposites with superior thermal stability and increased mechanical and barrier properties as compared to the neat PHB [53–56]. Among them, special attention has been paid to nanocellulose fillers due to their renewability, biodegradability, and superior mechanical properties.

3. Cellulose Nanocrystals

Cellulose is the nature's most abundant biopolymer and can be extracted from various sources such as plants, marine life, fungi, and bacteria [57]. Regardless of its source, the prime structural unit of cellulose is comprised of linear chains of D-glucose linked by repeating β-1,4-glycosidic bonds, followed by a 180° rotation for the next linkage [57,58]. Owing to its large network of intermolecular and intramolecular hydrogen bonds, cellulose is insoluble in water and most organic solvents. The degree of polymerization and molecular weight of cellulose are governed by both the cellulosic source and the methods employed to produce it [59]. Cellulose with nano-scale structural dimensions, referred to as nanocellulose, possesses high surface area, unique morphology, specific high strength and modulus, renewability, customizable surface chemistry, and good biocompatibility, offering myriad of opportunities for medical and engineering applications. As filler in polymers, nanocellulose has the advantage of being a stable material that cannot be melted during melt processing due to the high level of hydrogen bonding. Thus, nanocellulose can be used as a reinforcing agent for biopolymers, especially in the field of industrial packaging, where the melt processing techniques specific to thermoplastic polymers are intensively used [59].

Depending on the preparation methods and sources, nanocellulose can be classified into three categories, namely cellulose nanocrystals (CNCs), cellulose nanofibers (CNFs), and bacterial nanocellulose (BC). CNCs are one of the most important nanocelluloses that are produced at an industrial level using chemical treatments. Besides the wood and lignocellulosic fibers, a more convenient alternative from both an environmental and an economic point of view is represented by the agriculture along with food and beverage-processing waste and byproducts. Thus, CNCs are predominantly extracted from corn cobs [60], tea stalk [61], soy hulls [62], apple [63] and grape pomace [64], pineapple leaves [65], soybean [66], plum shells [67], barley [68] and garlic [69] straws, tomato peel waste [70], sugarcane bagasse [71], orange peel waste [72], or chili leftovers [73]. However, the exploitation of municipal waste and papermaking sludge may represent other alternative sources [58,74].

Different strong (sulfuric, hydrochloric and hydrobromic acids) and weak (phosphoric, citric and formic acids) acids were used for breaking the glycoside bonds in cellulose [75]. Acid hydrolysis involves the hydrolytic cleavage of the amorphous regions from the cellulose fibers, when the crystalline region domains are left behind. Reaction temperature and duration as well as the acid type and its concentration are the main parameters that determine the size and morphology of the isolated CNCs [76]. Thus, CNCs possess whisker or a short-rod-like morphology with uniform sizes ranging from 100 to 200 nm in length and 10 to 30 nm in diameter [77,78]. The use of weak acids leads to cellulose fibers with low crystallinity and fibrous morphology as a result of the low dissociation constant of these acids [76]. Despite being a simple and fast isolation method, sulfuric acid hydrolysis has the advantage of yielding CNCs with higher crystallinity degree (over 90%) and also leads to the sulfate esterification of the CNCs surface, which enhances CNCs' phase stability in aqueous medium [79]. However, sulfate esterification decreases their thermal stability in the case of thermal treatments. Moreover, CNCs' almost perfect crystalline structure ensures high mechanical properties such as tensile modulus and tensile strength, which are crucial for further applications [80]. It is worth mentioning that when incorporated in the polymer matrices, CNCs develop a network-like formation, upgrading the polymer's gas barrier and migration properties to a greater extent than the nanoclays or carbon-based materials [8]. However, the strong hydrophilic property of CNCs raises compatibility and dispersibility issues when combined with other polymers, especially with highly hydrophobic ones. Thus, extensive research studies have been conducted for the surface treatment of CNCs in order to overcome these problems [80].

CNFs, also known as nanofibrillar cellulose and cellulose nanofibers, possess hierarchical structures made up of interconnected fibrils with diameters ranging from 1 to 100 nm and an aspect ratio higher than 15 [81,82]. Top-down mechanical disintegration methods such as grinding, cryocrushing, high-intensity ultrasonication, and high-pressure homogenization are usually employed for the CNFs' isolation. Through these techniques, dilute suspensions of cellulose fibers are subjected to high shear and impact forces, thus leading to mechanical cleavage along the longitudinal direction of the cellulosic source [78,81–83]. Specifically, the defibrillation methods yield nanostructures with both crystalline and amorphous regions. The high flexibility of CNFs is due precisely to the presence of the amorphous component. However, due to their large aspect ratio, the mechanical derived nanocelluloses are more susceptible to fiber agglomeration, which makes their further processing more challenging.

Alternatively, BC, named bacterial nanocellulose, microbial cellulose, or bio-cellulose, is produced through a bottom-up approach using different aerobic non-pathogenic bacteria [82–84]. Besides being the purest form of nanocellulose, BC shows an ultrafine network structure containing fibers with micrometers in length and 20–100 nm in diameter, high water holding capacity and flexibility, and high crystallinity. Its outstanding physical, structural, and mechanical features make BC an ideal candidate for biomedical applications.

Nanocelluloses, regardless of their source or preparation method, have been intensively studied as reinforcements in biopolymers for improving their mechanical and barrier properties [16,17,85,86]. The biocomposites from aliphatic polyesters and bacterial cellulose, as well as the nanocomposite materials from microfibrillated cellulose and hydrophilic or hydrophobic polymers, have been already reviewed [87,88]. However, much recent literatures on nanocellulose based materials are focused on biopolymers reinforced with CNCs. CNCs may be easily obtained both in labs and industrial facilities by employing less energy intensive processes as compared to CNFs. Due to the bio-origin of both PHB and CNCs and their subsequent biodegradation to carbon dioxide and water, the PHB/CNCs nanocomposites are of particular interest in the context of the world's transition towards a circular economy (Figure 1).

Figure 1. Circuit of biodegradable PHB/CNCs nanocomposites in the context of a circular economy.

4. PHB Nanocomposites with Cellulose Nanocrystals

4.1. Preparation Routes

Solvent casting, melt processing, and electrospinning are the most used methods for obtaining PHB or PHBV nanocomposites with cellulose nanocrystals [89–91]. From these, solution casting is used more frequently due to some advantages such as its simplicity, easy application in lab conditions, and the obtaining of a good dispersion of the cellulose nanocrystals in the polymer matrix. However, this method has the drawback that it uses toxic solvents that are difficult to be entirely removed from the samples. Moreover, the largely different character of cellulose and PHB, strongly hydrophilic vs. hydrophobic, and, therefore, the different solvents needed to dissolve the PHB and disperse the CNCs, make difficult the mixing of their solutions. Melt processing by using batch mixers, kneaders, or twin-screw extruders followed by injection or compression molding is advantageous because it can be easily transposed in industry, being an environmentally friendly method, which does not use dangerous solvents. However, the specialized equipment required for performing this process, the energy consumption needed for melting the PHB and for the mixing process, and the tendency of the cellulose nanocrystals to aggregate in the polymer melt are several disadvantages that must be overcome. In addition, PHB is sensitive to thermal degradation during melt processing [32,92]. Electrospinning has advantages related to easy scale-up, inexpensiveness and ability to incorporate various additives and polymers in the process, however it suffers from the same drawbacks as the solvent casting method, related to toxicity and difficult removal of the solvent. Additive manufacturing or reactive blending [93], processing under supercritical conditions or foaming [94] were also tried as new routes to obtain PHB nanocomposites with cellulose nanofillers.

4.1.1. Solution Casting

CNCs extracted from different sources (bamboo pulp, microcrystalline cellulose (MCC), or bleached pulp board) by sulfuric acid hydrolysis were used to obtain PHB/CNCs nanocomposites by solution casting using chloroform or dimethylformamide (DMF) as a solvent. Several methods were used to disperse the CNCs into the PHB solution: (i) a CNCs suspension in water was dispersed in acetone and then in chloroform through a sequential solvent exchange process consisting of a succession of dispersions and centrifugations, and further the CNCs dispersion in chloroform was mixed with a chloroform solution of PHB [89,95,96]; (ii) the CNCs were transferred from water to DMF using a solvent exchange method and then mixed with the PHB solution in DMF [77]; (iii) the water suspension of CNCs was lyophilized, and then the CNCs were redispersed in DMF and finally mixed with the DMF solution of PHB [86,97–101]. The concentration of CNCs in PHB was maintained at low values of up to 6 wt% in all the reported works. At a CNCs concentration below 2 wt%, a homogeneous dispersion of the CNCs in the PHB matrix was obtained, while at higher contents, CNCs agglomerations were observed with the naked eye as small white dots distributed in the transparent PHB matrix [95]. Due to their ability to scatter light, the CNCs agglomerates led to a decrease in the transparency of the PHB/CNCs nanocomposites. FESEM images showed that the CNCs were homogeneously dispersed into the PHB matrix at a CNCs content of 1 wt% [95]. Similar results were reported by Zhang et al. [89] which prepared PHB/CNCs and PHB/CNFs nanocomposites with a nanocellulose content of 1, 3, and 5 wt%, using a solution casting method. The SEM images showed that the best dispersion of CNCs in the PHB matrix was obtained at a CNCs loading of 1 wt%, while at CNCs contents of 3 and 5 wt%, a high tendency of CNCs to form agglomerates was noticed. This was due to the strong hydrogen bonds formed between the CNCs nanoparticles at these higher concentrations [89]. The transparency tests revealed that the transmittance of the PHB/CNCs and PHB/CNFs nanocomposite films decreased with the increase of CNCs and CNFs concentration. This was attributed to the poor compatibility between the hydrophilic nanocellulose and the hydrophobic PHB matrix and to the formation of nanocellulose agglomerates, which prevent light transmission [89]. When PHB/CNCs nanocomposites with 2, 4 and 6 wt% CNCs were compared to a PHB/2 wt% BC nanocomposite, all of them being obtained by solution casting, the transparency tests showed that the transmittance of the nanocomposites was higher as compared to that of pure PHB [86]. The best transparency was observed for the PHB/CNCs films due to the good CNCs dispersion and favorable interactions between the CNCs and the PHB matrix [98].

Several methods were applied to improve the properties of PHB/CNCs nanocomposites obtained by solution casting, such as the addition of plasticizers or the use of PHBV copolymer instead of PHB. Different plasticizers were added to the PHB/CNCs nanocomposites during the solution mixing and casting process to improve their flexibility and processability [99,101]. For example, PHB/CNCs nanocomposites with a CNCs loading of 2 and 4 wt%, respectively, and 20 wt% glyceryl tributyrate (TB) or poly [di (ethylene glycol) adipate] (A) as plasticizers were prepared via solvent casting using DMF as a solvent. TB addition led to PHB/CNCs nanocomposites with a lower thermal stability than that of pure PHB due to its easy evaporation and increased mobility of the polymer chains, which facilitated the diffusion of the decomposition products. On the contrary, the addition of plasticizer A led to PHB/CNCs nanocomposites with higher thermal stability than neat PHB due to the high molecular weight of A plasticizer which prevents its migration from the nanocomposites. CNCs addition was shown to have a beneficial effect on the stiffness and barrier properties of the plasticized nanocomposites, especially in the case of TB containing nanocomposites. This was explained by the good dispersion of the CNCs in the TB plasticized-PHB matrix due to the good compatibility between TB and the PHB matrix and the favorable hydrogen-bonding interactions established between TB and CNCs [99]. Therefore, the addition of TB plasticizer enhanced the PHB—CNCs interactions and improved the dispersion of the CNCs [101]. The decrease in the contact angle (CA) value for the PHB/CNCs nanocomposites as compared to pure PHB, regardless of whether CNCs or BC was used as reinforcing agent and whether A or TB was used as plasticizer, indicated an increased hydrophilicity. This may be assigned to various causes: (i) the plasticizing effect of TB or A, which led to an increased mobility of the PHB chains facilitating the diffusion of water into the material, (ii) the existence of numerous hydrophilic groups on the CNCs surface, which might have increased the hydrophilicity of the material or (iii) the poor dispersion of BC in the PHB matrix, which left a higher mobility to the PHB chains, favoring the diffusion of water molecules inside the material. However, CA values did not decrease below 65°, except for the PHB/4 wt% CNCs nanocomposite (A plasticizer) [101]. The PHB/CNCs nanocomposite films, containing 20 wt% TB, were applied as coatings on a cellulose paperboard by compression molding [100]. The CNCs from the nanocomposite films increased the interaction between layers as a result of the hydrogen bonding interactions between the hydroxyl groups from their surface and the OH groups of the paperboard, leading to enhanced mechanical properties. In addition, the PHB/CNCs layer improved the barrier properties of the paperboard, which became more suitable for packaging applications [100].

PHB/CNCs nanocomposites containing 15 wt% PEG and a low amount of CNCs (up to 0.75 wt%) were prepared by a solvent casting method [102]. PEG was used as a plasticizer and compatibilizer based on its miscibility with PHB and its affinity to cellulose due to the formation of hydrogen bonds between the carbonyl (-C=O) groups of PEG and the hydroxyl (-OH) groups of CNCs. The CNCs were first dispersed in PEG, the CNCs surface being covered by a PEG layer. Indeed, the TEM images revealed a homogeneous dispersion of the CNCs in the PHB matrix at a CNCs content of 0.15 wt%. This confirmed the ability of PEG to act as a coupling agent between PHB and CNCs. Based on the ATR-FTIR spectra and the electron microscopy images of the PHB/PEG/CNCs nanocomposites, it was supposed that (i) for a CNCs s content up to 0.45 wt% nearly the entire surface of the CNCs was covered by PEG, so that all or almost all the interactions between the PHB and the CNCs occurred preferentially via their PEG coating; (ii) for a higher CNCs content, when the PEG/CNCs ratio was low, the amount of PEG was no longer enough to cover the entire surface of CNCs, so that the interactions between PEG and PHB decreased, becoming more likely that the CNCs interacted directly with the PHB matrix. This model was supported by the variation of the mechanical properties of the nanocomposites, up to a concentration of 0.45 wt% the PEG-coated CNCs showing no reinforcing effect [102].

The use of PHBV instead of PHB in the nanocomposites with CNCs was also tried as a method to improve the processability and flexibility of the nanocomposites [97,103]. Two methods were used to incorporate the CNCs in PHB. In one method [97], the gel-like CNCs, resulting from the sulfuric acid hydrolysis of microcrystalline cellulose (MCC), were freeze-dried. Then, the resulting powder was added to the PHBV solution in DMF, ultrasonicated and casted. Transparent films containing 1–5 wt% CNCs were thus formed [103]. In the second method [97], water suspensions with different concentrations of CNCs were added dropwise in DMF under stirring and, after the evaporation of water, PHBV was dissolved in the CNCs suspensions. PHBV/CNCs films with 0.5–4.6 wt% CNCs were obtained by solution casting, similar to the first method. The PHBV/CNCs films obtained by the two methods showed different thermal and mechanical properties [97,103]. Thus, a continuous decrease of the cold crystallization temperature (T_{cc}) with the increase of the CNCs concentration was observed in the nanocomposites obtained by the first method and a decrease in T_{cc} only at loadings lower than 2.3 wt% in the second case. A similar trend was noticed for the variation of the mechanical properties, which indicated a more homogeneous dispersion of the CNCs in the nanocomposites obtained by the first method and a worse dispersion of CNCs at CNCs loadings exceeding 2.3 wt% when the solvent exchange-solution casting method was employed. The CNCs agglomerations were clearly observed in the TEM image of the PHBV/4.6 wt% CNCs film (Figure 2) [97].

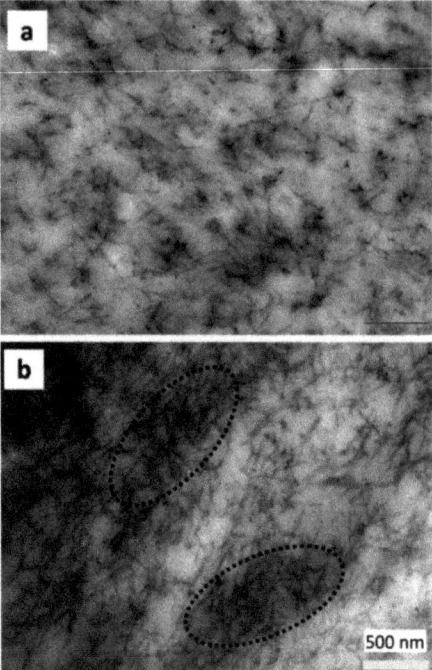

Figure 2. TEM images of the PHBV/CNCs nanocomposites with 2.3 wt% CNCs (**a**) and 4.6 wt% CNCs (**b**) Reprinted with permission from Ref. [97]. Copyright 2012 American Chemical Society.

4.1.2. Melt Processing

Melt processing may be considered as the most important method to obtain PHB/CNCs nanocomposites due to the eco-friendliness and good fitting to the industrial processing techniques. However, the incorporation of CNCs in the PHB melt may be a difficult task due to the high tendency of hydrophilic CNCs to agglomerate in a hydrophobic environment.

Chen et al. [104] used freeze-dried CNCs obtained by the sulfuric acid hydrolysis of MCC to prepare PHB/CNCs nanocomposites. A melt mixing method using a Haake Polylab Rheometer heated to 180 °C was employed to incorporate 2 wt% CNCs into the PHB and the resulted nanocomposite was subjected to crystallization studies to determine the CNCs effect on the PHB crystallization. Compared to MCC, CNCs had a higher influence on the spherulite morphology of PHB. CNCs acted as a heterogeneous nucleating agent causing an increase in the PHB crystallization rate simultaneously with a decrease in the energy barrier of PHB nucleation and in the folding surface free energy [104]. In addition, CNCs incorporation influenced the banded structure of the PHB spherulites, leading to a decrease in the average band space of the ring-banded spherulites. This was assigned to the increase in the crystallization rate of PHB in the presence of CNCs, which led to unbalanced stresses favoring the lamellae twist and the formation of ring-banded spherulites with reduced band space [104].

Jun et al. [90] used a PHBV matrix to prepare nanocomposites with two types of nanocelluloses (CNCs and CNFs) in different concentrations from 1 to 7 wt%. CNCs were obtained via the sulfuric acid hydrolysis of rice straws and CNFs resulted from the pressure-grinding of the cellulose extracted from the same source. For the preparation of nanocomposites, the PHBV powder was added to the nanocellulose suspensions under stirring and the mixtures were vacuum-dried for 24 h. A Haake co-rotating twin-screw extruder was used for melt compounding the mixtures. Both CNCs and CNFs showed a nucleation effect, accelerating PHBV crystallization and improving the Young's modulus of nanocomposites, the optimum mechanical properties being obtained at 1 wt% CNCs [90].

4.1.3. Other Methods

Electrospinning was used to obtain PHB/CNCs nanocomposite fibers with a content of CNCs of 5, 8, 12, 17, and 22 wt% [91]. CNCs were obtained by the sulfuric acid hydrolysis of alkali-treated and bleached corn husk [105]. A solvent exchange method was used to disperse the CNCs in achloroform/DMF mixture (90/10 volume ratio), which was employed as a solvent in the electrospinning process [91]. As revealed by SEM, the obtained PHB/CNCs nanocomposite fibers presented a uniform surface, without beads, regardless of the concentration of CNCs in the PHB matrix. A decrease of the PHB/CNCs nanocomposite fibers' diameter was observed with the increase of CNCs concentration in nanocomposites. This was attributed to the increase in the conductivity of the electrospinning solution with increasing CNCs loading as a result of the negatively charged sulfate ester groups formed on the CNCs surface during sulfuric acid hydrolysis [91].

PHB/CNCs nanocomposite foams with 2, 3 and 5 wt% CNCs, obtained via the sulfuric acid hydrolysis of pulp fibers, were prepared using a nonsolvent-induced phase separation (NIPS) method [106]. Chloroform was used as solvent while tetrahydrofuran (THF) or 1,4-dioxane (Diox) was used as nonsolvents. In the NIPS process, the addition of a nonsolvent reduced the polymer−solvent affinity leading to a phase-separated polymer solution with one phase rich in polymer, representing the backbone of the gel, which was penetrated by the polymer-poor phase (in the nonsolvent) [106]. When THF was used as nonsolvent, CNCs accelerated both the PHB crystallization and the nanocomposites gelation, showing a nucleating effect. In contrast, when Diox was used as nonsolvent, CNCs incorporation led to a decrease in both PHB crystallization and nanocomposites gelation rate. This was due to the better dispersion of the CNCs in Diox than in THF, preventing the movement of the PHB chains and delaying the crystals' growth. However, no significant differences between the degrees of crystallinity of the PHB/CNCs nanocomposites obtained using THF or Diox as nonsolvents were observed [106].

PHB/CNCs (4 wt%) nanocomposites obtained by solution casting using DMF as solvent were compression molded and further applied as a coating to cellulose paperboards, resulting in bilayer structures [107]. The PHB/paperboard ratio was varied between 5 and 20 wt%. The addition of CNCs improved the adhesion between the PHB layer and the cellulosic paperboard. The PHB/CNCs coatings decreased the water sensibility of the cellulose layer, leading to paperboard/PHB-CNCs bilayer composites suitable for packaging [107].

4.2. Methods Used to Improve the Compatibility in PHB/CNCs Nanocomposites

Different methods were applied to improve the compatibility between the strongly hydrophilic CNCs and the hydrophobic PHB matrix [108–111]. Dispersion agents and compatibilizers are one of the simplest and sometimes efficient additives for improving the compatibility in polymer nanocomposites with CNCs [112]. PEG is a hydrophilic polymer that is miscible with PHBV and, therefore, it was used as a compatibilizer in the

PHBV/CNCs nanocomposites [108]. PHBV/CNCs nanocomposites with CNCs contents of 2 and 5 wt% were prepared using solution casting or extrusion blending [108]. In the first method, CNCs were coated with PEG by dispersing the mixture of CNCs and PEG powders in DMF, and then the PHBV solution in DMF along with the CNCs/PEG suspension in DMF were mixed and casted. In the second method, the freeze-dried CNCs/PEG powder was premixed with PHBV and melt compounded using a co-rotating twin screw extruder followed by injection molding [108]. A very good dispersion of the CNCs in the PHBV matrix was obtained when the solvent casting method was used, also supported by the enhancement of the mechanical properties. However, despite the presence of the PEG compatibilizer, the CNCs could not be well dispersed in the PHBV during melt compounding, with effect on the mechanical properties, which decreased as compared to those of pure PHBV. A possible explanation for this behavior is that the high shear stress generated by the twin-screw removed the PEG coat from the CNCs surface and blended it with the PHBV matrix with which is compatible [108].

Another method proposed to improve the dispersion of the CNCs in PHBV and the compatibility between the two components is the chemical modification of CNCs by grafting PHBV onto their surface [110]. OH-terminated PHBV oligomers, prepared through transesterification using ethylene glycol in diglyme and dibutyltin dilaurate as a catalyst, were grafted on the surface of CNCs. The grafting reaction took place in anhydrous DMF using toluene diisocyanate (TDI) as a coupling agent. The ungrafted PHBV was removed by refluxing with chloroform [110]. PHBV-grafted CNCs (PHCNs) were used to obtain PHBV/PHCNs nanocomposites by solution casting, the content of modified CNCs ranging from 5 to 30 wt%. Most of the PHBV/PHCNs nanocomposite films showed a transparency similar to that of pure PHBV films. A decrease in the UV-vis transmittance with an increase in the PHCNs content was noticed; good results were obtained at PHCNs loadings up to 20 wt%, while a strong reduction in the transparency of the PHBV/PHCNs nanocomposites with 25 or 30 wt% PHCNs was observed. This was due to the formation of PHCNs agglomerates in the PHBV matrix at a higher content of modified CNCs [90]. The addition of PHCNs into PHBV led to a great increase in the mechanical properties for PHCNs contents of up to 20 wt%. This was due to the good adhesion between the components and the effective stress transfer at the PHBV-PHCNs interface [110].

CNCs grafted with polylactide (CNC-g-PLA) were prepared and used in PHB as a more compatible reinforcing agent [113]. CNCs resulted from the sulfuric acid hydrolysis of filter paper were grafted with polylactide by surface-initiated ring opening polymerization of L-lactide. The synthesis of CNC-g-PLA was carried out in 1-allyl-3-methylimidazolium chloride ionic liquid in the presence of catalytic amount of (dimethylamino)pyridine. Prior to the ring-opening polymerization of L-lactide, the CNCs were homogeneously acetylated. PHB nanocomposites with 2 wt% CNCs or CNC-g-PLA were obtained using a melt compounding method [113]. The calorimetric results showed a large influence of the CNCs treatment on the crystallization behavior of PHB. Untreated CNCs acted as a heterogeneous nucleating agent enhancing PHB crystallization. A different role was observed in the case of CNC-g-PLA which retarded the nucleation of PHB crystals and acted as an antinucleating agent during PHB crystallization [113]. Therefore, the PHB/CNCs nanocomposite exhibited a higher crystallization rate than neat PHB while the PHB/CNC-g-PLA nanocomposite presented a lower crystallization rate, showing that the crystallization behavior of PHB could be controlled by the CNCs' treatment [113].

A one pot acid hydrolysis/Fischer esterification process was used to obtain CNCs surface modified with butyric acid, lactic acid, and their mixture in the presence of 37% HCl [109]. To ensure a good dispersion of modified CNCs in PHBV, the CNCs_butyrate, CNCs_lactate and CNCs_butyrate_lactate nanofillers were subjected to a solvent exchange sequence, from water, to ethanol, acetone, and finally, to chloroform. The suspensions of modified CNCs in chloroform were then mixed with the solution of PHBV in chloroform/tetrachloroethane (50/50) and casted [109]. Due to the similarity between the chemical structures of the lactate and butyrate ester moieties grafted on the CNCs surface and that of the PHBV matrix, the adhesion between the polymer matrix and the reinforcing agents was considerably improved and a homogeneous dispersion of the modified CNCs in the PHBV nanocomposites was observed. Consequently, the PHBV/CNCs_butyrate, PHBV/CNCs_lactate and PHBV/CNCs_butyrate_lactate nanocomposites with 2 wt% CNCs showed a considerably improved transparency as compared to the nanocomposites containing unmodified CNCs [109]. However, the dynamical mechanical analysis results confirmed an improved interface only in the PHBV/CNCs_butyrate nanocomposites due to the similarity between the butyrate moieties attached on the CNCs surface and the PHBV matrix, which led to a better dispersion of the CNCs in the polymer matrix.

A mixture of PHB and poly(4-hydroxybutyrate), denoted as PHB/P4HB, was reinforced with surface hydrophobized CNCs, which were obtained by a double silanization process, as shown in Figure 3 [111].

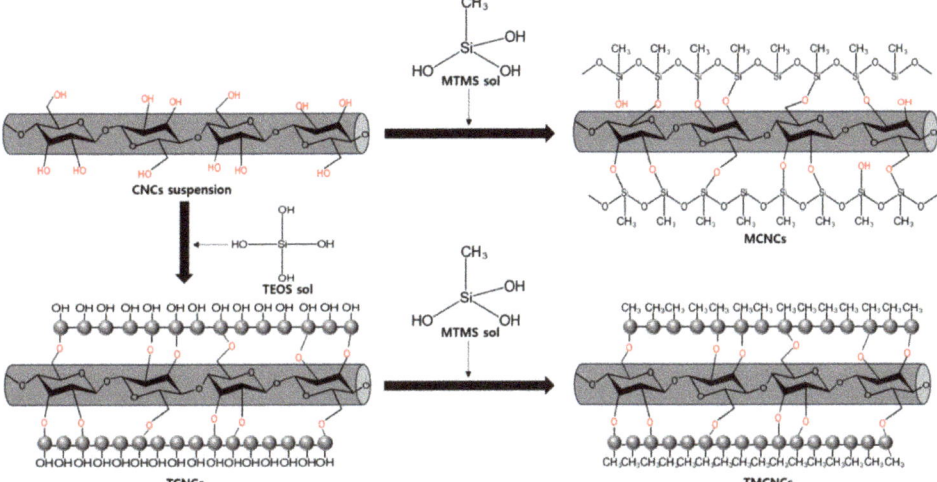

Figure 3. Schematic representation of the surface treatment of CNCs using a double silanization process [111].

To improve the compatibility between the CNCs and the hydrophobic matrix, the CNCs were modified with (i) methyltrimethoxysilane (MTMS) resulting CNCs with a hydrophobic surface (MCNCs), (ii) tetraethyl orthosilicate (TEOS) resulting CNCs with spherical SiO_2 nanoparticles on their surface (TCNCs) and (iii) TEOS then MTMS, resulting TCNCs with CH_3 ends (TMCNCs). The three types of modified CNCs were melt-compounded with PHB/P4HB by extrusion followed by compression molding resulting nanocomposite plates. MCNCs and TCNCs showed a low compatibility with PHB/P4HB and many aggregated nanocrystals were observed in the nanocomposites with 10 wt% modified celluloses. On the contrary, freeze-dried TMCNCs showed a homogenous dispersion in the PHB/P4HB matrix and no nanocrystals agglomeration [111].

4.3. Nanocomposites from PHB Blends and CNCs

The addition of a third polymer in PHB/CNCs nanocomposites was also used as a method to improve the compatibility and the properties of PHB/CNCs nanocomposites [114–117]. Based on the compatibility of poly(vinylacetate) (PVAc) and PEG with PHB or PHBV and their hydrophilic character, similar to that of CNCs, the two polymers were used as a third component in PHB/CNCs nanocomposites [114]. PHB/PVAc/CNCs, PHB/PEG/CNCs, PHBV/PVAc/CNCs, and PHBV/PEG/CNCs ternary nanocomposites with 2.4 or 4.8 wt% CNCs were prepared by melt mixing PHB or PHBV with PVAc/CNCs and PEG/CNCs masterbatches in a Haake double-screw mini-extruder. The masterbatches were prepared by dispersing the CNCs into a PVAc water emulsion or into a PEG solution in water, followed by solution casting as films and drying [114]. Due to the partial miscibility of PVAc or PEG with PHB and PHBV, they improved the dispersion of the CNCs into the polymer matrix. This was determined by the favorable hydrogen bonding interactions between the hydroxyl groups on the CNCs surface and the polar groups on the PVAc and PEG chains. The addition of a third polymer had as a result an improvement in the mechanical properties, more important in the case of the PVAc-containing nanocomposites. The plasticizing effect of PEG could be a cause of the lower improvement in the mechanical properties observed in the PEG-containing nanocomposites [114].

A combined solvent casting and melt processing technique was used to obtain a good dispersion of CNCs from pine cones in a PHB/poly(ε-caprolactorne) (PCL) blend [115]. CNCs were dispersed in chloroform and were added in a mixture of PHB/PCL (75:25) in chloroform under intense stirring. The solvent casted films were melt-compounded in a twin screw microextruder and compression molded. Nanocomposites with a content of CNCs of 3, 5 and 7 wt% were obtained by this method. In low amounts, CNCs enhanced the compatibility between the immiscible PHB and PCL and reduced their phase separation during the melt blending. This was due to the tendency of CNCs to locate at the PHB-PCL interface preventing the coalescence of the dispersed PCL phase in the continuous PHB matrix during the melt processing. The best dispersion of CNCs in the PHB/PCL matrix was obtained for a CNCs content of 3 wt%. In this case, the nanocomposite showed a significant increase in transparency that can be attributed to the good dispersion of CNCs and their compatibilization effect, which increased the PHB-PCL miscibility [115].

Poly(lactic acid) (PLA) and CNCs were added in a PHB plasticized with epoxidized canola oil (eCO) to improve its mechanical properties [116]. The PHB/PLA/eCO/CNCs nanocomposites with a PHB:PLA weight ratio of 3:1 and 5 wt% CNCs (related to the PHB/PLA amount) were obtained by melt-mixing using a conical twin-screw microextruder. The eCO green plasticizer was added to increase the flexibility and thermal stability of PHB, both important drawbacks of this biopolymer. The concomitant addition of PLA and eCO proved to be beneficial to both the elastic properties and thermal stability of nanocomposites [116].

A melt compounding masterbatch technique was applied to obtain PLA/PHB/CNCs nanocomposite films plasticized with a low content of acetyl tributyl citrate [117]. In addition to being an easily scalable and environmentally friendly technique, the melt compounding masterbatch technique ensured a good dispersion of the CNCs and plasticizer in the PLA/PHB blend. This technique consists in the fabrication of a PHB masterbatch, containing both the plasticizer and the CNCs, and its further dilution in PLA, both phases comprising melt compounding operations. The addition of 5 wt% CNCs to the plasticized PLA/PHB matrix led to an increase in the storage modulus due to the stiffening effect of the CNCs and the good dispersion of the high aspect ratio CNCs in the polymer matrix [117].

5. Influence of CNCs on the Properties of PHB/CNCs Nanocomposites

5.1. Thermal Properties

The crystallization behavior of PHB is very important because it determines many of its properties. PHB is characterized by a high crystallinity and a low crystallization rate which has as result the appearance of large spherulites. The addition of CNCs to PHB has a strong influence on its crystallization behavior and its crystalline structure. Thus, CNCs acted as nucleating agents, facilitating the PHB crystals growth and leading to the formation of greater lamellar thickness spherulites during the casting process [98]. This was proved by an increase in the melting temperature (T_m) of PHB following CNCs addition, observed in the differential scanning calorimetry (DSC) scans. A slight increase in the melting temperature of PHB, similar to that determined by the CNCs, was observed in nanocomposites with different nanocelluloses [86,99]. Thus, an increase of T_m with about 4 °C was recorded for PHB/CNFs nanocomposites with a CNFs loading of 5 wt% [89].

The nucleating effect of CNCs was also observed through the increase in the crystallization temperature (T_c) of PHB from around 60 °C for the pure PHB to approximately 83 °C for the PHB/CNCs nanocomposites [98]. The concentration of CNCs in PHB nanocomposites also influenced the T_c. An increase of T_c with about 18 °C was noticed for the PHB/CNCs nanocomposite with 1 wt% CNCs compared to neat PHB, showing the nucleating effect of CNCs that reduced the nucleation energy, increasing the PHB nucleating rate and facilitating its crystallization [89]. However, a lower increase of T_c with only 7.6 °C was observed at a CNCs loading of 5 wt% in nanocomposite. This was due to the CNCs tendency to form aggregates in the PHB matrix when incorporated in higher concentrations, which decreased their nucleating effect [89].

A low influence of CNCs on the degree of crystallinity of PHB was in general reported [86,98]. A slight increase in crystallinity (X_c) was noticed in PHB/CNCs nanocomposites with 5 wt% CNCs as compared to the pure PHB and the nanocomposites with lower content of CNCs [89]. The lower crystallinity of PHB/1 wt% CNCs nanocomposite was determined by the homogeneous dispersion of the rigid CNCs in the PHB matrix, which obstructed the segmental movement of the PHB chains, hindering the PHB crystallization [89].

The addition of a plasticizer in the PHB/CNCs nanocomposites may lead to an increase in the degree of crystallinity [100,101]. Thus, an increase in the X_c from 55.4% for the pure PHB to 62.7% for the PHB/CNCs nanocomposite containing 4 wt% CNCs and 20 wt% low molecular weight tributyrin as plasticizer was observed [101]. Therefore, the nucleating activity of CNCs was enhanced when TB was used as plasticizer. Similarly, the addition of 10 wt% epoxidized canola oil plasticizer in the PHB/PLA blend favored the increase of crystallinity, from 23.5% in the blend to 42.4% in the plasticized blend, although the addition of CNCs to the plasticized PHB/PLA blend did not lead to a further increase in the crystallinity degree [116]. This was supposed to be due to the poor dispersion of the CNCs in the PHB/PLA blend, which prevented the nucleation and crystallization.

The surface modification of CNCs also led to a reduction in the degree of crystallinity [110,113]. This was observed in the PHB nanocomposite with CNC-g-PLA vs. the nanocomposite containing untreated CNCs [113], as well as in the case of PHBV/PHCNs nanocomposites, where the CNCs grafted with PHBV were well dispersed in the PHBV matrix [110]. The treatment of CNCs ensured a good entanglement between the chains of the PHBV matrix and those grafted on the cellulose nanocrystals, as well as the formation of numerous hydrogen bonds interactions between the PHBV matrix and PHCNs. All these hindered the PHBV crystallization, leading to the formation of imperfect PHBV crystals and to a reduction in the crystallinity degree [110]. This idea was supported by the progressive increase in the glass transition temperature of the nanocomposites with the increase of PHCNs content that was assigned to the numerous hydrogen bonds and entanglements formation, which prevented the segmental movement of the PHBV chains, making the PHB/PHCNs nanocomposites more rigid as compared to the pristine PHBV [110].

The thermal stability of cellulose nanofillers strongly depends on the methods used to obtain them. Acid hydrolysis of cellulose using concentrated acids and, especially, sulfuric acid, is a harsh treatment leading to the degradation of the disordered amorphous regions and a strong decrease of the degree of polymerization of cellulose [118]. As a result, the good thermal stability of pure cellulose, which begins to decompose after 300 °C [119], may be strongly decreased, with an effect on the thermal stability of PHB. Different influences of CNCs on the thermal stability of PHB were reported [89,90,98]. Thus, a decrease in the thermal stability of PHB, corresponding to a reduction in the temperature of the maximum degradation rate (T_d) of around 20 °C, was observed after CNCs' addition [98]. This was considered as an effect of the sulfate groups on the CNCs surface, which induced and accelerated the scission of the PHB polymer chains. On the contrary, an improved thermal stability consisting of an increase in T_d from 286.5 °C for pure PHB to 291.7 °C for the nanocomposite with 5 wt% CNCs was observed [89]. A similar increase was reported for the PHBV/CNCs nanocomposites, from 278.6 °C for the pure PHBV to 284.3 °C for the nanocomposites with 7 wt% CNCs [90]. This was assigned to the hydrogen bonds formed between the carbonyl groups of PHB and the hydroxyl groups of CNCs, which slowed down the random chains breaking in PHB, improving its thermal stability [89]. Although no information on the thermal stability of the CNCs was disclosed in these two studies [89,98], the different processes used to obtain the CNCs and to incorporate them in the PHB matrix could influence these results. Thus, it is possible that the solvent exchange process [89] could contribute to a better thermal stability of the CNCs than the freeze drying process [98] along with the different solvents used in the solvent casting processes, requesting different conditions for their removal.

A different thermal stability of the PHB/CNCs nanocomposites depending on the concentration of CNCs was also reported [95]. Thus, a better thermal stability was observed for the nanocomposite with 3 wt% CNCs as compared to the pure PHB but a poor one for the nanocomposite with 5 wt% CNCs. Kinetic studies on the thermal degradation process of the nanocomposites showed that the rate of degradation decreased with an increase in the CNCs content up to about 3 wt% and increased with the further increase of CNCs concentration in nanocomposites [95]. This complex behavior was explained by the fact that, at CNCs concentration of 3 wt%, a good dispersion of the CNCs in the PHB was obtained. In this case, the formation of hydrogen bonds between the nanofiller and the polymer matrix hindered the movement of the chain segments of PHB, which impeded the diffusion of volatile products through the nanocomposites, enhancing their thermal stability. Conversely, the formation of CNCs aggregates in the nanocomposites with 5 wt% CNCs altered the thermal stability of PHB as they are believed to act as nucleation sites for the chain scission degradation of PHB at higher temperatures [95]. It was concluded that the thermal degradation mechanism of the PHB/CNCs nanocomposites followed first order reaction and phase boundary controlled reaction models with a random chain scission mechanism characteristic of most polyesters.

Although the addition of plasticizers generally decreased the thermal stability of PHB due to their high volatility, which reduced the onset degradation temperature [101], a favorable influence was observed after the addition of 15 wt% PEG to PHB/CNCs nanocomposites with low content of nanocrystals [102]. The highest increase in the T_d, of about 27 °C, was obtained for the PHB/PEG/CNCs nanocomposite with 0.45 wt% CNCs. This was assigned to the better adhesion between the CNCs and PHB matrix, ensured by the PEG coating which, at this CNCs concentration, covers almost the entire surface of the CNCs [102]. It has been shown that the presence in the nanocomposites of a high molecular weight poly(adipate diethylene) plasticizer instead of a low molecular weight one (TB), in a 20 wt% concentration relative to PHB, considerably increased the maximum degradation temperature compared to the pure PHB, PHB/CNCs nanocomposites and the nanocomposites containing the same amount of TB plasticizer [101].

Surface functionalization of CNCs had also a favorable influence on the thermal stability of PHBV [110]. In particular, CNCs grafted with PHBV (PHCNs) strongly increased the thermal stability of PHBV, an improvement in the T_d of about 35 °C being noticed at a concentration of PHCNs in PHBV of 20 wt%. This improvement was explained by the strong interactions established between the functionalized CNCs and the PHBV matrix as well as the entanglements formed between the PHBV chains of the polymer matrix and the PHBV chains attached to the CNCs surface. Due to the significant increase in the initial decomposition temperature, which approached 250 °C, all PHBV/PHCNs nanocomposites showed higher melt-processing windows as compared to the neat PHBV matrix [110]. However, only a slight increase of T_d was observed after the addition to PHBV of CNCs surface modified with butyric acid, lactic acid, and their mixture [109]. On the contrary, the incorporation of CNCs into the PHB/PCL blends led to a significant decrease in the thermal stability of PHB, probably due to the lower thermal stability of CNCs [115]. The most important thermal properties of PHB/CNCs nanocomposites are summarized in Table 1.

Table 1. Thermal and mechanical properties of selected PHB/CNCs nanocomposites.

Polymer Matrix	CNCs Loading (wt%)	Functionalization/ Additives	Processing Method	Thermal Properties					Mechanical Properties				References
				T_{max} (°C)	T_m (°C)	T_g (°C)	X_c (%)	Tensile Strength (MPa)	Young's Modulus (MPa)	Elongation at Break (%)	Sorage Modulus* (MPa)		
PHBV	2	-	Melt compounding	~280	168.3	-	55.6	36.34 ± 0.50	603.91 ± 22.77	15.35 ± 0.66	-	[90]
	5			~282	168.8	-	57.6	33.28 ± 0.57	645.62 ± 22.12	11.66 ± 0.67	-	
	7			284.3	171.4	-	59.0	32.01 ± 0.50	792.98 ± 23.56	10.04 ± 0.49	-	
PHB	2	L-lactide graft polymerization/-	Melt compounding	-	153.9/166.0	-	-	-	-	-	-	[113]
PHB	2.4	-/PVAc		293.49	~160/170	-	-	20 ± 2	1100 ± 95	17 ± 1	-	[114]
	4.8			294.10	~161/170	-	-	25 ± 2	1170 ± 50	16 ± 1	-	
PHB	2.4	-/PEG	Masterbatch melt compounding	294.53	~162/171	-	-	11 ± 1	1200 ± 69	2	-	
PHBV	2.4	-/PVAc		276.20	~150/160	-	-	25 ± 1	1320 ± 192	12 ± 1	-	
	4.8			286.49	~150/160	-	-	32 ± 2	1300 ± 243	51 ± 2	-	
PHBV	2.4	-/PEG		290.77	~153/162	-	-	13 ± 1	1270 ± 127	2	-	
PHB	3	-/PCL	Solvent casting and melt compounding	257.0	161.7/168.0	-	45.7	14.5 ± 0.7	902 ± 56	4.1 ± 0.4	-	[115]
	5			251.0	151.7/160.1	-	41.0	9.3 ± 1.4	1066 ± 65	1.8 ± 0.9	-	
PHBV	2	-/PEG	Solvent casting	-	~134/151	-	-	15.5	1100	7.1	~1900	[108]
	5			-	~132/150	-	-	26.1	1760	7.8	~2500	
PHBV	2	-/PEG	Melt compounding	-	-	-	-	~25.2	~1700	~4	-	
	5			-	-	-	-	~24.8	~1900	~3.4	-	
PHB	2	-	Solvent casting	-	171.2	-	47.7	-	-	-	-	[96]
	5			-	172.0	-	57.8	-	-	-	-	
PHBV	2	butyric acid/-	Solvent casting	277	171	-	50	-	-	-	4788	[109]
	5			285	170	-	44	-	-	-	3869	
PHBV	2	lactic acid/-		282	170	-	46	-	-	-	3283	
	2	butyric acid and lactic acid/-		275	170	-	51	-	-	-	3496	
PHBV	2	-	Solvent casting	-	~132/150	~-1.7	-	~35.5	~1850	-	~1100	[97]
	4.6			-	~124/142	~-4.2	-	~19	~1790	-	~1380	
PHB	2	-	Solvent casting	~288	160.6/170.8	-	53.6	29.7 ± 3.6	2400 ± 300	1.5 ± 0.2	-	[86]
	4			~290	160.5/170.4	-	53.7	29.2 ± 1.8	2500 ± 200	1.5 ± 0.2	-	

Table 1. Cont.

Polymer Matrix	CNCs Loading (wt%)	Functionalization/ Additives	Processing Method	Thermal Properties				Mechanical Properties				References
				T_{max} (°C)	T_m (°C)	T_g (°C)	X_c (%)	Tensile Strength (MPa)	Young's Modulus (MPa)	Elongation at Break (%)	Sorage Modulus * (MPa)	
PHB	2	-/glyceryl tributyrate (TB)	Solvent casting	~270	154.5/169.2	-	62.7	17 ± 2	1500 ± 100	1.9 ± 0.3	-	[99]
	4			~274	154.7/169.6	-	59.8	19 ± 2	1800 ± 100	1.6 ± 0.3	-	
PHB	2	-/poly[di(ethylene glycol) adipate] (A)	Solvent casting	~287.5	155.3/169.4	−13.6	57.1	15 ± 1	1200 ± 100	2.4 ± 0.5	-	
	4			~293	154.1/168.9	−13.9	56.8	14 ± 1	1100 ± 100	2.0 ± 0.4	-	
PHB	2	-	Solvent casting	~275	~162/172	-	55.4	~24.4	~1930	~1.77	-	[98]
	4			~271	~162/172	-	55.3	~25.3	~2050	~1.8	-	
PHB	0.15	-/PEG	Solvent casting	314	160	-	-	~17.1	~185	~180	-	[102]
	0.45			316	162	-	-	~17.5	~80	~300	-	
PHBV	5	PHBV grafting/-	Solvent casting	252.6	90.5/112.2/129.6	−16.9	-	~20	~300	~6.2	-	[110]
	20			279.5	99.5/121.3/136.7	−12.1	-	~33	~43	~7	-	
		-	Melt compounding	~308	176	-	39.6	28.1 ± 1.9	1518 ± 55	3.39 ± 0.18	1800	[116]
PHB	5	-/polylactic acid (PLA) and epoxidized canola oil (eCO)	Melt compounding	~317	171	-	43.4	24 ± 1	1604 ± 66	2.93 ± 0.16	1590	
P3HB/P4HB	5	tetraethyl orthosilicate (TEOS) and methyltrimethoxysilane (MTMS) double silanization/-	Melt compounding	-	158	~−17	-	19.2	5.52	301	-	[111]
PHB	3	-	Solvent casting	290.2	170.9	-	64.2	~25.8	~1890	~4.5	-	[89]
	5			291.7	171.0	-	67.1	~22.9	~2020	~2.6	-	
PHB	5	-	Electrospinning from solution	-	183.2	-	-	4.2 ± 0.3	4931.3 ± 245.4	19.4 ± 1.7	-	[91]
	8			-	184.7	-	-	4.4 ± 0.2	5120.1 ± 215.3	18.3 ± 1.2	-	
PHB	2	-	Solvent casting	-	-	-	-	57.1 ± 1.2	870 ± 20	6.5 ± 1.0	-	[95]
	5			-	-	-	-	45.2 ± 1.8	1410 ± 70	3.2 ± 0.8	-	

* Storage modulus value at room temperature.

5.2. Mechanical Properties

The reinforcing effect of CNCs in PHB nanocomposites is based on the remarkable mechanical properties of cellulose nanofillers [120,121]. The Young's modulus of a cellulose whisker calculated from the shift in the characteristic band at 1095 cm^{-1} in the Raman spectrum was 143 GPa [120], while the transverse Young's modulus of a nanocellulose film determined by PeakForce Atomic Force Microscopy was 19 GPa [121]. Similar to other biopolymers, the mechanical properties of PHB are weaker than those of most common thermoplastic polymers; therefore, the addition of cellulose reinforcements can bring significant gains in stiffness and other mechanical properties. However, the mismatch between the PHB hydrophobicity and the strong hydrophilicity of CNCs prevents the good dispersion of CNCs in the polymer matrix and reduces the efficiency of the load transfer at the polymer–fiber interface. As a result, the addition of CNCs in PHB did not always lead to a strong improvement in strength and stiffness [90,106,107].

The addition of CNCs in PHB generally determined a continuous increase in the Young's modulus with an increase in the CNCs concentration in nanocomposites [89,98]. This effect was assigned on the one hand to the rigid nature of the CNCs, which potentiated the PHB stiffness and increased the degree of crystallinity of PHB when present in higher amounts in the PHB matrix [89] and, on the other hand, to the homogeneous dispersion of the CNCs in the PHB matrix due to the possible hydrogen bonding interactions between CNCs and PHB [86]. A different effect of CNCs on the tensile strength of PHB/CNCs nanocomposites was reported, depending on the dispersion of the CNCs in the PHB matrix. A continuous improvement in the tensile strength of PHB with increasing CNCs content between 2 and 6 wt% was observed in nanocomposites with a homogeneous dispersion of the CNCs [98] while a decrease in the tensile strength with increasing CNCs content from 1 to 5 wt% was achieved for the nanocomposites where CNCs agglomerates were pointed out by SEM [89]. This may be due to the fact that CNCs agglomerates acted as stress concentrators from where the cracks were initiated. The presence of rigid CNCs in higher concentration in the PHB matrix and possible fibers agglomeration were also reflected in a decrease of elongation at break [89,98]. Curiously, an increase of the elongation at break was observed in the nanocomposite with 1 wt% CNCs [89]. At low CNCs loadings, the homogeneous dispersion of the CNCs in the PHB matrix led to a restriction of the segmental movement of the PHB chains and impeded the PHB crystallization causing an increase in the elongation at break [89].

A different mechanical behavior was reported for the PHBV/CNCs nanocomposites [90,97]. In nanocomposites containing 0.5–4.6 wt% CNCs, an increase in the tensile strength and Young's modulus with an increase in the CNCs concentration, followed by a leveling off from a content of CNCs of 2.3 wt%, was noticed [97]. Therefore, the good dispersion of CNCs in the PHBV matrix at loadings of up to 2.3 wt% ensured a large interface area between PHBV and CNCs and the enhancement of the mechanical properties, while the CNCs agglomerates, which were formed at higher concentrations of CNCs, were probably the main cause of the mechanical properties' decline after the 2.3 wt% concentration threshold [97]. In PHBV/CNCs nanocomposites with 1–7 wt% CNCs obtained by melt compounding [90], the increase in the concentration of CNCs determined a gradual increase in the Young's modulus, along with a gradual decrease in the tensile strength and elongation at break. The decrease in the tensile strength highlighted the lack of adhesion between CNCs and PHBV, while the drop in elongation showed the reduced deformability at PHBV/CNCs interface [90].

The influence of CNCs in large amounts, from 5 to 22 wt%, on the mechanical properties of PHB/CNCs nanocomposite fibers prepared via electrospinning highlighted the presence of a more favorable concentration of CNCs for which the improvement of the mechanical properties of the nanocomposite fibers was maximum [91]. Accordingly, the largest improvement in tensile strength (by 13%) and elastic modulus (by 6%) was observed for a content of CNCs of 8 wt%. The formation of CNCs agglomerates which acted as stress concentrators and the weaker PHB-CNCs interface were considered responsible for the decrease in the mechanical properties of the nanocomposite fibers with high CNCs content [91].

The addition of a plasticizer together with CNCs was also tested as a method to improve both the flexibility-processability and the stiffness of PHB [99,102]. However, in most cases, no important changes or a decrease in the mechanical properties was obtained due to the diametrically opposite effects of the plasticizer and the reinforcing agent: while the reinforcing agent improves the PHB's stiffness, increasing its elastic modulus and tensile strength and reducing its elongation at break, the plasticizer increases the elongation at break at the expense of the stiffness. A small variation of the Young's modulus, along with a slight reduction in the tensile strength and strain at break, were observed for the PHB/TB/CNCs nanocomposites containing 20 wt% glyceryl tributyrate and 2 or 4 wt% CNCs as compared to neat PHB [99]. The highest values of Young's modulus and tensile strength were obtained for the PHB/TB/CNCs composite with 4 wt% CNCs, showing that TB may enhance CNCs dispersion in the PHB matrix. This was due to the good compatibility between TB and PHB and the hydrogen bonding interactions between TB and CNCs. However, using a high molecular weight plasticizer such as poly [di (ethylene glycol) adipate] instead of TB in these nanocomposites had a detrimental effect on the mechanical properties, probably due to the lower miscibility between this plasticizer and the PHB matrix which didn't allow a good dispersion of CNCs in PHB [99]. When the PHB/TB/CNCs nanocomposite film was used to obtain PHB/TB/CNCs-paperboard bilayer composites with a PHB content of 15 wt% related to the paperboard weight, 20 wt% TB plasticizer and 4 wt% CNCs relative to the PHB weight, an improvement in the tensile strength, Young's modulus, and elongation at break compared to neat PHB was obtained [100]. In addition, peeling tests were undertaken to assess the adhesive strength between layers, the paperboard and the PHB/TB/CNCs film. During these tests, the PHB/TB/CNCs layer was peeled off from the paperboard surface with broken cardboard fibers adhering to its surface. This indicated that the PHB/TB/CNCs-paperboard adhesion was higher than resistance of the fibers in the paperboard due to the PHB penetration between the paperboard fibers during the compression molding [100].

A special effect was reported in the case of PHB/PEG/CNCs nanocomposites containing PEG plasticizer (15 wt%) and low amounts of CNCs (0.022–0.75 wt%) [102]. The addition of CNCs in PHB/PEG led to a significant reduction in both tensile strength (from 30 MPa to 22 MPa) and elastic modulus (from 1180 MPa to 537 MPa) of PHB, simultaneously with an increase in the elongation at break. The most important increase of flexibility was noticed in the nanocomposites with 0.45 wt% CNCs, where the total surface of CNCs was entirely covered by PEG and a 25-fold increase in the elongation at break was obtained [102]. Therefore, the CNCs slippage and the alignment of the PHB macromolecules during the plastic deformation can activate the shear flow of the polymer matrix, so that the materials can withstand greater deformations without a rapid catastrophic fracture. As a result, CNCs may act as a reinforcing agent only for higher amounts of CNCs in PHB/PEG [102]. Similarly, a decrease in the tensile strength of the PHB/PLA/eCO/CNCs nanocomposites of about 23% and a slight increase of the Young's modulus of about 10% as compared to the PHB/PLA blend were observed after the addition of epoxidized canola oil (10 wt%) as a plasticizer and CNCs (5 wt%) as a reinforcing agent in a PHB/PLA (3:1) blend (Figure 4) [116]. The small efficiency of the CNCs on the mechanical properties of the plasticized blend was attributed to the inhomogeneous dispersion of CNCs in the presence of the eCO plasticizer and the formation of CNCs agglomerates which initiated the rupture of the material [116].

When PHBV-grafted CNCs were used as reinforcing agents in PHBV-based nanocomposites, a strong enhancement of the mechanical properties of PHBV, consisting in an increase in the tensile strength, Young's modulus, and elongation at break, was reported over the entire concentration range [110]. This increase was attributed to the surface modification of CNCs, which ensured a good adhesion between PHCNs and PHBV both due to the entanglements between the chains of the PHBV matrix and the PHBV chains grafted on the CNCs surface, and the formation of hydrogen bonding interactions between the PHBV matrix and PHCNs. The largest improvements in the tensile strength, Young's modulus, and elongation at break, by 113%, 95%, and 17%, were obtained for the PHBV/PHCNs nanocomposite with 20 wt% PHCNs [110]. A less pronounced reinforcing effect on the mechanical properties of PHBV was obtained when the CNCs were surface modified with butyric acid, lactic acid, or their mixture [109]. In this case, an improved storage modulus at temperatures higher than the T_g of PHBV was reported only for the nanocomposites containing butyrate-modified CNCs, where the improved interface between PHBV and CNCs_butyrate led to increased T_g and tan δ values [109]. The shift of the glass transition and the improved dumping showed a better stress transfer between the PHBV matrix and CNCs_butyrate determined by the good dispersion of CNCs_butyrate in the PHBV matrix, and to the similarity between the butyrate moieties attached on the CNCs surface and the PHBV matrix [109].

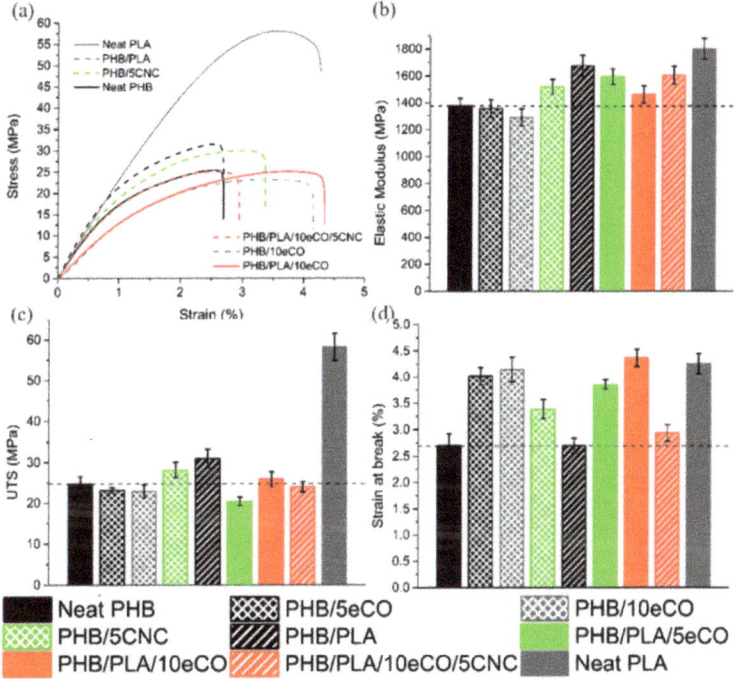

Figure 4. (a) Stress–strain curves of PHB blends and nanocomposites (b) Elastic modulus (c) Tensile strength at break, and (d) Elongation at break for neat PHB, neat PLA, PHB/5eCO and PHB/10eCO, PHB/PLA, PHB/PLA/5eCO and PHB/PLA/10eCO blends and nanocomposites (PHB/CNCs and PHB/PLA/10eCO/CNCs) [116].

A different effect of CNCs on the mechanical properties of a PHB/P4HB mixture was reported after the surface treatment of CNCs with silanes [111]. The addition of 5 and 10 wt% freeze-dried double-silanized CNCs (TMCNCs) in PHB/P4HB determined a strong increase of the elongation at break along with a gradual decrease of the tensile strength and a decrease of the Young's modulus only for the higher content of TMCNCs in nanocomposites. The strong increase of the flexibility was due to the action of TMCNCs as a plasticizer and nucleating agent, promoting the formation of smaller spherulites and enhancing the dispersibility of the surface treated CNCs in the polymer matrix [111]. The most important mechanical properties of PHB/CNCs nanocomposites are summarized in Table 1.

5.3. Barrier Properties

The barrier properties against water, light, gases or volatiles are essential when the PHB/CNCs nanocomposites are intended for food packaging application. Compared to other biopolymers, PHB shows better barrier properties due to its hydrophobicity, high crystallinity, and stereo-regularity [10], and the addition of CNCs may further improve these properties. Migration studies using nonpolar and polar food simulants were carried out on PHB/CNCs (1–5 wt%) nanocomposite films obtained by solvent casting [96] in order to determine if these substances can pass by diffusion or adsorption to a potential food product that was packaged using these nanocomposite films. The PHB/CNCs nanocomposites were subjected to migration studies according to the Commission Regulation EU No 10/2011: (i) in ethanol 10% (v/v) at 40 °C for 10 days and (ii) in isooctane at 20 °C for 2 days. When ethanol was used as a food simulant, the overall migration of the PHB/CNCs nanocomposites decreased as compared to the pristine PHB at low CNCs loadings, of 1 and 2 wt%. However, at CNCs loadings greater than 3 wt%, the migration increased significantly, reaching a maximum value for the nanocomposites having 5 wt% CNCs. At low CNCs contents, the strong interactions between the CNCs and PHB, which allowed a good dispersion of the CNCs in the PHB matrix and led to restrictions of the movement of the polymer chains, hindered the migration in ethanol. At higher CNCs contents, CNCs tended to agglomerate, and the adhesion between CNCs and PHB was considerably decreased, which left free way for the migration of the nanocomposites' components in the simulant. The same allure of the migration levels as in the case of ethanol was observed when isooctane was used as food simulant, although the overall migration was much reduced [96]. The maximum migration values of 175 µg/kg in ethanol and 40 µg/kg in isooctane, obtained for the nanocomposites with 5 wt% CNCs, remained, however, far below the standard migration limits of 60 mg/kg established by the EU current legislation. Therefore, the migration tests showed that the PHB/CNCs nanocomposites are safe for using as food packaging materials [96].

When the PHB/CNCs nanocomposites are intended for packaging fresh produce, a high barrier to oxygen is important [10]. It was observed that the incorporation of low CNCs loadings, of 1 and 2 wt%, into the PHB matrix led to a significant reduction in the oxygen transmission rate (OTR), with 46% and 65% [96]. This was ascribed to the good PHB/CNCs interface established as a result of the hydrogen bonding between CNCs and PHB and the formation of a CNCs network inside the polymer matrix, which reduced the permeation of oxygen through the PHB/CNCs nanocomposites [96]. Moreover, a three-times reduction of the OTR was obtained in PHB/CNCs nanocomposites containing 3 wt% CNCs [89]. A schematic representation of the tortuous gas diffusion path in the case of a PHB/CNCs nanocomposite compared to pristine PHB is shown in Figure 5.

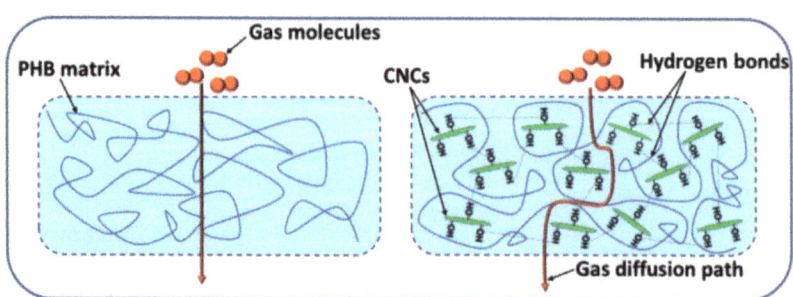

Figure 5. A schematic representation of the tortuous gas diffusion path in the case of a PHB/CNCs nanocomposite compared to pristine PHB.

Good water vapor barrier properties are required for the PHB/CNCs nanocomposites intended to be used in the food packaging industry in order to prevent premature loss of flavor or spoilage of the food products [89]. PHB is characterized by a higher water vapor permeability (WVP) as compared to that of common polymers used in packaging, which may raise problems when baked products are packaged [10]. However, the addition of CNCs in PHB reduced its WVP [86,89,98]. Accordingly, a three-times reduction in WVP was reported for PHB/3 wt% CNCs nanocomposite as compared to neat PHB, although no further decrease was observed for a higher content of CNCs in the nanocomposites [89]. In the case of the PHB/CNCs nanocomposites obtained by solvent casting, a similar decrease in the WVP was reported after the addition of 2 wt% CNCs in PHB and no variation in the WVP value for higher CNCs amounts up to 6 wt% [98]. The decrease of the water vapor permeation through the nanocomposites films was determined by the CNCs, which acted as physical barriers hindering the diffusion of water molecules [89]. The differences observed in the variations of the WVP with the amount of CNCs may be determined by the good dispersion vs. agglomeration of CNCs, the adhesion to the PHB matrix, porosity of the films, and other factors [98]. Moreover, it was observed that the addition of a plasticizer (20 wt% TB) increased the water vapors permeation through the PHB/CNCs films, thus reducing their ability to protect the packaged food from humidity during shelf life [99]. In such nanocomposites, the plasticizer increased the free volume in the polymer, favoring water vapors diffusion and reducing the moisture barrier properties of the nanocomposites, while the CNCs acted as a barrier, increasing the tortuosity of the water vapor diffusion path in the materials [99] (Figure 5). A significant reduction in the water vapor permeability of the cellulose paperboard, with around two orders of magnitude, was observed after coating it with plasticized PHB/CNCs [100]. This was assigned to the CNCs homogeneously dispersed in the PHB matrix, acting as physical barriers to the diffusion of water vapor molecules.

An important characteristic of the PHB/CNCs nanocomposites intended for the food-packaging industry is their transparency, which allows seeing the packaging content. The transmittance of the films determined by UV-visible spectroscopy is an important characteristic showing not only the dispersion of CNCs in the PHB matrix but also the compatibility of the phases in the composite material [89]. In general, well dispersed CNCs should not significantly influence the transparency of the PHB. Indeed, in low concentration, CNCs did not change the transmittance of the PHB film; however, from a content of 5 wt% CNCs [89] or 2 wt% CNCs [96] in the nanocomposites their transparency declined. The surface treated CNCs allowed the addition of higher contents of CNCs in PHBV without a significant decrease in the UV-visible transmittance, [110] or even an increase in transparency compared to neat PHBV [109]. An increase of transparency was also reported after the addition of CNCs in the PHB/PCL blend [115].

5.4. Biodegradability

PHB is biodegradable in soil and water according to following ISO and ASTM biodegradation standards: ISO 14855 (controlled industrial and home composting conditions), ISO 15985 (anaerobic digestion), ISO 17556 (soil biodegradation test), ISO 14851 (biodegradation in aerobic, aquatic conditions), ISO 14853 (aqueous anaerobic biodegradation test) and ASTM D6691 (marine biodegradation) [122].

It was observed that coating cellulose paperboard with PHB/CNCs nanocomposites did not significantly change the biodegradation rate of paperboard under composting conditions [100]. Thus, the plasticized PHB/CNCs-cellulose paperboard bilayer composites showed a biodegradation behavior quite similar to that of the neat paperboard, a disintegration degree higher than 90% being achieved six weeks after they were incubated for biodegradation under composting condition. However, in the fifth week of the biodegradation test, a certain decrease in the biodegradation rate was observed for the plasticized PHB/CNCs-paperboard composites as compared to the uncoated paperboard. This was determined by the increased crystallinity degree of PHB and the formation of higher ordered

PHB crystals following the CNCs and TB addition, which were much more resistant to biodegradation than the amorphous regions in the material [100]. Similarly, the addition of CNCs in PHB did not affect the good biodegradability of PHB [101]. Thus, the PHB/CNCs nanocomposites were disintegrated in a high proportion (90%) in about 21 days under composting conditions. At a low CNCs loading of 2 wt%, the PHB/CNCs nanocomposite containing TB plasticizer showed a lower degradation rate compared to pristine PHB in the first 14 days. This was probably due to the stiffening effect of the CNCs, which hindered the diffusion of water into the material, slowing down the attack of microorganisms, together with the higher PHB crystallinity, determined by the addition of the CNCs. As the CNCs loading increased to 4 wt%, the nanocomposites presented a much higher degradation rate, even in the first 14 days. This was due to the increased hydrophilicity and water absorption of the nanocomposites at a higher CNCs content, which favored the degradation of PHB by hydrolysis. Overall, in the case of PHB/CNCs nanocomposites where TB was used as a plasticizer, the effect of CNCs on biodegradation was stronger than that of the plasticizer. However, the type of plasticizer affected the biodegradation rate. In the case of PHB/CNCs nanocomposites containing a high molecular weight plasticizer, a rather low degradation rate was noticed, the nanocomposites having a similar biodegradation behavior with PHB. Therefore, in this case, the CNCs incorporation did not lead to an acceleration of the material biodegradation, despite their hydrophilicity [101]. When the CNCs were added in a PHB/PCL blend, the biodegradability of PHB/PCL/CNCs nanocomposites tested under composting conditions was also influenced by the presence of CNCs and their concentration [115]. It was observed that an increase in the CNCs content led to an increase in the biodegradation rate of the PHB/PCL/CNCs nanocomposites as compared to the PHB/PCL blend and neat PHB. This was assigned both to the increased hydrophilicity as a result of the CNCs addition, which promoted and accelerated the hydrolytic degradation of PHB, and to the decrease in the degree of crystallinity of PHB as a result of PCL and CNCs addition [115].

5.5. Biocompatibility

As combinations of two biocompatible materials, the PHB/CNCs nanocomposites may be useful for biomedical applications. Choi et al. obtained PHB/CNCs nanocomposite foams with different contents of CNCs by employing a nonsolvent-induced phase separation technique [106]. Regardless of whether THF or Diox was used as a nonsolvent, cytotoxicity tests showed that the resulted PHB/CNCs nanocomposites showed sufficiently high cell viability (more than 80%) for at least 4 days, which render them safe for potential biomedical application [106]. Furthermore, an enhanced cell adhesion was observed after CNCs addition to PHBV [110]. The cytocompatibility of the PHBV/PHCNs nanocomposites was evaluated by culturing human MG-63 cells on their surface. It was found that the addition of PHBV-grafted CNCs (PHCNs) in PHBV increased the MG-63 cells growth. Moreover, an increase in the number of living cells from the PHBV/PHCNs nanocomposites surface was observed as the content of PHCNs in the PHBV/PHCNs nanocomposites increased (Figure 6). This was explained by the higher hydrophilicity of the nanocomposites as compared to the neat PHBV which promoted cells attachment and proliferation. Therefore, the addition of PHBV-grafted CNCs improved cell−matrix interactions and the adhesion of MG-63 cell on the surface of nanocomposites [110].

The PHB/CNCs electrospun mats were also characterized by in vitro cytotoxicity tests using a direct contact test with L-929 cells and MTT assay [91]. The PHB/CNCs nanocomposite mats with a CNCs content of 22 wt% were proven to be noncytotoxic, supporting the viability and proliferation of the L-929 fibroblasts cells. In addition, a significant increase in the water absorption capacity of PHB, which is known to be a hydrophobic material, was observed with increasing the CNCs content in the PHB/CNCs nanocomposites mats. This was due to the presence of the numerous hydroxyl (-OH) groups on the CNCs surface, which were able to establish hydrogen bonding interactions with water molecules. Consequently, an improvement in the percentage of water absorption of around 311% was observed in the case of the PHB/22 wt% CNCs nanocomposite mats as compared to the neat PHB mats. Therefore, PHB/CNCs nanocomposites mats are more suitable as materials for scaffolds or other biomedical applications as compared to the pure PHB [91].

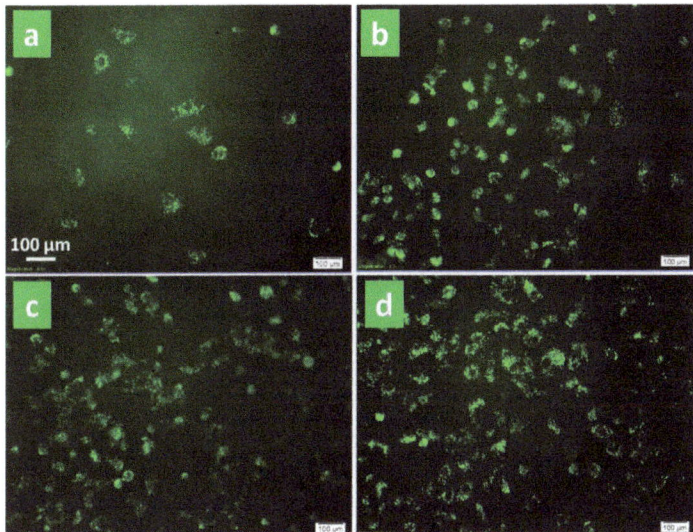

Figure 6. Fluorescence images showing the attachment of MG-63 on neat PHBV (**a**) and PHBV/PHCNs nanocomposites with 10% (**b**), 20% (**c**), and 30% (**d**) PHCNs. Reprinted with permission from Ref. [110]. Copyright 2014 American Chemical Society.

6. Applications of PHB/CNCs Nanocomposites

Due to their biodegradability and improved thermal and mechanical features, PHB/CNCs bionanocomposites have a great market potential, especially in the medical and packaging sectors (Figure 7). In addition to the "green" character, the incorporation of CNCs derived from cheap natural resources in PHB can reduce the production costs, one of the important barriers to the widespread marketing of PHB. When intended for use as packaging materials, PHB/CNCs nanocomposites must satisfy certain 'essentialrequirements' like easy processing, transparency, high elongation at break and good barrier properties. Moreover, when packages reach the end of their life, the waste must ultimately decompose into mainly carbon dioxide, biomass, and water in order to have a minimal impact on the environment [123]. A major sector of food packaging is represented by the short-life packaging where biodegradability represents a compulsory characteristic. Showing a rapid biodegradation, which is in general enhanced by the addition of CNCs, PHB nanocomposites are suitable biomaterials for short-life packaging. Besides the enhanced flexibility and processability, the food packages should also show specific barrier properties. It was observed that PHB/PCL/3 wt% CNCs nanocomposites obtained through solvent casting followed by extrusion and compression molding showed balanced mechanical properties and good UV barrier properties, along with high transparency and degradation rate [115]. Therefore, the nanocomposite with 3 wt% CNCs is promising for packaging application. Luzi et al. studied the suitability of PLA/PHB/CNCs films plasticized with an oligomeric lactic acid and containing carvacrol, a natural antimicrobial additive, for food packaging [124]. The extruded films showed a good Young's modulus and an excellent elongation at break (430%), along with a reduction of the oxygen transmission rate following CNCs incorporation. Moreover, all formulations disintegrated in less than 17 days under composting conditions, showing a weight loss higher than 80% [124].

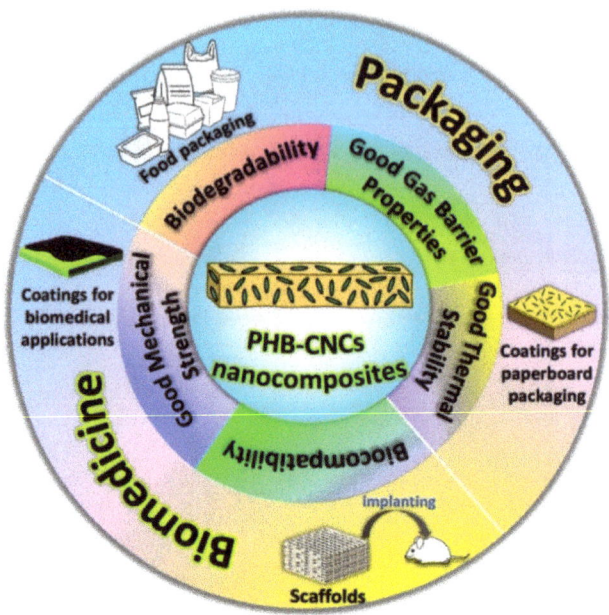

Figure 7. Schematic representation of PHB/CNCs nanocomposites' applications.

Although the addition of a plasticizer is needed for a better processability and flexibility of the PHB/CNCs films intended for packaging application, their addition can negatively influence the mechanical and barrier properties. Seoane et al. studied the possibility of using PHB/CNCs nanocomposites, plasticized with tributyrin or poly (diethylene adipate), as biodegradable packaging materials [101]. The investigation of the PHB/CNCs nanocomposites' biodegradability was performed under simulated composting conditions according to the ISO 20200 standard. The addition of small molecular weight plasticizers improved the dispersion of CNCs in the PHB matrix. Moreover, all the prepared PHB/CNCs nanocomposites showed a degree of disintegration of about 90% in 21–28 days after the beginning of the disintegration tests, which, together with their promising mechanical and thermal properties recommend them, as suitable candidates for the packaging industry [101].

Besides their use as films for packaging, the PHB/CNCs nanocomposites can also serve as coatings for paperboard or other hydrophilic substrates [100]. Thus, plasticized PHB/CNCs nanocomposites were used as layers for paperboard in order to improve its mechanical properties and to protect it against moisture. The new biodegradable and low cost composite materials were characterized in terms of barrier properties and disintegration in composting conditions for the evaluation of these materials as packaging. Although reduced after the addition of CNCs, the water vapor permeability of the paperboard coated by PHB/CNCs was still higher with about one order of magnitude compared to that of the same paperboard coated by polyethylene. However, the paperboard coated by PHB/CNCs achieved a much higher disintegration level (80%) than the neat paperboard (55%) after 35 days in composting conditions, which strongly support these composites as cheap packaging materials [100].

The application of PHB and PHB/CNCs in biomedicine has a great future. For the application in the biomedical field, the PHB/CNCs nanocomposites must possess two essential characteristics, namely biocompatibility (both PHB and CNCs are inherently biocompatible) and nontoxicity. However, each biomedical application may demand supplementary characteristics such as bioactivity, biodegradability, superior mechanical strength, or a certain degree of flexibility that may require the addition of other specific components to the PHB/CNCs formulations. In contrast to PLA, PGA, or PLGA, largely used in medical applications, the biodegradation of PHB and its copolymers in living tissues does not cause medium acidification [125]. The predominant degradation product of PHB, 3-hydroxybutyric acid, is much weaker than lactic acid, the main biodegradation product of PLA and PLGA; therefore, PHB and its nanocomposites do not cause chronic tissue irritation due to a decrease of pH, which is an extremely serious problem associated with PLA implants [125]. Up to now, only in vitro studies on PHB/CNCs nanocomposites were done [91], which demonstrated their lack of toxicity. However, in vivo studies are necessary for a more rapid application of PHB/CNCs nanocomposites as scaffolds and dressings.

Besides these applications, the PHB/CNCs nanocomposites may be useful in other fields such as encapsulation of medicine or fertilizers, and even electronics and other engineering applications.

7. Conclusions and Future Perspectives

The addition of CNCs in PHB is an efficient tool for designing fully biobased and biodegradable PHB/CNCs nanocomposites with improved properties. PHB/CNCs bio-nanocomposites proved to be suitable as replacements for fossils fuel-based materials because they have many properties comparable to that of the latter and, additionally, an insignificant carbon-footprint during processing and disposal. Even though PHB/CNCs nanocomposites have experienced a rapid advance in many fields and, especially, in packaging, there are still some scientific and technological challenges that remain open. One of them is the large improvement in interfacial adhesion and CNCs dispersion in the PHB polymer required for a strong enhancement of properties. Therefore, CNCs need to be efficiently functionalized in order to serve as effective reinforcing agents in PHB. In this respect, several approaches have been made, by using dispersing agents, compatibilizers, or grafting polymers compatible with PHB or PHBV on the CNCs' surface. However, it should be taken into account that all the surface treatments applied to CNCs will increase the costs of the final material. Therefore, as a future outlook, there is still a need for further advancements in the treatment of CNCs and in the methods used for obtaining PHB/CNCs based nanocomposites with finely dispersed CNCs. Concerning the applications of PHB/CNCs nanocomposites, more attention needs to be paid to the specific studies such as barrier properties, migration, toxicity, and biocompatibility, especially in vivo. The deepening of such specific tests would contribute to a great extent to a clearer outline of the behavior of these materials in terms of their application. In summary, although there are some barriers that need to be overcome, recent and continuous advances in the scalability of the industrial production of PHB/CNCs nanocomposites offer great promise and potential for their utilization in the packaging and biomedical sectors, while contributing to a sustainable future and economy.

Author Contributions: Conceptualization, A.N.F. and D.M.P.; writing—original draft preparation, C.D.U. and S.B.; writing—review and editing, A.N.F. and D.M.P.; project administration, A.N.F. All authors have read and agreed to the published version of the manuscript.

Funding: This research was funded by the Romanian Ministry of Research, Innovation, and Digitization, grant number 67TE/2020.

Data Availability Statement: Not applicable.

Acknowledgments: This work was supported by a grant from the Romanian Ministry of Research, Innovation, and Digitization, CNCS/CCCDI—UEFISCDI, project number 67TE/2020, within PNCDI III.

Conflicts of Interest: The authors declare no conflict of interest.

References

1. Ncube, L.K.; Ude, A.U.; Ogunmuyiwa, E.N.; Zulkifli, R.; Beas, I.N. Environmental impact of food packaging materials: A review of contemporary development from conventional plastics to polylactic acid based materials. *Materials* **2020**, *13*, 4994. [CrossRef]
2. Katiyar, V. *Bio-Based Plastics for Food Packaging Applications Paperback*; Smithers Rapra Technology: Shrewsbury, UK, 2017; ISBN 978-1911088363.
3. Coppola, G.; Gaudio, M.T.; Lopresto, C.G.; Calabro, V.; Curcio, S.; Chakrabordy, S. Bioplastic from renewable biomass: A facile solution for a greener environment. *Earth Syst. Environ.* **2021**, *5*, 231–251. [CrossRef]
4. Kurtela, A.; Antolović, N. The problem of plastic waste and microplastics in the seas and oceans: Impact on marine organisms. *Croat. J. Fish.* **2019**, *77*, 51–56. [CrossRef]
5. Sirohi, R.; Pandey, J.P.; Gaur, V.K.; Gnansounou, E.; Sindhu, R. Critical overview of biomass feedstocks as sustainable substrates for the production of polyhydroxybutyrate (PHB). *Bioresour. Technol.* **2020**, *311*, 123536. [CrossRef]
6. Lee, J.; Park, H.J.; Moon, M.; Lee, J.S.; Min, K. Recent progress and challenges in microbial polyhydroxybutyrate (PHB) production from CO_2 as a sustainable feedstock: A state-of-the-art review. *Bioresour. Technol.* **2021**, *339*, 125616. [CrossRef]
7. Yeo, J.C.C.; Muiruri, J.K.; Thitsartarn, W.; Li, Z.; He, C. Recent advances in the development of biodegradable PHB-based toughening materials: Approaches, advantages and applications. *Mater. Sci. Eng. C* **2018**, *92*, 1092–1116. [CrossRef]
8. Briassoulis, D.; Tserotas, P.; Athanasoulia, I.-G. Alternative optimization routes for improving the performance of poly (3-hydroxybutyrate) (PHB) based plastics. *J. Clean. Prod.* **2021**, *318*, 128555. [CrossRef]
9. Maximize Market Research PVT. LTD. Available online: https://www.maximizemarketresearch.com/market-report/global-polyhydroxy-butyrate-market/64542/ (accessed on 13 April 2022).
10. Popa, M.S.; Frone, A.N.; Panaitescu, D.M. Polyhydroxybutyrate blends: A solution for biodegradable packaging? *Int. J. Biol. Macromol.* **2022**, *207*, 263–277. [CrossRef]
11. Rhim, J.W.; Park, H.M.; Ha, C.S. Bio-nanocomposites for food packaging applications. *Prog. Polym. Sci.* **2013**, *38*, 1629–1652. [CrossRef]
12. Liu, Y.; Ahmed, S.; Sameen, D.E.; Wang, Y.; Lu, R.; Dai, J.; Li, S.; Qin, W. A review of cellulose and its derivatives in biopolymer-based for food packaging application. *Trends Food Sci. Technol.* **2021**, *112*, 532–546. [CrossRef]
13. Khairnar, Y.; Hansora, D.; Hazra, C.; Kundu, D.; Tayde, S.; Tonde, S.; Naik, J.; Chatterjee, A. Cellulose bionanocomposites for sustainable planet and people: A global snapshot of preparation, properties, and applications. *Carbohydr. Polym. Technol. Appl.* **2021**, *2*, 100065. [CrossRef]
14. Panaitescu, D.M.; Vizireanu, S.; Stoian, S.A.; Nicolae, C.-A.; Gabor, A.R.; Damian, C.M.; Trusca, R.; Carpen, L.G.; Dinescu, G. Poly(3-hydroxybutyrate) modified by plasma and TEMPO-oxidized celluloses. *Polymers* **2020**, *12*, 1510. [CrossRef]
15. Anjana; Raturi, G.; Shree, S.; Sharma, A.; Panesar, P.S.; Goswami, S. Recent approaches for enhanced production of microbial polyhydroxybutyrate: Preparation of biocomposites and applications. *Int. J. Biol. Macromol.* **2021**, *182*, 1650–1669. [CrossRef]
16. Panaitescu, D.M.; Nicolae, C.A.; Gabor, A.R.; Trusca, R. Thermal and mechanical properties of poly(3-hydroxybutyrate) reinforced with cellulose fibers from wood waste. *Ind. Crops Prod.* **2020**, *145*, 112071. [CrossRef]
17. Panaitescu, D.M.; Frone, A.N.; Chiulan, I.; Nicolae, C.A.; Trusca, R.; Ghiurea, M.; Gabor, A.R.; Mihailescu, M.; Casarica, A.; Lupescu, I. Role of bacterial cellulose and poly (3-hydroxyhexanoate-co-3-hydroxyoctanoate) in poly (3-hydroxybutyrate) blends and composites. *Cellulose* **2018**, *25*, 5569–5591. [CrossRef]
18. Popa, M.S.; Frone, A.N.; Radu, I.C.; Stanescu, P.O.; Truşcă, R.; Rădiţoiu, V.; Nicolae, C.A.; Gabor, A.R.; Panaitescu, D.M. Microfibrillated Cellulose Grafted with Metacrylic Acid as a Modifier in Poly(3-hydroxybutyrate). *Polymers* **2021**, *13*, 3970. [CrossRef]
19. Oprea, M.; Panaitescu, D.M.; Nicolae, C.A.; Gabor, A.R.; Frone, A.N.; Raditoiu, V.; Trusca, R.; Casarica, A. Nanocomposites from functionalized bacterial cellulose and poly(3-hydroxybutyrate-co-3-hydroxyvalerate). *Polym. Degrad. Stab.* **2020**, *179*, 109203. [CrossRef]
20. Thirumala, M.; Vishnuvardhan, R.S.; Mahmood, S.K. Production and characterization of PHB from two novel strains of *Bacillus* spp. isolated from soil and activated sludge. *J. Ind. Microbiol.* **2010**, *37*, 271–278. [CrossRef]
21. Surendran, A.; Lakshmanan, M.; Chee, J.Y.; Sulaiman, A.M.; Thuoc, D.V.; Sudesh, K. Can polyhydroxyalkanoates be produced efficiently from waste plant and animal oils? *Front. Bioeng. Biotechnol.* **2020**, *8*, 169. [CrossRef]
22. Tang, X.; Chen, E.Y.-X. Chemical synthesis of perfectly isotactic and high melting bacterial poly(3-hydroxybutyrate) from bio-sourced racemic cyclic diolide. *Nat. Commun.* **2018**, *9*, 2345. [CrossRef]
23. Portugal-Nunes, D.J.; Pawar, S.S.; Lidén, G.; Gorwa-Grauslund, M.F. Effect of nitrogen availability on the poly-3-d-hydroxybutyrate accumulation by engineered *Saccharomyces cerevisiae*. *AMB Express* **2017**, *7*, 35. [CrossRef]
24. Dilkes-Hoffman, L.S.; Lant, P.A.; Laycock, B.; Pratt, S. The rate of biodegradation of PHA bioplastics in the marine environment: A meta-study. *Mar. Pollut. Bull.* **2019**, *142*, 15–24. [CrossRef]
25. Barkoula, N.M.; Garkhail, S.K.; Peijs, T. Biodegradable composites based on flax/polyhydroxybutyrate and its copolymer with hydroxyvalerate. *Ind. Crops Prod.* **2010**, *31*, 34–42. [CrossRef]
26. McAdam, B.; Brennan Fournet, M.; McDonald, P.; Mojicevic, M. Production of polyhydroxybutyrate (PHB) and factors impacting its chemical and mechanical characteristics. *Polymers* **2020**, *12*, 2908. [CrossRef]
27. Santos, A.; Dalla Valentina, L.V.O.; Schulz, A.; Duarte, M.A.T. From obtaining to degradation of PHB: Material properties. Part I. *Ing. Cienc.* **2017**, *13*, 269–298. [CrossRef]

28. Raza, Z.A.; Khalil, S.; Abid, S. Recent progress in development and chemical modification of poly(hydroxybutyrate)-based blends for potential medical applications. *Int. J. Biol. Macromol.* **2020**, *160*, 77–100. [CrossRef]
29. de Carvalho, A.J.; das Graças Silva-Valenzuela, M.; Wang, S.H.; Valenzuela-Diaz, F.R. Biodegradable nanocomposite microcapsules for controlled release of urea. *Polymers* **2021**, *13*, 722. [CrossRef]
30. Râpă, M.; Zaharia, C.; Stănescu, P.O.; Cășărică, A.; Matei, E.; Predescu, A.M.; Pantilimon, M.C.; Vidu, R.; Predescu, C.; Cioflan, H. In vitro degradation of PHB/bacterial cellulose biocomposite scaffolds. *Int. J. Polym. Sci.* **2021**, 3820364. [CrossRef]
31. Bugnicourt, E.; Cinelli, P.; Lazzeri, A.; Alvarez, V. Polyhydroxyalkanoate (PHA): Review of synthesis, characteristics, processing and potential applications in packaging. *Express Polym. Lett.* **2014**, *8*, 791–808. [CrossRef]
32. Ma, P.; Cai, X.; Chen, M.; Dong, W.; Lemstra, P.J. Partially bio-based thermoplastic elastomers by physical blending of poly(hydroxyalkanoate)s and poly(ethylene-co-vinyl acetate). *Express Polym. Lett.* **2014**, *8*, 517–527. [CrossRef]
33. Leroy, E.; Petit, I.; Audic, J.L.; Colomines, G.; Deterre, R. Rheological characterization of a thermally unstable bioplastic in injection molding conditions. *Polym. Degrad. Stab.* **2021**, *97*, 1915–1921. [CrossRef]
34. Cesário, M.T.; Raposo, R.S.; de Almeida, M.C.M.D.; van Keulen, F.; Ferreira, B.S.; da Fonseca, M.M. Enhanced bioproduction of poly-3-hydroxybutyrate from wheat straw lignocellulosic hydrolysates. *New Biotechnol.* **2014**, *31*, 104–113. [CrossRef] [PubMed]
35. Bedade, D.K.; Edson, C.B.; Gross, R.A. Emergent approaches to efficient and sustainable polyhydroxyalkanoate production. *Molecules* **2021**, *26*, 3463. [CrossRef] [PubMed]
36. Reddy, M.V.; Mawatari, Y.; Onodera, R.; Nakamura, Y.; Yajima, Y.; Chang, Y.C. Bacterial conversion of waste into polyhydroxybutyrate (PHB): A new approach of bio-circular economy for treating waste and energy generation. *Bioresour. Technol. Rep.* **2019**, *113*, 456–460. [CrossRef]
37. Sayyed, R.Z.; Shaikh, S.S.; Wani, S.J.; Rehman, M.T.; Al Ajmi, M.F.; Haque, S.; El Enshasy, H.A. Production of biodegradable polymer from agro-wastes in *Alcaligenes* sp. and *Pseudomonas* sp. *Molecules* **2021**, *26*, 2443. [CrossRef]
38. Soto, L.R.; Byrne, E.; van Niel, E.W.J.; Sayed, M.; Villanueva, C.C.; Hatti-Kaul, R. Hydrogen and polyhydroxybutyrate production from wheat straw hydrolysate using Caldicellulosiruptor species and Ralstoniaeutropha in a coupled process. *Bioresour. Technol.* **2019**, *272*, 259–266. [CrossRef]
39. Van Thuoc, D.; Chung, N.T.; Hatti-Kaul, R. Polyhydroxyalkanoate production from rice straw hydrolysate obtained by alkaline pretreatment and enzymatic hydrolysis using *Bacillus* strains isolated from decomposing straw. *Bioresour. Bioprocess.* **2021**, *8*, 98. [CrossRef]
40. Pagliano, G.; Gugliucci, W.; Torrieri, E.; Piccolo, A.; Cangemi, S.; Di Giuseppe, F.A.; Robertiello, A.; Faraco, V.; Pepe, O.; Ventorino, V. Polyhydroxyalkanoates (PHAs) from dairy wastewater effluent: Bacterial accumulation, structural characterization and physical properties. *Chem. Biol. Technol. Agric.* **2020**, *7*, 29. [CrossRef]
41. Dalsasso, R.R.; Pavan, F.P.; Bordignon, S.E.; de Aragão, G.M.F.; Poletto, P. Polyhydroxybutyrate (PHB) production by *Cupriavidus necator* from sugarcane vinasse and molasses as mixed substrate. *Process. Biochem.* **2019**, *85*, 12–18. [CrossRef]
42. Hassan, M.A.; Bakhiet, E.K.; Hussein, H.R.; Ali, S.G. Statistical optimization studies for polyhydroxybutyrate (PHB) production by novel Bacillus subtilis using agricultural and industrial wastes. *Int. J. Environ. Sci. Technol.* **2018**, *16*, 3497–3512. [CrossRef]
43. Gonzalez, A.; Iriarte, M.; Iriondo, P.J.; Iruin, J.J. Miscibility and carbon dioxide transport properties of blends of bacterial poly(3-hydroxybutyrate) and a poly(vinylidene chloride-co-acrylonitrile) copolymer. *Polymer* **2002**, *43*, 6205–6211. [CrossRef]
44. Eraslana, K.; Aversab, C.; Nofarac, M.; Barlettab, M.; Gisariod, A.; Salehiyane, R.; Goksu, Y.A. Poly(3-hydroxybutyrate-co-3-hydroxyhexanoate) (PHBH): Synthesis, properties, and applications—A review. *Eur. Polym. J.* **2022**, *67*, 111044. [CrossRef]
45. Jian, Y.; Zhu, Y. Poly 3-hydroxybutyrate 4-hydroxybutyrate (P34HB) as a potential polymer for drug-eluting coatings on metal coronary stents. *Polymers* **2022**, *14*, 994. [CrossRef] [PubMed]
46. Naser, A.Z.; Deiab, I.; Darras, B.M. Poly(lactic acid) (PLA) and polyhydroxyalkanoates (PHAs), green alternatives to petroleum-based plastics: A review. *RSC Adv.* **2021**, *11*, 17151–17196. [CrossRef] [PubMed]
47. Ibrahim, M.I.; Alsafadi, D.; Alamry, K.A.; Hussein, M.A. Properties and applications of poly(3-hydroxybutyrate-co-3-hydroxyvalerate) biocomposites. *J. Polym. Environ.* **2021**, *29*, 1010–1030. [CrossRef]
48. Nosal, H.; Moser, K.; Warzała, M.; Holzer, A.; Stańczyk, D.; Sabura, E. Selected fatty acids esters as potential PHB-V bioplasticizers: Effect on mechanical properties of the polymer. *J. Polym. Environ.* **2012**, *29*, 38–53. [CrossRef]
49. Frone, A.N.; Nicolae, C.A.; Eremia, M.C.; Tofan, V.; Ghiurea, M.; Chiulan, I.; Radu, E.; Damian, C.M.; Panaitescu, D.M. Low molecular weight and polymeric modifiers as toughening agents in poly(3-hydroxybutyrate) films. *Polymers* **2020**, *12*, 2446. [CrossRef]
50. Garcia-Garcia, D.; Ferri, J.M.; Boronat, T.; Lopez-Martinez, J.; Balart, R. Processing and characterization of binary poly(hydroxybutyrate) (PHB) and poly(caprolactone) (PCL) blends with improved impact properties. *Polym. Bull.* **2016**, *73*, 3333–3350. [CrossRef]
51. Jakić, M.; Vrandečić, N.S.; Erceg, M. Thermal degradation of poly(3-hydroxybutyrate)/poly(ethylene oxide) blends: Thermogravimetric and kinetic analysis. *Eur. Polym. J.* **2016**, *81*, 376–385. [CrossRef]
52. Ma, P.; Hristova-Bogaerds, D.G.; Lemstra, P.J.; Zhang, Y.; Wang, S. Toughening of PHBV/PBS and PHB/PBS blends via in situ compatibilization using dicumyl peroxide as a free-radical grafting initiator. *Macromol. Mater. Eng.* **2012**, *297*, 402–410. [CrossRef]
53. Vale Iulianelli, G.C.V.; dos David, G.S.; dos Santos, T.N.; Sebastião, P.J.O.; Tavares, M.I.B. Influence of TiO_2 nanoparticle on the thermal, morphological and molecular characteristics of PHB matrix. *Polym. Test.* **2018**, *65*, 156–162. [CrossRef]

54. Silva, M.B.R.; Tavares, M.I.B.; Junior, A.W.M.; Neto, R.P.C. Evaluation of intermolecular interactions in the PHB/ZnO nanostructured materials. *J. Nanosci. Nanotechnol.* **2016**, *16*, 7606–7610. [CrossRef]
55. Liao, H.-T.; Wu, C.-S. Poly(3-hydroxybutyrate)/multi-walled carbon nanotubes nanocomposites: Preparation and characterizations. *Des. Monomers Polym.* **2013**, *16*, 99–107. [CrossRef]
56. Jandas, P.J.; Prabakaran, K.; Kumar, R.; Mohanty, S.; Nayak, S.K. Eco-friendly poly(hydroxybutyrate) nanocomposites: Preparation and characterization. *J. Polym. Res.* **2021**, *28*, 285. [CrossRef]
57. Kaur, J.; Sengupta, P.; Mukhopadhyay, S. Critical review of bioadsorption on modified cellulose and removal of divalent heavy metals (Cd, Pb, and Cu). *Ind. Eng. Chem. Res.* **2022**, *61*, 1921–1954. [CrossRef]
58. Gan, I.; Chow, W.S. Antimicrobial poly(lactic acid)/cellulose bionanocomposite for foodpackaging application: A review. *Food Packag. Shelf Life* **2018**, *17*, 150–161. [CrossRef]
59. Tanpichai, S.; Boonmahitthisud, A.; Soykeabkaew, N.; Ongthip, L. Review of the recent developments in all-cellulose nanocomposites: Properties and applications. *Carbohydr. Polym.* **2022**, *286*, 119192. [CrossRef]
60. Louis, A.C.F.; Venkatachalam, S. Energy efficient process for valorization of corn cob as a source for nanocrystalline cellulose and hemicellulose production. *Int. J. Biol. Macromol.* **2020**, *163*, 260–269. [CrossRef]
61. Guo, Y.; Zhang, Y.; Zheng, D.; Li, M.; Yue, J. Isolation and characterization of nanocellulose crystals via acid hydrolysis from agricultural waste-tea stalk. *Int. J. Biol. Macromol.* **2020**, *163*, 927–933. [CrossRef]
62. Neto, W.P.F.; Silvério, H.A.; Dantas, N.O.; Pasquini, D. Extraction and characterization of cellulose nanocrystals from agro-industrial residue–soy hulls. *Ind. Crop. Prod.* **2013**, *42*, 480–488. [CrossRef]
63. Melikoglu, A.Y.; Bilek, S.E.; Cesur, S. Optimum alkaline treatment parameters for the extraction of cellulose and production of cellulose nanocrystals from apple pomace. *Carbohydr. Polym.* **2019**, *215*, 330–337. [CrossRef] [PubMed]
64. Coelho, C.C.S.; Michelin, M.; Cerqueira, M.A.; Gonçalves, C.; Tonon, R.V.; Pastrana, L.M.; Freitas-Silva, O.; Vicente, A.A.; Cabral, L.M.C.; Teixeira, J.A. Cellulose nanocrystals from grape pomace: Production, properties and cytotoxicity assessment. *Carbohydr. Polym.* **2018**, *192*, 327–336. [CrossRef] [PubMed]
65. Prado, K.S.; Spinacé, M.A.S. Isolation and characterization of cellulose nanocrystals from pineapple crown waste and their potential uses. *Int. J. Biol. Macromol.* **2019**, *122*, 410–416. [CrossRef] [PubMed]
66. Souza, A.G.; Santos, D.F.; Ferreira, R.R.; Pinto, V.Z.; Rosa, D.S. Innovative process for obtaining modified nanocellulose from soybean straw. *Int. J. Biol. Macromol.* **2020**, *165*, 1803–1812. [CrossRef]
67. Frone, A.N.; Chiulan, I.; Panaitescu, D.M.; Nicolae, C.A.; Ghiurea, M.; Galan, A.-M. Isolation of cellulose nanocrystals from plum seed shells, structural and morphological characterization. *Mater. Lett.* **2017**, *194*, 160–163. [CrossRef]
68. Fortunati, E.; Benincasa, P.; Balestra, G.M.; Luzi, F.; Mazzaglia, A.; Del Buono, D.; Puglia, D.; Torre, L. Revalorization of barley straw and husk as precursors for cellulose nanocrystals extraction and their effect on PVA_CH nanocomposites. *Ind. Crops Prod.* **2016**, *92*, 201–217. [CrossRef]
69. Kallel, F.; Bettaieb, F.; Khiari, R.; García, A.; Bras, J.; Chaabouni, E.S. Isolation and structural characterization of cellulose nanocrystals extracted from garlic straw residues. *Ind. Crops Prod.* **2016**, *87*, 287–296. [CrossRef]
70. Jiang, F.; Hsieh, Y.-L. Cellulose nanocrystal isolation from tomato peels and assembled nanofibers. *Carbohydr. Polym.* **2015**, *122*, 60–68. [CrossRef]
71. Ferreira, F.V.; Mariano, M.; Rabelo, S.C.; Gouveia, R.F.; Lona, L.M.F. Isolation and surface modification of cellulose nanocrystals from sugarcane bagasse waste: From a micro- to a nano-scale view. *Appl. Surf. Sci.* **2018**, *436*, 1113–1122. [CrossRef]
72. Hideno, A.; Abe, K.; Yano, H. Preparation using pectinase and characterization of nanofibers from orange peel waste in juice factories. *J. Food Sci.* **2014**, *79*, N1218–N1224. [CrossRef]
73. Nagalakshmaiah, M.; Mortha, G.; Dufresne, A. Structural investigation of cellulose nanocrystals extracted from chili leftover and their reinforcement in cariflex-IR rubber latex. *Carbohydr. Polym.* **2016**, *136*, 945–954. [CrossRef] [PubMed]
74. Pennells, J.; Godwin, I.D.; Amiralian, N.; Martin, D.J. Trends in the production of cellulose nanofibers from non-wood sources. *Cellulose* **2020**, *27*, 575–593. [CrossRef]
75. Dhali, K.; Ghasemlou, M.; Daver, F.; Cass, P.; Adhikari, B. A review of nanocellulose as a new material towards environmental sustainability. *Sci. Total Environ.* **2021**, *775*, 145871. [CrossRef] [PubMed]
76. Salimi, S.; Sotudeh-Gharebagh, R.; Zarghami, R.; Chan, S.Y.; Yuen, K.H. Production of nanocellulose and its applications in drug delivery: A critical review. *ACS Sustain. Chem. Eng.* **2019**, *7*, 15800–15827. [CrossRef]
77. Pradhan, D.; Jaiswal, A.K.; Jaiswal, S. Emerging technologies for the production of nanocellulose from lignocellulosic biomass. *Carbohydr. Polym.* **2022**, *285*, 119258. [CrossRef]
78. Wang, L.; Li, K.; Copenhaver, K.; Mackay, S.; Lamm, M.E.; Zhao, X.; Dixon, B.; Wang, J.; Han, Y.; Neivandt, D.; et al. Review on nonconventional fibrillation methods of producing cellulose nanofibrils and their applications. *Biomacromolecules* **2021**, *22*, 4037–4059. [CrossRef]
79. Noremylia, M.B.; Hassan, M.Z.; Ismail, Z. Recent advancement in isolation, processing, characterization and applications of emerging nanocellulose: A review. *Int. J. Biol. Macromol.* **2022**, *206*, 954–976. [CrossRef]
80. Trache, D.; Hussin, M.H.; Haafiz, M.K.M.; Thakur, V.K. Recent progress in cellulose nanocrystals: Sources and production. *Nanoscale* **2017**, *9*, 1763–1786. [CrossRef]
81. Frone, A.N.; Panaitescu, D.M.; Donescu, D.; Spataru, C.I.; Radovici, C.; Trusca, R.; Somoghi, R. Preparation and characterization of PVA composites with cellulose nanofibers obtained by ultrasonication. *Bioresources* **2011**, *6*, 487–512. [CrossRef]

82. Mohammad Taib, M.N.A.; Hamidon, T.S.; Garba, Z.N.; Trache, D.; Uyama, H.; Hussin, M.H. Recent progress in cellulose-based composites towards flame retardancy applications. *Polymer* **2022**, *244*, 124677. [CrossRef]
83. Teodoro, K.B.R.; Sanfelice, R.C.; Migliorini, F.L.; Pavinatto, A.; Facure, M.H.M.; Correa, D.S. A review on the role and performance of cellulose nanomaterials in sensors. *ACS Sens.* **2021**, *6*, 2473–2496. [CrossRef] [PubMed]
84. Vatansever, E.; Arslan, D.; Nofar, M. Polylactide cellulose-based nanocomposites. *Int. J. Biol. Macromol.* **2019**, *137*, 912–938. [CrossRef] [PubMed]
85. Panaitescu, D.M.; Lupescu, I.; Frone, A.N.; Chiulan, I.; Nicolae, C.A.; Tofan, V.; Stefaniu, A.; Somoghi, R.; Trusca, R. Medium chain-length polyhydroxyalkanoate copolymer modified by bacterial cellulose for medical devices. *Biomacromolecules* **2017**, *18*, 3222–3232. [CrossRef] [PubMed]
86. Seoane, I.T.; Cerrutti, P.; Vazquez, A.; Manfredi, L.B.; Cyras, V.P. Polyhydroxybutyrate-based nanocomposites with cellulose nanocrystals and bacterial cellulose. *J. Polym. Environ.* **2017**, *25*, 586–598. [CrossRef]
87. Panaitescu, D.M.; Frone, A.N.; Chiulan, I. Nanostructured biocomposites from aliphatic polyesters and bacterial cellulose. *Ind. Crops Prod.* **2016**, *93*, 251–266. [CrossRef]
88. Siró, I.; Plackett, D. Microfibrillated cellulose and new nanocomposite materials: A review. *Cellulose* **2010**, *17*, 459–494. [CrossRef]
89. Zhang, B.; Huang, C.; Zhao, H.; Wang, J.; Yin, C.; Zhang, L.; Zhao, Y. Effects of cellulose nanocrystals and cellulose nanofibers on the structure and properties of polyhydroxybutyrate nanocomposites. *Polymers* **2019**, *11*, 2063. [CrossRef] [PubMed]
90. Jun, D.; Guomin, Z.; Mingzhu, P.; Leilei, Z.; Dagang, L.; Rui, Z. Crystallization and mechanical properties of reinforced PHBV composites using melt compounding: Effect of CNCs and CNFs. *Carbohydr. Polym.* **2017**, *168*, 255–262. [CrossRef]
91. Kampeerapappun, P. The electrospunpolyhydroxybutyrate fibers reinforced with cellulose nanocrystals: Morphology and properties. *J. Appl. Polym. Sci.* **2016**, *133*, 43273. [CrossRef]
92. Panaitescu, D.M.; Popa, M.S.; Raditoiu, V.; Frone, A.N.; Sacarescu, L.; Gabor, A.R.; Nicolae, C.A.; Teodorescu, M. Effect of calcium stearate as a lubricant and catalyst on the thermal degradation of poly(3-hydroxybutyrate). *Int. J. Biol. Macromol.* **2021**, *190*, 780–791. [CrossRef]
93. Frone, A.N.; Batalu, D.; Chiulan, I.; Oprea, M.; Gabor, A.R.; Nicolae, C.A.; Raditoiu, V.; Trusca, R.; Panaitescu, D.M. Morpho-structural, thermal and mechanical properties of PLA/PHB/cellulose biodegradable nanocomposites obtained by compression molding, extrusion, and 3D printing. *Nanomaterials* **2020**, *10*, 51. [CrossRef] [PubMed]
94. Panaitescu, D.M.; Trusca, R.; Gabor, A.R.; Nicolae, C.A.; Casarica, A. Biocomposite foams based on polyhydroxyalkanoate and nanocellulose: Morphological and thermo-mechanical characterization. *Int. J. Biol. Macromol.* **2020**, *164*, 1867–1878. [CrossRef] [PubMed]
95. Dhar, P.; Vangala, S.P.; Tiwari, P.K.; Kumar, A.; Katiyar, V. Thermal degradation kinetics of poly (3-hydroxybutyrate)/cellulose nano- crystals based nanobiocomposite. *J. Thermodyn. Catal.* **2014**, *5*, 1000134.
96. Dhar, P.; Bhardwaj, U.; Kumar, A.; Katiyar, V. Poly (3-hydroxybutyrate)/cellulose nanocrystal films for food packaging applications: Barrier and migration studies. *Polym. Eng. Sci.* **2015**, *55*, 2388–2395. [CrossRef]
97. Ten, E.; Bahr, D.F.; Li, B.; Jiang, L.; Wolcott, M.P. Effects of cellulose nanowhiskers on mechanical, dielectric, and rheological properties of poly(3-hydroxybutyrate-co-3-hydroxyvalerate)/cellulose nanowhisker composites. *Ind. Eng. Chem. Res.* **2012**, *51*, 2941–2951. [CrossRef]
98. Seoane, I.T.; Fortunati, E.; Puglia, D.; Cyras, V.P.; Manfredi, L.B. Development and characterization of bionanocomposites based on poly(3-hydroxybutyrate) and cellulose nanocrystals for packaging applications. *Polym. Int.* **2016**, *65*, 1046–1053. [CrossRef]
99. Seoane, I.T.; Cerrutti, P.; Vazquez, A.; Cyras, V.P.; Manfredi, L.B. Ternary nanocomposites based on plasticized poly(3-hydroxybutyrate) and nanocellulose. *Polym. Bull.* **2019**, *76*, 967–988. [CrossRef]
100. Seoane, I.T.; Luzi, F.; Puglia, D.; Cyras, V.P.; Manfredi, L.B. Enhancement of paperboard performance as packaging material by layering with plasticized polyhydroxybutyrate/nanocellulose coatings. *J. Appl. Polym. Sci.* **2018**, *135*, 46872. [CrossRef]
101. Seoane, I.T.; Manfredi, L.B.; Cyras, V.P.; Torre, L.; Fortunati, E.; Puglia, D. Effect of cellulose nanocrystals and bacterial cellulose on disintegrability in composting conditions of plasticized PHB nanocomposites. *Polymers* **2017**, *9*, 561. [CrossRef]
102. de O. Patrício, P.S.; Pereira, F.V.; dos Santos, M.C.; de Souza, P.P.; Roa, J.P.B.; Orefice, R.L. Increasing the elongation at break of polyhydroxybutyrate biopolymer: Effect of cellulose nanowhiskers on mechanical and thermal properties. *J. Appl. Polym. Sci.* **2013**, *127*, 3613–3621. [CrossRef]
103. Ten, E.; Turtle, J.; Bahr, D.; Jiang, L.; Wolcott, M. Thermal and mechanical properties of poly(3-hydroxybutyrate-co-3-hydroxyvalerate)/cellulose nanowhiskers composites. *Polymer* **2010**, *51*, 2652–2660. [CrossRef]
104. Chen, J.; Xu, C.; Wu, D.; Pan, K.; Qian, A.; Sha, Y.; Wang, L.; Tong, W. Insights into the nucleation role of cellulose crystals during crystallization of poly(β-hydroxybutyrate). *Carbohydr. Polym.* **2015**, *134*, 508–515. [CrossRef] [PubMed]
105. Kampeerapappun, P. Extraction and characterization of cellulose nanocrystals produced by acid hydrolysis from corn husk. *J. Met. Mater. Miner.* **2015**, *25*, 19–26.
106. Choi, J.; Kang, J.; Yun, S.I. Nanofibrous foams of poly(3-hydroxybutyrate)/cellulose nanocrystal composite fabricated using nonsolvent-induced phase separation. *Langmuir* **2021**, *37*, 1173–1182. [CrossRef] [PubMed]
107. Seoane, I.T.; Manfredi, L.B.; Cyras, V.P. Bilayer biocomposites based on coated cellulose paperboard with films of polyhydroxybutyrate/cellulose nanocrystals. *Cellulose* **2018**, *25*, 2419–2434. [CrossRef]
108. Jiang, L.; Morelius, E.; Zhang, J.; Wolcott, M.; Holbery, J. Study of the poly(3-hydroxybutyrate-co-3-hydroxyvalerate)/cellulose nanowhisker composites prepared by solution casting and melt processing. *J. Compos. Mater.* **2008**, *42*, 2629–2645. [CrossRef]

109. Magnani, C.; Idström, A.; Nordstierna, L.; Müller, A.J.; Dubois, P.; Raquez, J.M.; Lo Re, G. Interphase design of cellulose nanocrystals/poly(hydroxybutyrate-ran-valerate) bionanocomposites for mechanical and thermal properties tuning. *Biomacromolecules* **2020**, *21*, 1892–1901. [CrossRef] [PubMed]
110. Yu, H.Y.; Qin, Z.Y.; Yan, C.F.; Yao, J.M. Green nanocomposites based on functionalized cellulose nanocrystals: A study on the relationship between interfacial interaction and property enhancement. *ACS Sustain. Chem. Eng.* **2014**, *2*, 875–886. [CrossRef]
111. Jo, J.; Kim, H.; Jeong, S.-Y.; Park, C.; Hwang, H.S.; Koo, B. Changes in mechanical properties of polyhydroxyalkanoate with double silanized cellulose nanocrystals using different organosiloxanes. *Nanomaterials* **2021**, *11*, 1542. [CrossRef]
112. Oksman, K.; Mathew, A.P.; Bondeson, D.; Kvien, I. Manufacturing process of polylactic acid (PLA)—cellulose whiskers nanocomposites. *Compos. Sci. Technol.* **2006**, *66*, 2776–2784. [CrossRef]
113. Chen, J.; Wu, D.; Tam, K.C.; Pan, K.; Zheng, Z. Effect of surface modification of cellulose nanocrystal on nonisothermal crystallization of poly(β-hydroxybutyrate) composites. *Carbohydr. Polym.* **2017**, *157*, 1821–1829. [CrossRef] [PubMed]
114. Pracella, M.; Mura, C.; Galli, G. Polyhydroxyalkanoate nanocomposites with cellulose nanocrystals as biodegradable coating and packaging materials. *ACS Appl. Nano Mater.* **2021**, *4*, 260–270. [CrossRef]
115. Garcia-Garcia, D.; Lopez-Martinez, J.; Balart, R.; Strömberg, E.; Moriana, R. Reinforcing capability of cellulose nanocrystals obtained from pine cones in a biodegradable poly(3-hydroxybutyrate)/poly(ε-caprolactone) (PHB/PCL) thermoplastic blend. *Eur. Polym. J.* **2018**, *104*, 10–18. [CrossRef]
116. Lopera-Valle, A.; Caputo, J.V.; Leão, R.; Sauvageau, D.; Luz, S.M.; Elias, A. Influence of Epoxidized Canola Oil (eCO) and Cellulose Nanocrystals (CNCs) on the Mechanical and Thermal Properties of Polyhydroxybutyrate (PHB)—Poly(lactic acid) (PLA) Blends. *Polymers* **2019**, *11*, 933. [CrossRef] [PubMed]
117. Frone, A.N.; Ghiurea, M.; Nicolae, C.A.; Gabor, A.R.; Badila, S.; Panaitescu, D.M. Poly(lactic acid)/Poly(3-hydroxybutyrate) Biocomposites with Differently Treated Cellulose Fibers. *Molecules* **2022**, *27*, 2390. [CrossRef]
118. Vanderfleet, O.M.; Reid, M.S.; Bras, J.; Heux, L.; Godoy-Vargas, J.; Panga, M.K.R.; Cranston, E.D. Insight into thermal stability of cellulose nanocrystals from new hydrolysis methods with acid blends. *Cellulose* **2019**, *26*, 507–528. [CrossRef]
119. Kian, L.K.; Jawaid, M.; Ariffin, H.; Alothman, O.Y. Isolation and characterization of microcrystalline cellulose from roselle fibers. *Int. J. Biol. Macromol.* **2017**, *103*, 931–940. [CrossRef]
120. Sturcova, A.; Davies, G.R.; Eichhorn, S.J. Elastic modulus and stress-transfer properties of tunicate cellulose whiskers. *Biomacromolecules* **2005**, *6*, 1055–1061. [CrossRef]
121. Panaitescu, D.M.; Frone, A.N.; Ghiurea, M.; Chiulan, I. Influence of storage conditions on starch/PVA films containing cellulose nanofibers. *Ind. Crops Prod.* **2015**, *70*, 170–177. [CrossRef]
122. Narancic, T.; Verstichel, S.; Chaganti, S.R.; Morales-Gamez, L.; Kenny, S.T.; De Wilde, B.; Padamati, R.B.; O'Connor, K.E. Biodegradable plastic blends create new possibilities for end-of-life management of plastics but they are not a panacea for plastic pollution. *Environ. Sci. Technol.* **2018**, *52*, 10441–10452. [CrossRef]
123. Kadja, G.T.M.; Ilmi, M.M.; Azhari, N.J.; Khalil, M.; Fajar, A.T.N.; Subagjo; Makertihartha, I.G.B.N.; Gunawan, M.L.; Rasrendra, C.B.; Wenten, I.G. Recent advances on the nanoporous catalysts for the generation of renewable fuels. *J. Mater. Res. Technol.* **2022**, *17*, 3277–3336. [CrossRef]
124. Luzi, F.; Dominici, F.; Armentano, I.; Fortunati, E.; Burgos, N.; Fiori, S.; Jiménez, A.; Kenny, J.M.; Torre, L. Combined effect of cellulose nanocrystals, carvacrol and oligomeric lactic acid in PLA_PHB polymeric films. *Carbohydr. Polym.* **2019**, *223*, 115131. [CrossRef] [PubMed]
125. Bonartsev, A.P.; Bonartseva, G.A.; Reshetov, I.V.; Kirpichnikov, M.P.; Shaitan, K.V. Application of Polyhydroxyalkanoates in Medicine and the Biological Activity of Natural Poly(3-Hydroxybutyrate). *Acta Nat.* **2019**, *11*, 4–16. [CrossRef] [PubMed]

Review

Nanocelluloses: Sources, Pretreatment, Isolations, Modification, and Its Application as the Drug Carriers

Valentino Bervia Lunardi [1], Felycia Edi Soetaredjo [1,2], Jindrayani Nyoo Putro [1], Shella Permatasari Santoso [1,2], Maria Yuliana [1], Jaka Sunarso [3], Yi-Hsu Ju [4,5] and Suryadi Ismadji [1,*]

[1] Department of Chemical Engineering, Widya Mandala Surabaya Catholic University, Kalijudan 37, Surabaya 60114, Indonesia; valentinolunardi70@gmail.com (V.B.L.); felyciae@yahoo.com (F.E.S.); jindranyoo@yahoo.com (J.N.P.); shella_p5@yahoo.com (S.P.S.); mariayuliana@ukwms.ac.id (M.Y.)
[2] Department of Chemical Engineering, National Taiwan University of Science and Technology, No. 43, Section 4, Keelung Rd, Da'an District, Taipei City 10607, Taiwan
[3] Research Centre for Sustainable Technologies, Faculty of Engineering, Computing and Science, Swinburne University of Technology, Kuching 93350, Sarawak, Malaysia; jsunarso@swinburne.edu.my
[4] Graduate Institute of Applied Science, National Taiwan University of Science and Technology, No. 43, Section 4, Keelung Rd, Da'an District, Taipei City 10607, Taiwan; yhju@mail.ntust.edu.tw
[5] Taiwan Building Technology Center, National Taiwan University of Science and Technology, No. 43, Section 4, Keelung Rd, Da'an District, Taipei City 10607, Taiwan
* Correspondence: suryadiismadji@yahoo.com; Tel.: +62-31-389-1264

Abstract: The 'Back-to-nature' concept has currently been adopted intensively in various industries, especially the pharmaceutical industry. In the past few decades, the overuse of synthetic chemicals has caused severe damage to the environment and ecosystem. One class of natural materials developed to substitute artificial chemicals in the pharmaceutical industries is the natural polymers, including cellulose and its derivatives. The development of nanocelluloses as nanocarriers in drug delivery systems has reached an advanced stage. Cellulose nanofiber (CNF), nanocrystal cellulose (NCC), and bacterial nanocellulose (BC) are the most common nanocellulose used as nanocarriers in drug delivery systems. Modification and functionalization using various processes and chemicals have been carried out to increase the adsorption and drug delivery performance of nanocellulose. Nanocellulose may be attached to the drug by physical interaction or chemical functionalization for covalent drug binding. Current development of nanocarrier formulations such as surfactant nanocellulose, ultra-lightweight porous materials, hydrogel, polyelectrolytes, and inorganic hybridizations has advanced to enable the construction of stimuli-responsive and specific recognition characteristics. Thus, an opportunity has emerged to develop a new generation of nanocellulose-based carriers that can modulate the drug conveyance for diverse drug characteristics. This review provides insights into selecting appropriate nanocellulose-based hybrid materials and the available modification routes to achieve satisfactory carrier performance and briefly discusses the essential criteria to achieve high-quality nanocellulose.

Keywords: drug delivery; drug release; functionalization; nanocellulose

1. Introduction

Drug delivery technology (DDT) is a cutting-edge applied science for delivering drugs to specific targets. This technology regulates the absorption and release of therapeutic drugs via various drug carriers to the desired organs, including subcellular organs, tissues, and cells, to improve human health [1]. DDT has advanced rapidly in the past few decades, enabled by various discoveries in various fields, including pharmaceutical, materials, and biomedical sciences. DDT development aims to improve therapeutic drugs' pharmacological activity and overcome various disadvantages of conventional therapeutic drugs such as drug agglomeration, biodistribution deficiency, low bioavailability, limited solubility, and insufficient selectivity to prevent the concurrent effects of therapeutic drugs.

The majority of research studies on drug delivery technology revolve around developing materials suitable for drug delivery with desirable characteristics such as high drug adsorption capacity, targeted drug administration, controlled release, biocompatibility, and non-immunogenic and non-toxic effects that optimize therapeutic efficacy and eliminates side effects [2]. Many engineered nanomaterials have been studied for drug delivery applications [3]. Some nanomaterials have recently been undergoing development and clinical investigation; however, each nanomaterial has its various characteristics and limitations, challenging the researcher in creating a suitable drug delivery system.

Natural-based polymers have drawn considerable attention as suitable biomaterials for numerous applications in drug delivery systems. Various nature-based polymers such as polysaccharides (cellulose, chitosan, hyaluronic acid, pectins, alginate, cellulose ethers), proteins (silk fibroin and collagen), and peptides have been identified as promising biomaterials for drug delivery systems given their biocompatibility, processability, and characteristics (e.g., nanoparticles, hydrogels, aerogels, tablets, and so on) that can be regulated by modifying various polymer functional groups such as amino groups, carboxyl groups, and hydroxyl groups [4]. The current development of these mentioned various polysaccharides, proteins, and peptides for drug delivery systems have been well-reviewed elsewhere [4–7].

Several natural polymers have been shown to have a higher affinity for cell receptors and modulate cellular processes such as adhesion, migration, and proliferation. These advantages make these natural polymers attractive for effective and high-efficiency drug delivery systems [8]. They can also be degraded in the presence of in vivo enzymes, which ensures their ability to create responsive local delivery systems. However, only polysaccharides and proteins have been extensively studied in drug delivery systems (DDS). These natural polymers have unique characteristics in each tissue and have identical characteristics in the extracellular skeleton. These characteristics support these natural polymers' utilization as drug carriers with insignificant invasive features [9–11].

Cellulose is the most abundant and commonly found natural polymer [12]. Its annual production is estimated at more than $7.5 \cdot 10^{10}$ tons [13]. As a promising fuel and chemical precursor, cellulose has been widely utilized in various industries such as textile, pulp, paper, composite, and pharmaceutical excipients [2]. However, the development of cellulose-based materials as a direct molecule controller for drug adsorption and release had not been evaluated until the discovery of nanocellulose, which became a turning point for using carbohydrate-based nanomaterials in the field of drug delivery [14,15].

As illustrated in Figure 1, the publication on nanocellulose for biomedical engineering applications increases every year, especially for drug delivery applications. The increase in the number of publications on the utilization of nanocellulose for drug delivery systems is a strong indication of the potential application of this material in the future. The rapid development of nanotechnology and materials science has brought about nanocellulose as a potential drug carrier because of its extraordinary physicochemical and biological characteristics. Nanocellulose has a large surface-area-to-volume ratio, thus enabling more significant adsorption and therapeutic drug-binding capacity than other materials. With these properties, nanocellulose can facilitate drug release mechanisms and allocate drug delivery precisely to the target to drastically reduce drug consumption, leading to improved drug delivery system effectiveness [16,17]. Nanocellulose additionally exhibits other attractive characteristics such as stiffness, high mechanical strength, biocompatibility, low toxicity, lightweight, tunable surface chemistry, and renewability [11,18], which are desirable for the design of advanced drug delivery system.

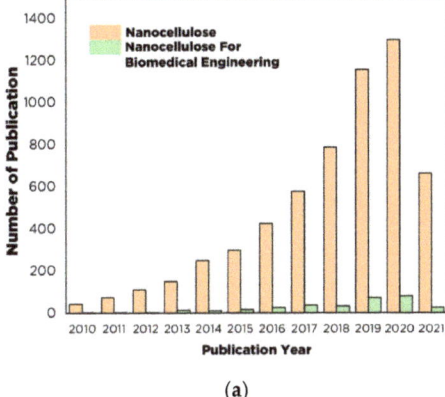 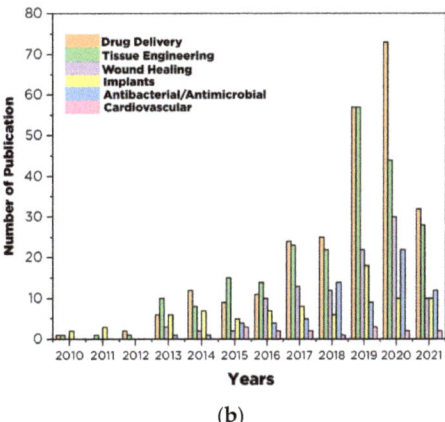

Figure 1. The number of publications in the area of nanocellulose and nanocellulose for biomedical engineering indexed by Scopus from 2010-until recent (10 June 2021) (**a**); data representation of annual publication of nanocellulose in various categories of biomedical engineering within the last decades (**b**); data analysis performed on Scopus using the terms nanocellulose and nanocellulose for "x" (x refer to biomedical engineering, drug delivery, tissue engineering, wound healing, implants, Antibacterial/antimicrobial, and cardiovascular).

Nanocellulose can be utilized as either carrier or excipient for broad application in drug delivery systems such as microparticles, tablets, hydrogels, aerogels, regulating nanoparticles, and membrane drug delivery systems [19]. Nanocellulose has been manufactured on the laboratory and industrial scale, i.e., ranging from 140 g day^{-1} to 50 ton year^{-1} in three different forms as nanocrystalline cellulose (NCC), nanofiber cellulose (NFC), and bacterial nanocellulose (BNC) [20]. Several recent research and review articles have comprehensively overviewed the process, extraction, characterization, and applications of nanocellulose and their modified structures in drug delivery systems [12,17,21–24].

The drug binding and the release time of nanocellulose-based drugs vary depending on the nanocellulose configuration, therapeutical ingredient's activity, the production method, and the modification [25,26]. Therefore, nanocellulose is a promising carrier for various drug delivery systems such as oral administration, ophthalmic drug delivery, intratumoral administration, transdermal drug delivery, topical administration, and local drug delivery.

This review provides a comprehensive overview of the preparation procedures of nanocellulose and the various effects on drug formulation and delivery. Three types of nanocelluloses and a brief description of their synthesis processes are discussed at the beginning of this review. Subsequently, the effects of raw materials and the synthesis process on the characteristics of the resultant nanocellulose are discussed. This is then followed by the application of nanocellulose to various drug delivery systems.

2. Conversion of Cellulose into Nanocellulose and Its Characteristic

Cellulose is the most abundant natural polymer globally and is a renewable source and essential raw material for various industries. Cellulose is a crucial constituent compound for plants, marine animals, algae, fungi, bacteria, and amoebae [12]. In 1838, French chemist Anselme Payen discovered and isolated cellulose from plant fibers using nitric acid and determined its chemical structure. The primary sources of cellulose are plant fibers with a high cellulose content, such as cotton (containing more than 90% cellulose content) [27] and wood (up to 50% cellulose). Other compounds such as hemicellulose, lignin, pectin, and wax are also present; they can be recovered during the separation process.

Recently, various agricultural wastes with high cellulose content were explored as a source of cellulose, such as oil palm empty fruit bunches (OPEFB) [28], palm and banana fronds, passionfruit peel waste [29], bagasse, wheat straw, rice straw, bamboo stalks, hemp bark, potato tubers, mulberry bark, hemp avicel, and sugar beets [30]. Cellulose derived from these non-plant precursors can have a molecular structure similar to that of plant cellulose. However, the main difference is that much less hemicellulose or lignin is present in these non-plant-based precursors; higher cellulose content with much lower impurities can be obtained from these precursors.

In terms of chemical structure, cellulose is composed of a linear homopolysaccharide consist of β-D-glucopyranose units entirely condensed and bonded through β-1,4-glycoside linkages (Figure 2). The structure foundation of the cellulose network is arranged by a chain glucose dimer comprising two anhydrous glucoses (AG) defined as cellobiose [31] (Figure 2). The raw material or the pretreatment (chemical or mechanical) of cellulose may affect the cellulose chain length and thus lead to molecular weight variation. The number of AG units in each chain is known as the polymerization degree (PD). The value of PD for cellulose powder varies from 100 to 300 units and around 26,500 for cellulose pulp [32]. The PD value for cellulose from cotton is 15,000, and wood is approximately 10,000 [33].

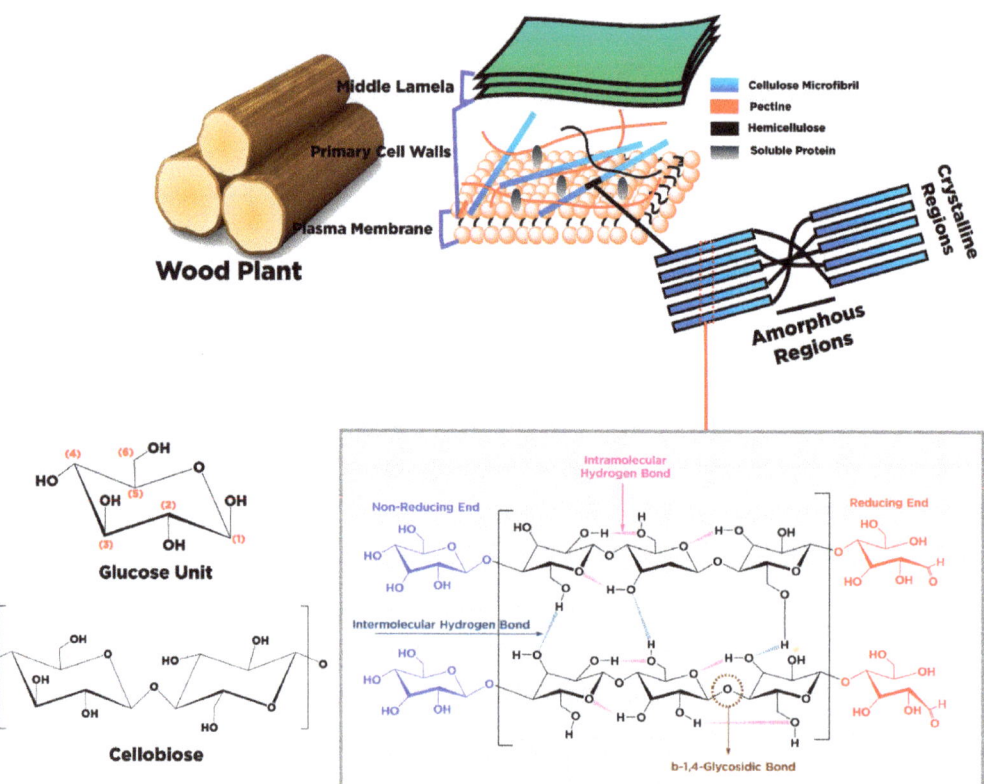

Figure 2. Schematic of cellulose production from wood plant and structural chemistry of exhibiting arrangement betwixt individual fibers.

Each cellulose monomer contains three reactive hydroxyl groups in the repeating chemical structure of the β-D-glucopyranose unit. In the same chain, these hydroxyl groups can make hydrogen bonds with the adjacent β-D-glucopyranose units. At different chain locations, the bonds present are intramolecular and intermolecular hydrogen bonds responsible for the crystal arrangement, determining the cellulose's physical characteristics. Based on molecular orientation and hydrogen network between molecules and intramolecular, cellulose is classified into different types, i.e., I, II, III, III_I, III_{II}, IV_I, and IV_{II}. For details about the classification of cellulose, the reader can refer to the work of Moon et al. [34]. Some of the cellulose characteristics are mainly represented by hydrogen linkage coordination [35,36].

Structurally, the cellulose is a linear chain polymer with a rod-like configuration, aided by the glucose residues' equatorial conformation that is intensely aggregated together with the lateral size 3–5 nm [36]. Primary chains of cellulose, especially polysaccharide chains, are found on the secondary walls of plants arranged in a parallel configuration. The cellulose's basic fibers have a cross-sectional diameter between 10–450 nm with a length of several micrometers that depend on the diversity of material sources [37]. Moreover, the elementary fibrils were arranged into large pack units called microfibrils, further foregathered into fibrils [13]. There are regions within the cellulose fibrils where the cellulose chains are organized into a highly crystalline structure with a length of 50–150 nm and disordered amorphous regions with 25–50 nm [34]. The cellulose chains construct the crystalline regions through Van der Waals forces, strong intra- and intermolecular hydrogen linkage, and β-1,4-glycosidic bonds. In contrast, amorphous regions are built up through the deficiency of hydrogen bonds in the crystalline region. The crystalline and amorphous regions in cellulose may vary depending on various sources.

The crystalline constituent within cellulose fibers can be refined through various chemical treatments by destructing and removing the disordered amorphous or paracrystalline regions. The purified crystal fragments with particle sizes on the nanometer scale are called nanocrystalline cellulose (NCC) (Figure 3). Different shapes of NCC are present such as needle and elongated rod-like shape or spindle-like shape with high stiffness of crystalline fragments [38], which are reported as cellulose whisker [39], nanowhisker [40], nanorod [41], and spherical nanocrystal [42].

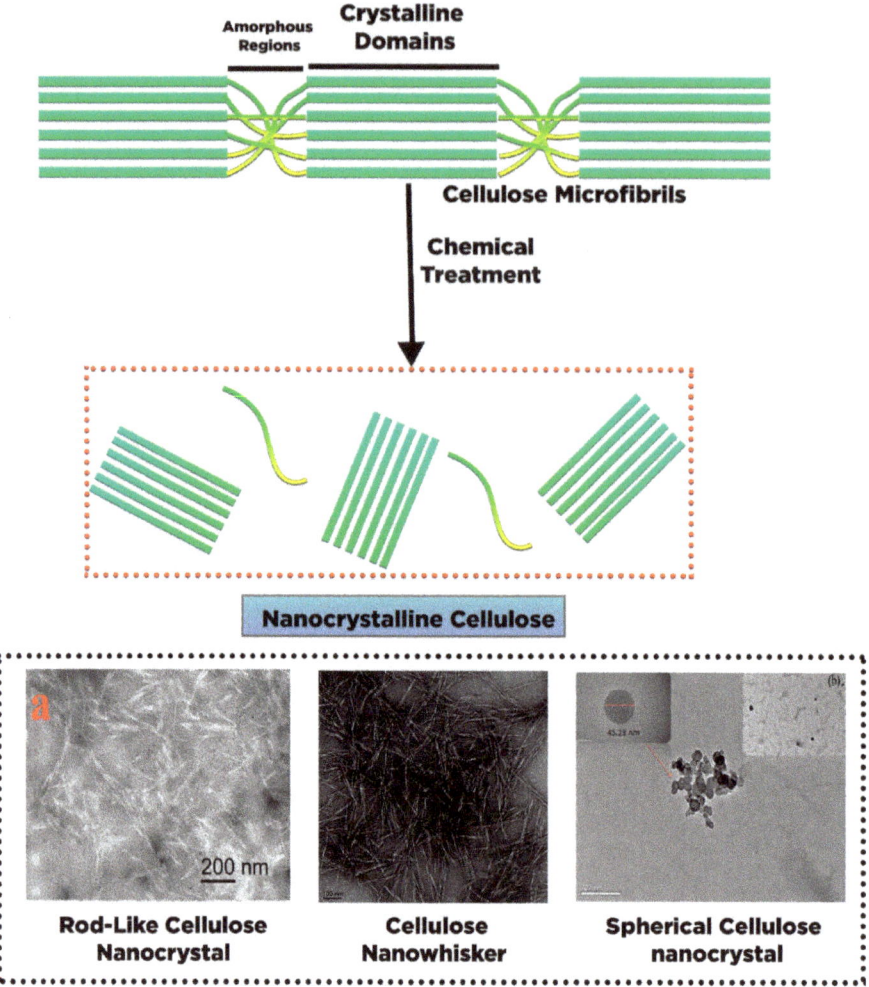

Figure 3. Schematic representation nanocrystalline cellulose fabrication by chemical treatment ((**a**) transmission electron microscopy (TEM) images of rod-like cellulose nanocrystals [38], reprinted with permission; transmission electron microscopy (TEM) images of cellulose nano whisker reprinted with permission from [25]. Copyright © 2019 Elsevier B.V.; (**b**) transmission electron microscopy (TEM) images of spherical cellulose nanocrystal reprinted with permission from [43]. Copyright © 2018 Elsevier B.V.).

A top-down process has been applied for NCC production in which a large unit of cellulose fibers (cm) is disintegrated through chemical or mechanical treatment into small units of nanocellulose (nm) [44]. NCC's chemical structure is constructed by intra- and intermolecular hydrogen linkage of cellulose macromolecules with a high crystallinity value varying from 54 to 88% [45]. NCC's particle size depends on the origin of the cellulose sources, with the diameter and length typically varying between 5 and 30 nm and between 100 and 500 nm, respectively [46]. Thus, NCCs have become an attractive candidate as drug carriers, given their outstanding physical and chemical properties [21,47,48].

Cellulose nanofiber (CNF), also known as cellulose nanofibril, micro-fibrillated cellulose, nano-fibrillar cellulose, nano-fibrillated cellulose, or cellulose microfibril, has a similar molecule structure to NCCs with nano-size particles. Similar to NCC, CNF can also be produced from various cellulose sources. However, the morphology and crystallinity of NCC and CNF are the unique features that differentiate these two cellulose-based compounds. CNFs have long, flexible cellulose chains of amorphous and crystalline regions isolated from cellulose fibrils through mechanical treatment (Figure 4) [46]. The diameter of CNFs varies from 1 to 100 nm, while their length varies between 500 and 2000 nm. The dimension of CNFs molecules is strongly influenced by mechanical treatment and defibrillation [49].

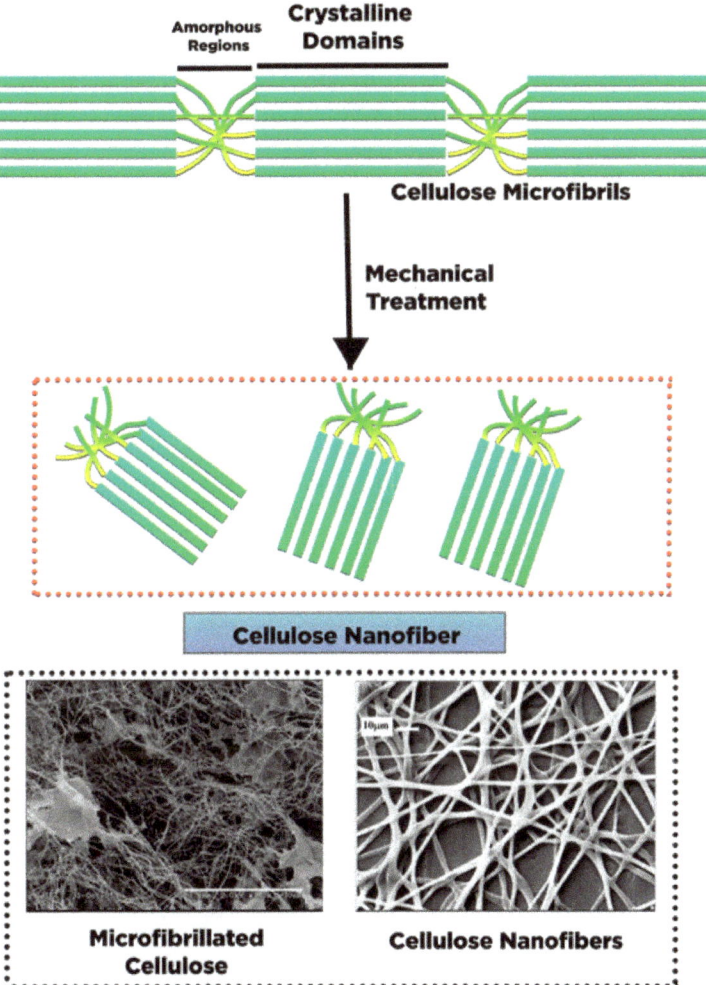

Figure 4. Schematic representation of cellulose nanofibers fabrication by mechanical treatment (scanning electron microscopy (SEM) images of micro fibrillated cellulose reprinted with permission from ref. [50]; Copyright © 2007 Elsevier Ltd.; scanning electron microscopy (SEM) images of cellulose nanofibers reprinted with permission from ref. [51]. Copyright © 2006 Elsevier Ltd.).

NCC has high crystalline cellulose purity, resulting in a rigid structure, whereas the CNF structure consists of irregular amorphous parts, with some parts exhibiting a high degree of crystallinity. The amorphous regions in CNF control the structure flexibility of nanocellulose [52]. Figure 4 presents an illustration of CNF extracted from cellulose fragments via mechanical defibrillation. The exerted force fractures the cellulose fibrils along its longitudinal axis [34]. Compared with NCC, CNF exhibits unique properties such as extended length with excellent aspect proportion (length to diameter), superlative surface area, hydrophilicity, biocompatibility, and adjustable characteristic through surface modification [53].

Microbial cellulose (MC), bacterial nanocellulose (BC), and bio-cellulose (BC) have been used as the other terms for bacteria cellulose (BC). In contrast to NCC and CNF, BC's structure comprises sugars with low molecular weight. Many bacteria strains have been used to generate BC as an extracellular metabolic product, such as *Gluconacetobacter, Sarcina, Aerobacteria, Escheria, Achromobacter, Rhizobium, Rhodobacter, Azotobacter,* and *Agrobacterium* [54,55]. However, only *Gluconacetobacter xylinus* has been commercially utilized to produce BC on an industrial scale [27]. The bacteria strains are commonly incubated in nutrient-rich aqueous media and produce BC on the upper layer (interface with air) as an exopolysaccharide. In this case, the β-D-glucopyranose units are initially present during the growth of cellulose molecules within the bacterial cell. The elementary fibril is released across the pores of the cellulose surface, which was further arranged and crystallized into microfibrils with twisting ribbons shape followed by pellicle formation (Figure 5) [56]. The fabricated BC comprises a nanofibers framework with a diameter of 20–100 nm with a length of several micrometers and a large surface area composed mainly of water (99%) [57].

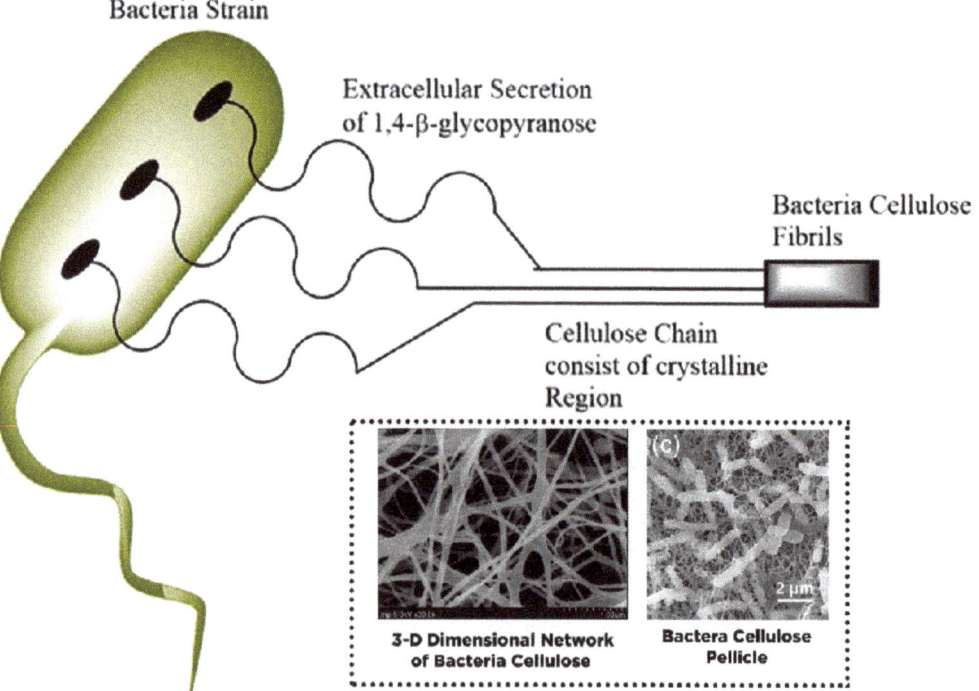

Figure 5. Schematic production of bacteria cellulose through extracellular secretion (scanning electron microscopy (SEM) images of 3-D dimensional network of bacteria cellulose [58], reprinted with permission; (c) scanning electron microscopy (TEM) images bacteria cellulose pellicle, reprinted with permission from ref. [59]. Copyright © 2019 Elsevier Ltd.)

In terms of chemical composition, BC is indistinguishable from plant-based nanocellulose (e.g., NCC and CNF). However, BC has higher crystallinity (up to 84–89%) with fewer amorphous regions than NCC and CNF. Moreover, BC contains fewer impurities and contaminants such as hemicellulose, lignin, and pectin, mainly found in plant-based nanocellulose. BC is a biocompatible material with non-cytotoxicity and non-genotoxicity for biomedical applications, especially drug delivery [60]. BC synthesis does not involve a complicated process such as mechanical and chemical treatment to cleave the hemicellulose or lignin within the lignocellulosic biomass, thereby allowing high cellulose purity.

BC's properties can be modulated by various techniques such as substrate manipulation, culture condition and operation parameter, and proper bacterial strain selection [17,54]. In contrast to NCC and CNF, BC provides unique characteristics such as high crystallinity of nanocellulose (84–88%) and polymerization grade, high water uptake capacity (exceeding 100 times of its weight), large surface area (high aspect proportion of fiber), outstanding tensile strength (Young modulus of 15–18 GPa), flexibility, foldability, moldability, mechanical stability, and high porosity [60]. A summary of the characteristics of various types of nanocellulose is listed in Table 1.

Table 1. Summary of the characteristics of various types of nanocelluloses.

Types / Parameter	Nanocrystalline Cellulose (NCC)	Cellulose Nanofibers (CNF)	Bacterial Cellulose
Common names	Cellulose whisker, cellulose nanowhisker, cellulose nanowire, and cellulose nanorod or spherical cellulose nanocrystals	Cellulose nanofibril, micro fibrillated cellulose, Nanofibrillar cellulose, Nanofibrillated cellulose, and cellulose microfibril	Microbial cellulose (MC), bacterial nanocellulose (BC), and bio-cellulose (BC)
Morphological structure	Needles like shape, elongated rod-like shape, and spindle shape	Smooth, extended, and flexible chain	Twisted ribbons like shape
Structure of Nanocellulose	Crystalline domains	amorphous and crystalline domains	Crystalline domains
Chain Length	≥500	500–15,000	4000–10,000
Crystallinity (%)	54–88	-	84–88
Other Impurities and contaminant	Possible to contain hemicellulose, lignin, and pectin	Possible to contain hemicellulose, lignin, and pectin	Contain no hemicellulose, lignin, and pectin
Size (Length and Diameter)	Diameter: 5–30 nm and Length: 100–500 nm	Diameter: 1–100 nm and Length: 500–2000 nm	Diameter 20–100 nm and several micrometric lengths
Process System	Top-down system	Top-down system	Bottom-up system
Tensile strength (Gpa)	7.5–7.7 [34]	13	0.2–0.3
Modulus Young (Gpa)	110–220 [45]	Approximately 15	18–20 [60]
Density (gr/cm^3)	1.6 [61]	1.42	1.1
Characteristics	Homogenous nanorod form, exceptional aspect ratio (length to diameter), appreciable specific surface area (SSA), biocompatibility, liquid crystalline attribute, inferior breaking expansion, high young's modulus, hydrophilicity, outstanding mechanical stiffness, tunable surface characteristic due to the reactive hydroxyl group and low density	Extended length with excellent aspect proportion (length to diameter), superlative surface area, hydrophilicity, biocompatibility and adjustable characteristic through surface modification afforded by high extensive of hydroxyl groups in CNF.	High crystallinity of nanocellulose (84–88%) and polymerization grade, high water uptake capacity (exceeding 100 times of its weight), remarkable surface area (high aspect proportion of fiber), outstanding tensile strength (young modulus 15–18 Gpa), and flexibility, foldability, moldability, mechanical stability, highly biocompatible material, non-cytotoxic, un-genotoxic and high porosity

Based on the previous discussion, cellulose can be subjected to a mechanical, biological, and chemical treatment to produce three different NCs, i.e., nanocrystalline cellulose, cellulose nanofibrils, and biological cellulose. They are classified based on various aspects such as morphology, particle size, crystallinity, nanocellulose structure, extraction techniques, and cellulose sources [56]. Moreover, other important factors such as interfibrillar arrangement, microfibril inclination, chemical constituent, cell dimension, and defects can also vary depending on the cellulose sources [62]. Among all the mentioned characteristics, mechanical strength is essential in the drug delivery field [63]. As summarized in

Table 1, NCC possesses a high modulus young, up to 220 GPa, which is higher than glass (86 GPa) [61] and kevlar KM2 fiber (88 GPa) [45]. Furthermore, the mechanical stiffness of NCC can reach up to 7.7 GPa, which is higher than 302 stainless steel (3.88 GPa) [45] and kevlar KM2 fiber (1.28 Gpa) [45].

3. Sources and Pretreatment of Raw Materials for Nanocellulose Productions

In general, the production of nanocellulose (NC) consists of three steps: (1) Finding the suitable sources, (2) raw material pretreatment, and (3) NC extraction. The raw material's source and type influence the physical and chemical properties and the NC product's yield. Currently, most nanocellulose sources utilize high-quality biomass such as cotton, wood pulp, and dissolving pulp, which comprises the high cellulose content. However, in response to recent essential issues, such as the depletion of non-renewable energy and increasing global temperature, the researchers realized the development of waste-based biomass as a feedstock for the production of nanocellulose. Various types of biomass waste, including forest residues, algae, agricultural, and industrial by-products, appear as potential raw materials for nanocellulose production. In terms of chemical composition, each category of biomass waste is primarily composed of cellulose, lignin, hemicellulose, pectin, and other minor substances with different physical and chemical characteristics [64]. Agricultural and forest residues have similarities in their chemical composition, but lignin composition in agricultural waste is significantly high, while the cellulose content in forest residues is higher than in agricultural waste [64,65].

Among all of the waste-based cellulose sources, the nanocellulose extraction from industrial waste seems more complex since the chemical and structural composition of feedstock is variable and crucially depends on the residue types. The various impurities (e.g., hemicellulose, lignin, wax, and pectin) act as a structural barrier that hinders the accessibility to the cellulose material for the extraction process [22]. Therefore, pretreatment is necessary to remove the cellulose framework's impurities, permitting the aperture of the material framework to expedite cellulose microstructure access. Moreover, removing impurities is also beneficial to reduce the consumption energy of mechanical treatment for cellulose disintegration [66]. Another objective of raw material pretreatment is to regulate the biomass structure and size and overcome the plant cell wall recalcitrance.

The pretreatment is generally divided into four categories such as physical (milling, grinding, microwave, ultrasound, etc.), chemical (dilute acid, mild alkali, TEMPO mediated oxidation, organosolv, and ionic liquid), biological (fungi, bacterial, and archaeal), and physicochemical (steam explosion, liquid hot water, wet oxidation, etc.) [67]. The effectiveness of the biomass pretreatment process depends on pH, temperature, type of catalyst, and pretreatment time. Selecting the appropriate pretreatment would allow avoiding the structure disintegration or loss of cellulose, ensuring low cost, and minimizing energy use to reduce toxic and hazardous waste [68].

The chemical pretreatment process is considered the most efficient and economically feasible for the disintegration of biomass with low pretreatment severity. However, chemical pretreatment is non-environmentally friendly and requires a wastewater treatment process [69]. Physical pretreatment is environmentally friendly and scarcely generates hazardous or toxic substances, but the major disadvantage lies in its high energy consumption, which is generally higher than chemical treatment [70]. Biological treatment is widely known as an eco-friendly process, operates under mild conditions, and consumes a lower energy amount. However, long pretreatment duration, low conversion, and carbohydrate loss tendency throughout pretreatment remain the main challenges of biological pretreatment by the microorganism [71]. Physicochemical pretreatment using a combination of chemicals and high temperature or pressure in extreme conditions can effectively escalate biomass degradation. Nevertheless, high energy input is required, which translates to high operation costs for this method. Proper pretreatment of cellulosic fibers can improve the hydroxyl group's accessibility, inner surface enhancement, crystallinity alteration, and fracture of the intra and inter hydrogen bonds of cellulose, leading to the increased fibers reactivity [72]. Detailed pretreatment of cellulose-based raw materials has been thoroughly discussed elsewhere [73].

The integrated pretreatment strategy of lignocellulosic waste biomass comprising two or more pretreatment stages increases the pretreatment process's effectiveness, product characteristics, and versatility of composition in extracted cellulose. An additional process that adds more steps to cellulose purification is highly undesirable [74]. For instance, de Carvalho Benini [75] performed alkaline treatment coupled with multiple stages of bleaching pretreatment followed by sequential dilute acid hydrolysis to increase the efficiency of impurities removal (e.g., starch, hemicellulose, and lignin/pectins) from the cellulose framework. Similarly, Wijaya et al. [29] combined alkaline and bleaching treatment to obtain higher purified cellulose from passion fruit peel. In a different study, Maciel et al. [76] obtained the soluble and insoluble lignin after alkaline treatment reached 60 and 75%, respectively. The summary of the pre-treatment strategy of waste-based nanocellulose sources is presented in Table 2.

Table 2. Summary of waste-based sources for nanocellulose production and its characteristic.

Waste Residue Sources	Nanocellulose Isolation Technique		Nanocellulose Characteristics	References
	Pretreatment	Treatment		
WASTE BASED FOREST RESIDUE				
Birch and Spruce sawdust	Hot water treatment and subsequent delignification; TEMPO oxidation	Mechanical defibrillation	CNF σ = 171,6 MPa; E = 6.4 Gpa;	[77]
Medium-density fiberboard	Soxhlet extraction (Ethanol and toluene), NaOH, and recurrent bleaching	Acid hydrolysis (H_2SO_4)	NCC L:164.7 nm; W: 6.7 nm; CrI (%): 71	[78]
Eucalyptus sawdust	Hot water treatment, alkaline delignification, O_2 residual delignification, TEMPO-mediated Oxidation	High pressure homogenization	CNF D_{avg}: 41.0 nm; SSA: 60 m^2/g; Y (%) = 60	[79]
Pinecone biomass	Alkali treatment followed with acidification ($NaClO_2$:CH_3COOH)	Mechanical grinding.	CNF σ: 273 MPa; E: 17 GPa; CrI (%): 70%; D: 5–20 nm.	[80]
Logging residues	Alkaline and bleaching pretreatment	Acid hydrolysis (H_2SO_4)	NCC L/D > 10; CrI (%): 86–93; TS (°C): 208.4–211	[81]
Bamboo log chips	Pretreatment with glycerol; and screw extrusion	Mechanical refining/Milling treatment assisted by H_2SO_4 (0.15%) as a catalyst	CNF D: 20–80 nm; CrI (%): 52.7%; Y (%): 77.2	[82]
WASTE BASED ALGAE RESIDUE				
Cladophorales	-	TEMPO Oxidation;	CNF W: 80 nm; SSA: 77 m^2/g CrI (%): 93%; D: 80 nm; Excellent mechanical and rheological characteristics	[83]
Red algae	-	Acid hydrolysis (H_2SO_4)	NCC L: 432 nm; W = 28.6 nm; L/D: 15.1; CrI (%): 69.5; Yield: 20.5%; TS (°C): 220 °C	[84]
Green Seaweed *Ulva lactuca*	Methanol pretreatment (Soxhlet extraction) followed by bleaching, alkaline pretreatment, and neutralization	Acid hydrolysis (H_2SO_4)	NCC CrI (%): 83; TS (°C): 225 °C	[85]
Industrial kelp (*Laminaria japonica*) waste	Two stages of bleaching pretreatment (Chlorine dioxide followed with hydrogen peroxide)	Acid hydrolysis (H_2SO_4)	NCC L: 100–500 nm; D = 20–50 nm; L/W: 5–20; Yield: 52.3%; TS (°C): 240 °C	[86]

Table 2. Cont.

Waste Residue Sources	Nanocellulose Isolation Technique		Nanocellulose Characteristics	References
	Pretreatment	Treatment		
Dealginate kelp residue From Giant Kelp (Calrose variety)	Na_2CO_3 (2% wt) treatment, residual sodium alginate extraction by NaOH (2% wt); Ultrasonic irradiation; $NaClO_2$ (0.7% wt) buffer solution bleaching treatment and delignification	Acid hydrolysis (H_2SO_4)	NCC L: 100–500 nm; D = 20–50 nm; L/W: 30–70; CrI (%): 74.5; TS (°C): 120–180 °C; l = 120–480 nm	[87]
Chaetomorpha antennina	Bleaching method	Acid hydrolysis (HCl) followed with Ultrasonic irradiation	CNF E = 0.9 Gpa; CrI (%): 85.02; Y = 34.09%; TS (°C) = 200–370 °C	[88]
Gelidium sesquipedale	Soxhlet Extraction (Ethanol: Toluene) Bleaching treatment, delignification (5% KOH solution)	Acid hydrolysis (H_2SO_4) followed with neutralization (NaOH)	NCC L: 467–1650 nm; D = 18–29 nm; L/W: ~40; CrI (%): ~70%;	[89]
Gelidium elegansred	Alkali and bleaching pretreatment	Acid hydrolysis (H_2SO4)	NCC L: 547.3 nm; D = 21.8 nm; L/W: 25; CrI (%): 73%; TS (°C): 334 °C	[90]
WASTE BASED AGRICULTURAL RESIDUE				
Waste sugarcane bagasse	Acidification and alkaline pretreatment	Acid hydrolysis (H_2SO_4)	NCC L: 170 nm; D = 35 nm; h = 70–90 nm; CrI (%): 93%; TS (°C): 249–345 °C	[91]
Jute dried stalks	Alkali treatment followed by steam explosion; sodium chlorite bleaching	Acid hydrolysis (oxalic acid) followed by steam explosion.	CNF L: few micrometers D = 50 nm; CrI (%): 82.2%; E: 138 Gpa; TS (°C): 250–400 °C	[92]
Coconut husk	Ultrasonic-aided solvent submersion. Delignification and Bleaching Pretreatment, followed by TEMPO-mediated Oxidation (TEMPO/NaClO/$NaClO_2$; pH = 4.8)	Ultrasonication	CNF L: 150–350; D = 2–10 nm; CrI (%): 56.3%; TS (°C): 190–380 °C	[93]
Citrus waste	Alkaline and Bleaching Pretreatment	Enzymatic hydrolysis and ultrasonication	CNF L: 458 nm; W: 10.3 nm; D_{avg} = 10 nm; L/W: 47; CrI (%): 55%; TS (°C): 190–380 °C	[94]
Raw rice husk	Size Reduction, Soxhlet extraction (toluene and ethanol); Acidification ($NaClO_2$ and CH_3COOH); and delignification (5% KOH)	High pressure homogenization and high-intensity ultrasonication processes (500 W, 40 min).	CNF L: 1800 nm; W: 10 nm; CrI (%): 77.5%; L/D > 180; TS (°C): 323 °C	[95]
Corn cobs	-	One pot synthesis via mechanochemical esterification	CNF σ = 110–125 MPa; E = 5.5 Gpa; D: 1.5–2.8 nm	[96]
Kenaf bast fiber	Delignification and three stage of bleaching pre-treatments	Mechanical grinder	CNF D: 1.2–34 nm; CrI (%): 82.52%; Y (%) 60.25; TS (°C): 200–400	[97]
Passion Fruit Peels	Alkaline and bleaching pretreatment	Acid hydrolysis (H_2SO_4) followed with ultrasonication	NCC L: 103–173.5 nm; CrI (%): 77.96%; TS (°C): 303.4; Y (%): 58.1	[29]
WASTE BASED INDUSTRIAL BY PRODUCT				
Olive industry solid waste	Pretreatment including pulping and bleaching	Acid hydrolysis (H_2SO_4)	NCC	[98]
Lime residues	Autoclaving pretreatment	High shear and high-pressure homogenization	CNF D: 5–28 nm; CrI (%): 44–46	[99]

Table 2. Cont.

Waste Residue Sources	Nanocellulose Isolation Technique		Nanocellulose Characteristics	References
	Pretreatment	Treatment		
Recycled Tetra Pak Food Packaging Wastes	Delignification and bleaching pretreatment	Acid hydrolysis (H_2SO_4) followed with ultrasonication	NCC L: 127–258 nm; D: 11.4–14 nm; L/D: 10; CrI (%): 94.8%; TS (°C): 204	[100]
Waste paper	Deinking method and alkaline pretreatment	Acid hydrolysis (H_2SO_4) followed with ultrasonication	NCC L: 271 nm	[101]
Discarded cigarette filters	Ethanol extraction, alkaline pretreatment, and bleaching pretreatment,	Acid hydrolysis (H_2SO_4) followed with ultrasonication	NCC L: 143 nm; W: 8 nm; CrI (%): 96.77%; Y (%): 29.4	[102]
Recycled Paper Mill Sludge	Ozonation pretreatment	Acid hydrolysis (Maleic acid)	NCC L: 2431 nm; W: 165 nm; L/D: 16.7 CrI (%): 77%; Y (%): 0.8	[103]
Citrus Pulp of Floater (CPF)	Alkaline and bleaching pretreatment with autoclave	Enzymatic hydrolysis	n.d CrI (%):60	[104]
Sweet lime pulp waste	Blending and acid hydrolysis (H_2SO_4)	*Komagataeibacter europaeus* SGP37 incubated in static intermittent fed-batch cultivation	BNC Y(g/L): CrI (%):89.6; TS (°C): 348	[105]

Abbreviation: D: Diameter; L: Length; W: Width; TS: Thermal Stability; Y: Yield; L/D: Aspect Ratio; CrI: Crystallinity Index; l: Lateral size; σ: Tensile strength; E: Young Modulus.

4. Isolation of Nanocellulose

4.1. Isolation of Nano-Fibrillated Cellulose (NFC)

Regardless of its cellulose sources, NFC is mainly fabricated from cellulose pulp through mechanical treatment by breaking down the linkage of interfibrillar hydrogen [106]. The exerted mechanical force triggers the cracking phenomenon to form a critical tension center in fibrous substances. The development of NFC from fibrous material requires intense mechanical treatment with or without pretreatment. However, fibrous material's mechanical disintegration may cause pulp clogging, causing the fiber to agglomerate and require high energy to break it down. Thus, another pretreatment is required to overcome this problem.

Several pretreatments have been introduced before the primary mechanical treatment to diminish the polymerization degree and debilitate the hydrogen linkage. These pretreatments include mechanical refining, alkaline hydrolysis, solvent-assisted pretreatment, organic acid hydrolysis, 2,6,6-tetramethylpiperidine-1-oxyl (TEMPO)-mediated oxidation, enzymatic disintegration, periodate-chlorite oxidation, oxidative sulfonation, cationization, ionic liquid, carboxymethylation, deep eutectic solvents, and acetylation [17].

The earliest production of NFC was reported by Turbak et al. [107] and Herrick et al. [108]. They isolated NFC from wood via high-pressure homogenization (HPH). HPH exerted a mechanical force on cellulose fibrils driven by crushing, shear, and cavitational forces in which cellulose pulp is transferred into the chamber through a small nozzle to enable particle size reduction to the nanoscale of the cellulose fibrils [72]. Currently, the HPH is the most commonly utilized method for NFC production on an industrial and laboratory scale, given its simplicity, high efficiency, and lack of organic solvent requirements [109]. Furthermore, HPH enables high conversion of cellulose material toward CNF. High energy, high pressure, and long duration of the HPH process may also escalate the fibrillation degree. However, the difficulty of cleaning the equipment due to the blockage in the homogenizer valve is the major drawback of the HPH method [110]. Different processes have also been developed to produce CNF, such as micro-fluidization, micro-grinding, cryo-crushing, ultrasonication, mechanical refining, radiation, ball milling, blending, extrusion, steam explosion, and aqueous counter collision [111].

4.2. Isolation of Cellulose Nanocrystal (NCC)

According to the previous discussion, the main difference between NCC and CNF lies in their structure, in which CNF comprises amorphous and crystalline regions while NCC has high crystalline purity in cellulose regions. Therefore, the primary step in isolating NCC is to break down the disordered amorphous or paracrystalline regions that integrate the crystalline regions within cellulose fibrils. Initially, an NCC suspension was produced in 1949 from lignocellulosic biomass through an integrated alkaline and bleaching pretreatment and acid hydrolysis [13]. Acid hydrolysis remains the paramount process for NCC extraction. The crystalline part in cellulose fibers is not hydrolyzed because it has a high resistance to acids, although acids can easily hydrolyze the amorphous regions [112]. In this method, sulfuric acid (H_2SO_4), hydrochloric acid (HCl), hydrobromic acid (HBr), and phosphoric acid (H_3PO_4) have been extensively employed as the acid component to breakdown the amorphous region of cellulose.

Following acid hydrolysis, the remaining free acid molecules and other impurities should be removed by diluting and washing with water using centrifugation and dialysis processes. Moreover, specific mechanical treatment like sonication may be needed to stabilize the NCC particles in uniform suspensions. However, the high tendency of corrosion, low recuperation rate, and high acid wastewater produced due to the high amount of water for the washing process for nanocellulose suspension neutralization become the significant drawbacks of the acid hydrolysis process [46]. To avoid excessive equipment corrosion and environmental issue, various nanocellulose isolation processes have been developed, such as extraction using ionic liquids, TEMPO oxidation, enzymatic, and others. Various researchers have carried out the combination and integration of various isolation processes to increase the isolation process's efficiency, such as enzymatic hydrolysis with TEMPO oxidation and enzymatic hydrolysis with ultrasonication [113]. Chemical treatment is crucial for NCC isolation, while mechanical treatment is the vital stage for CNF production.

4.3. Isolation of Bacteria Cellulose (BC)

The selection of strains of microorganisms is a very crucial factor in the synthesis of BC. There are currently two main methods that have been used for BC production, i.e., static fermentation and submerged fermentation [54]. Static fermentation has been widely employed as an extracellular-based production route. In the static fermentation, a 3D network of gelatinous pellicles with high water content formed during the interspersing and intertwining of the ribbons structure form of BC, reaching a particular thickness corresponding to longer incubation time and causing the entrapment of bacteria cells and its further inactivity. The static fermentation produces BC with excellent crystallinity and mechanical strength, although prolonged cultivation and low productivity limit their industrial utilization.

Furthermore, the BC layer's uneven thickness is produced due to the exposure of bacteria to uncertain conditions (nutrient, oxygen level, and cell distribution) throughout the growth cycle. Fed-batch strategies and submerged fermentation involving aeration and agitation fermentation have been introduced to overcome static fermentation's significant drawbacks. Submerged fermentation leads to higher BC productivity than static fermentation, which has been extensively utilized commercially. The cultivated bacteria are adequately exposed to oxygen, thereby generating a high yield of BC in the shape of small granules or pellets during aerated fermentation [114]. Moreover, agitation in the fermentation would result in a more homogeneous BC and oxygen evenly distributed to bacterial cells. However, the produced BC has lower crystallinity and mechanical strength than static fermentation [115].

Several submerged fermentation issues such as the advancement of cellulose non-production strains [116], irregular shapes of BC granules or pellets, and physical characteristic modification of BC remain challenging for the researcher to overcome. In addition, excessive-high rotation speed and hydrostatic stresses may promote gluconic acid production by bacteria due to the accumulation of self-protection metabolism [117]. Several factors such as bacterial strains, fermentation medium carbon sources, growth condition, and its characteristic and yield should be evaluated carefully to choose the most suitable BC synthesis process selection approach. The summary of the recent studies of BC production is given in Table 3.

Table 3. Recent study of bacteria cellulose production.

Bacteria Cultivation	Source of Carbon and Its Concentration	Culture Medium	Fermentation Conditions	Yield (g/L)	References
Komagataeib acter xylinus K2G30 (UMCC 2756)	Glucose	GY Broth	Static; 28 °C; 9 days	6.17 ± 0.02	[118]
	Mannitol			8.77 ± 0.04	
	Xylitol			1.36 ± 0.05	
Komagataeibacter rhaeticus PG 2	Glycerol	Hestrin–Schramm (HS) liquid media	Static; 28 °C; 15 days	~6.9	[119]
	Glucose			~4.05	
	Sorbitol and Mannitol			~1.65–3.41	
Komagataeibacter xylinus B12068	Glucose	Hestrin–Schramm (HS) liquid media	Static; 30 °C; 7 days	~2.2	[120]
	Sucrose			~1.6	
	Galactose			~1.4	
	Maltose and Mannitol			~0.1–0.2	
Komagataeibacter medellinensis	Glucose	Standard Hestrin–Schramm (HS) Medium	Static; 28 °C; 8 days	2.80	[121]
	Sucrose			1.68	
	Fructose			0.38	
Gluconacetobacter xylinus (PTCC 1734)	Date syrup	Yamanaka	150 rpm; 28 °C; 7 days,	~1.15	[122]
	glucose			~0.85	
	mannitol,			~1.4	
	sucrose			~1.45	
	food-grade sucrose			~0.7	
	Date syrup	Hestrin–Schramm		~0.65	
	glucose			~0.7	
	mannitol,			~1.05	
	sucrose			~1.5	
	food-grade sucrose			~1.1	
	Date syrup	Zhou		~0.9	
	glucose			~1	
	mannitol,			~1.85	
	sucrose			~1.65	
	food-grade sucrose			~1.15	

5. Surface Chemistry of Nanocellulose for Drug Delivery

Biocompatibility, biodegradability, and drug carrier capability to confine, control, and localize the drug release towards the target sites are desirable for nano-drug carrier formulation. The ability of nano-drug carriers to transport the drug and specify the sites for targeted drug release is influenced by the particle size, the surface charge, modification, and hydrophobicity. These aspects govern the nano-drug carrier interface with the plasma membrane and its diffusion across the physiological drug barrier [123]. Most NCs exhibit high specific surface area and negative interface charges as potential drug carriers, making them suitable as hydrophilic drug carriers. Therefore, the NCs' surface can be attached to the desired drug [124]. However, pristine NC cannot be used effectively as a drug carrier given its limited water solubility, moisture sensitivity, thermal instability, and lack of stability in various buffer solutions. Even though the pH adjustment of the environment can enhance the dispersibility of NCs, the scattering examination divulged the aggregation tendency of NCs, which translates to the colloidal instability of NCs. The size reduction obtained by converting cellulose into NC provides an exponential improvement of hydrogen bonding that triggers the NC aggregation. This limitation can be made worse by the drug coordination, which is exposed on the NC exterior, consequently altering the dispersibility and solubility [125]. Therefore, various surface modification and pretreatment fiber methodologies have been developed to overcome limitations and advance specific characteristics [126].

From a structural perspective, the three hydroxyl groups in each cellulose monomer are the most prominent characteristic that makes the NC surface reactive. The reactivity of hydroxyl groups influences the surface modification of anhydroglucose units. It was reported that in the molecular framework of cellulose, the hydroxyl group at the sixth position behaves as primary alcohol with a reactivity ten times larger than the other hydroxyl groups, while the hydroxyl group at the second position has two-fold higher reactivity than that in the third position, both of which serve as secondary alcohols. This phenomenon manifests from the steric hindrance of each hydroxyl group, in which the hydroxyl group at the sixth position attached to the carbon atom that is connected to only one alkyl groups while the carbon atom that carries the hydroxyl groups in the second and third positions bonded to two alkyl groups [127]. Regarding the surface receptiveness of NC's hydroxyl groups, the addition of solvent and reactant may alter the group's receptiveness in diverse positions. De la Motte et al. [128] modified NCC through cationic epoxide 2,3-epoxypropyltrimethyl ammonium chloride (EPTMAC) by spray technique. It was revealed that the hydroxyl bunch receptiveness of cationic modified NC follows the order of $OH-C_6 = OH-C_2 > OH-C_3$, which was validated through nuclear magnetic resonance (NMR).

Nanocellulose surface modification for drug delivery was developed by modulating the NC hydroxyl groups. In general, the main objective of nanocellulose surface modification is to incorporate new functional groups or drug components into the nanocellulose framework to escalate the degree of substitution and the efficacy of material grafting without altering the structure, morphology, and crystallinity of nanocellulose [129]. Several processes have been developed for the surface modification of NC, either by physical or chemical processes, presented in more detail in the following sections.

5.1. Functionalization of Nanocellulose through Physical Technique

Several physical techniques such as surface defibrillation, irradiation, electric current, and electric discharge have been developed to modify and functionalize nanocellulose surfaces for diverse applications [130]. Surface defibrillation disintegrates cellulose into elementary fibrils by exerting mechanical force using various devices such as ultra-refining, a high-pressure homogenizer, a grinder, a microfluidizer, and spray-drying. In nanocellulose functionalization, the combination of nanocellulose and drug entities can be subjected to surface defibrillation to modify the morphology of nanocellulose and construct a new matrix system with a tight fiber network.

Microparticles from BC with fibrillar structure morphology have been prepared by spray-drying technique. An ultra-refining-assisted method was also conducted to construct bacteria cellulose nanofiber (BCNF) with various sizes and shapes. The coating of BCNF with mannitol (MN), maltodextrin (MF), and hydroxypropylmethylcellulose (HPCM) were also carried out at various ratios to study the drug release characteristics. The addition of such coating matrices exhibits benefits towards the spray-drying process and drug carrier ability, i.e., superior protection of drug confinement, decreased droplet adhesion on the drying chamber, and improved powder performance. As a result, the BC-microparticles can successfully enhance the drug uptake capacity and sustain the drug release of diclofenac sodium (hydrophilic) and caffeine (lipophilic) [131].

As a recent advanced method, irradiation exerted high energy, which modifies the cellulose exterior. For example, the radiated gamma energy can generate reactive intermediates comprising ions and free radicals that provoke reaction pathways such as cross-linking, scission degradation, oxidation, and polymer and molecule grafting. The presence of irradiation beams, such as microwave and electron, accelerates the polymer growth. UV-irradiation has also been developed to improve the reaction rate to allow pre-synthesized grafted polymer formation on the nanocellulose surface [132]. Recently, this method has been developed to induce polymer grafting and polymer growth on nanocellulose surfaces.

Plasma treatment is considered an environmentally friendly method to achieve nanocellulose surface functionalization by utilizing plasma ionized gas without altering its characteristics. Researchers have widely applied this method for various modifications, such as increasing material–cell interaction, introducing the surface of NC with hydrophobicity or hydrophilicity characteristics, and incorporating chitosan towards cellulose substrates. For instance, Kusano et al. [133] modified CNF by utilizing dielectric-based plasma discharge treatment, leading to the formation of many carboxyl groups, carbonyl groups, and oxygen-containing groups on the surface of nanocellulose [133]. Moreover, assisted ultrasonic irradiation combined with plasma discharge treatment can refine the wetting and oxidation of the nanofibers coating. Plasma treatment is an attractive route for surface functionalization of nanocellulose given its benefits such as non-polluting, fast-modification, and simple chemical treatments compared to the conventional modification method.

5.2. Functionalization through Chemical Synthesis of Nanocellulose

Chemical treatments use reactive chemical species for nanocellulose formation through cellulosic framework disintegration. As mentioned in the previous section, acid hydrolysis has been extensively exploited as the primary process for CNF and NCC isolation from the cellulosic fiber. The strong acidic environment leads to the disintegration of amorphous regions that act as structural defects in the cellulosic framework, facilitating nanoparticle production. Other chemical processes, such as TEMPO-based oxidation and APS oxidation, are also used in the CNF and NCC synthesis. The schematic mechanisms of acid-based hydrolysis and oxidation processes are presented in Figure 6. The summary of chemical modification of nanocellulose is tabulated in Table 4.

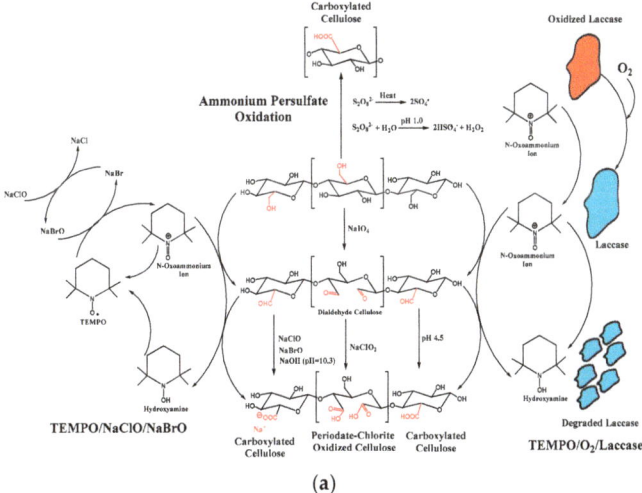

Figure 6. Cont.

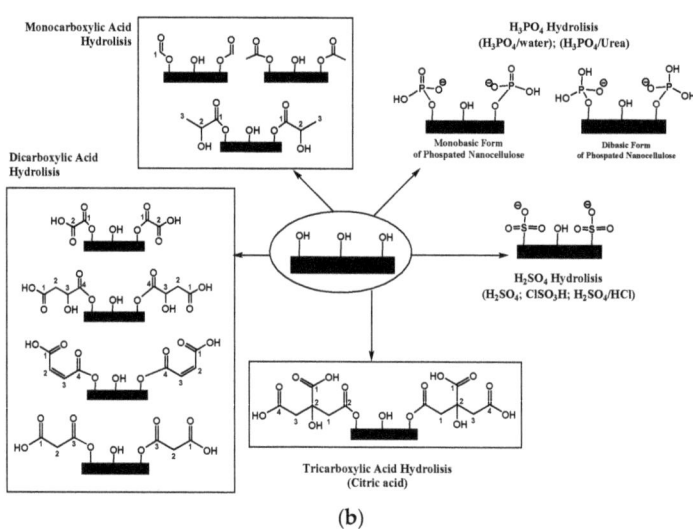

Figure 6. Simplified mechanisms of chemical synthesis nanocellulose; (**a**) acid-based chemical modification; (**b**) oxidation based chemical modification.

Table 4. The influence of chemical functionalization on morphological nanocellulose.

Methods	Reagents	Aided Reagents	Operation Parameter	Sources of Cellulose	Mechanical Technique	Yield (%)	Morphology (nm)	CI(%)	Zeta Potential (mV)	Surface Charge Density (mmol/g)	Ref.
Mineral Acids	H_2SO_4	-	52% H_2SO_4 50 °C; 60 min	Passion Fruit Peels	Ultrasonication	58.1	NCC L: 103–173.5	77.96	−25	-	[29]
		-	63% H_2SO_4 50 °C; 90 min	Microcrystalline Cellulose	Ultrasonication	30%	NCC L:250; W: 16	-	−46.1	-	[134]
		-	-	Filter Paper	-	-	NCC W: 22	85	-	-SO_3H (0.0985)	[135]
	H_2SO_4/HCl	-	H_2SO_4:HCl:H_2O (3:1:6); Ultrasonic 50 hZ; 10 h	Microcrystalline Cellulose	Ultrasonication	-	S-CNC (D:10–180 nm)	-	-	-	[136]
	ClSO$_3$H (Post-sulfonation)	-	ClSO$_3$H in 50 mL DMF; RT; 2 h	Sulfated NCC	Ultrasonication	79.31	NCC L:152; W: 22.7; h: 5.0	88%	−66.1	-SO_3H 0.409	[137]
	H_3PO_4	-	73.9% H_3PO_4; 100 °C; 90 min	Filter Paper	Blending (15 min)	76–80	NCC L:316; W: 31;	81	-	-PO_3 (0.0108)	[135]
		-	10.7 M H_3PO_4; 100 °C; 30 min				NCC	83	−27	-PO_3 (0.4352)	
		-	10.7 M H_3PO_4; 100 °C; 30 min	Cellulose Biotethanol Residue	Homogenizer (10 times)		CNF L: 2500 nm	81	−23	-PO_3 (0.018)	[138]
	H_3PO_4 in molten Urea	-	10.7 M H_3PO_4; 150 °C; 30 min				NCC L: 610 nm	83	−34	-PO_3 (1.038)	
		-					CNF L: 330–480 nm	86	−24	-PO_3 (1.173)	
	HCl	-	2.5 M HCl; 105 °C; 40 min	Filter Paper	Blending (40 min)	-	NCC W: 20	79%	-	-	[135]
Organic Acids	Acetic Acid	H_2SO_4	80 °C; 3 h	Bleached eucalyptus kraft pulp	-	81	NCC L: 264; W: 16	80	−33	-SO_3H (0.015)	[139]
		HCl	105 °C; 9 h	Cotton	Blending (20 min)	30	NCC L: 269; W: 45	-	-	-	[140]
	Formic Acid	6M HCl	80 °C; 4 h	Microcrystalline Cellulose	-	-	NCC L: 236; W: 25	88	−1.7	Formate (0.4)	[141]
		0.015 M FeCl$_3$	90 °C; 6 h	Bleached eucalyptus kraft pulp	-	75	NCC L:594	75	−6.53	Formate	[142]
	Lactic Acid	HCl	150 °C; 3 h	Cotton	Blending (20 min)	-	NCC L: 200; W = 20	80	-	Lactate	[143]

Table 4. Cont.

Methods	Reagents	Aided Reagents	Operation Parameter	Sources of Cellulose	Mechanical Technique	Yield (%)	Morphology (nm)	CI(%)	Zeta Potential (mV)	Surface Charge Density (mmol/g)	Ref.
	Butyric Acid	0.027 M HCl	105 °C; 9 h	Cotton	Blending (20 min)	20	NCC L: 226; W = 34	-	-	Butyrate	[140]
	Maleic Acid (MA)	-	70% MA; 100 °C; 45 min	Bleached eucalyptus kraft pulp	-	12%	NCC	-	−33	-COOH (0.29)	[144]
		-	60% MA; 120 °C; 2 h	Bleached eucalyptus kraft pulp	Microfluidizer (120 mPa; 5 passes)	3%	L: 329.9; h = 15.9	-	−46.9	-COOH (0.368)	[145]
						84%	CNF h: 13.4	-	−45.2	-COOH (0.059)	
	Oxalic Acid (OA)	-	8.75% OA; 110 °C; 15 min	Filter paper	Sonication (60 min)	93.77	NCC L: 150–200; W: 5–20	-	−36	-COOH, 0.29	[146]
		-	70% OA; 100 °C; 1 h	Bleached eucalyptus kraft pulp	-	24.7	NCC	80	−42.5	-COOH	[144]
		-	30% OA; 100° C; 30 min	Celery	Sonication (18 min)	76.8	CNF h: 5.5	49	−32.9	-COOH	[147]
	Malonic Acid	-	80% wt of Malonic Acid; 140 °C; 3 h			5%		-	-	-COOH	
		0.025 M HCl				19.8%		75	-	-COOH	
	Malic acid	-	80% wt of Malic Acid; 140 °C; 3 h	Ramie Cellulose	Blending (5 min)	3.4%	NCC L: ~220; W: ~12	-	-	-COOH, (1.617)	[148]
		0.05 M HCl				20%		78	-	-COOH	
	Citric Acid	-	80% wt of Citric Acid; 140 °C; 3 h			5.1		-	-	-COOH	
		0.05 M HCl				20.5		78	-	-COOH, (1.884)	
		-	80% wt of Citric Acid; 100 °C; 4 h	Bleached Baggase Pulp	Ultrasonication	32	NCC, L: 251; W: 21	78	−122.9	-COOH, 0.65	[149]
						-	CNF, L: 654; W: 32	69	190.3	-COOH, 0.3	
Oxidation Treatment	TEMPO/NaCl /NaBr	-	TEMPO (0.094 mmol)-NaBr (1.57 mmol)- NaClO (1.24 M); 10 °C; 45 min	Nanocrystalline Cellulose	Ultrasonication	-	NCC, L: 100; W: 5–20	80%	-	-	[150]
		-	TEMPO (0.1 mmol mmol)-NaBr (1 mmol)- NaClO (5 mmol/g cellulose); Ambient condition; 1.5 h	HBKP	Ultrasonication	-	CNF	85%	-	-COOH; -CHO (1.191)	[151]
	TEMPO/O$_2$/Laccase	-	50 mM TEMPO, 5 U mL−1 laccase; 96 h	HBKP	Ultrasonication	-	CNF, L: > 100; W: 4–8	-	-	-COOH; -CHO (0.837)	
	Sequential Periodate-Chlorite Oxidation	1 M Acetic Acid (2)	(1). 46 mmol NaIO$_4$; 50 °C;4.5 h followed by (2). 12 g NaClO$_2$; 50 °C; 40 h	Hardwood Pulp	Homogenizer (5 passes; 80 MPa)	-	CNF, L: 95.8; W: 2.72	-	−128	-COOH (2.0)	[152]
	APS Oxidation	-	1 M APS; 75 °C; 16 h	Cotton Linters	-	34.4	CNF, L: 95.8; W: 2.72	63.8	-	-COOH (0.16); -SO$_3$ (0.98)	[153]

In general, NCC isolation comprises exposing pure cellulose material under strong acid hydrolysis with strictly controlled operating parameters such as temperature, agitation, time, and concentration of chemical species. As mentioned earlier, various chemical reagents such as H_2SO_4, HCl, HBr, and H_3PO_4 have been utilized as cellulosic disintegrators. The selection of acid reagents has the most crucial role in determining drug carrier characteristics and synthesis pathways for incorporation or grafting through chemical/physical modification for particular functional groups. The amorphous decomposition using HCl and HBr is not widely adopted because they provide low dispersion stability of NCC and increase the agglomeration tendency of NCC in an aqueous suspension. H_2SO_4 and H_3PO_4, on the other hand, exhibit better performance as a hydrolyzing agent because the chemical moieties can be attached to the hydroxyl group of NCC during the reaction to isolate charged surface of NCC for subsequent incorporation of phosphate or sulfate functional groups. The new functional group incorporation causes the spontaneous dispersibility of NCCs in an aqueous environment due to the colloidal stability restoration through electrostatic repulsion refinement, which is the preferred characteristic of drug carriers.

A subsequent treatment of H_2SO_4 followed by HCl synthesis has been utilized to control the sulfate moieties on the NC surfaces. The as-synthesized particle had a similar particle size to those particles directly acquired from acid hydrolysis. Nevertheless, the surface charge density can be adjusted on the hydroxyl groups exploited by sulfate groups [49]. Lin and Dufresne [137] proposed a strategy of inaugurating progressive sulfate group content on NCCs surface through the modulation ratio of reactants and post-sulfonation (chlorosulfonic acid) and desulfonation conditions. They also evaluated the impact of sulfonation degree on the morphology, dimension, physical characteristic, and surface chemistry of modified NCCs. Diverse zeta potential ranged from -7 mV to -66 mV and approximately 0.0563 mmol/g–1.554 mol/g of sulfonation degree was acquired. Therefore, it is indicated that the zeta potential of nanocellulose is mainly controlled by the sulfonation degree of nanocellulose itself [137].

Wijaya et al. [29] successfully isolated NCC through sulfuric acid hydrolysis of bleached passionfruit peels waste fiber by adjusting the acid concentration, hydrolysis time, and reaction temperature. The NCC was used for tetracycline hydrochloride adsorption through electrostatic and Van der Waals interaction. The adsorption isotherm was correlated using Langmuir and Freundlich isotherm models. With pH environment adjustment, the adsorption affinity of the drug can be altered to control the uptake and sustained release of drugs [29].

(2,2,6,6-tetramethylpiperidine-1-oxyl)-mediated (or TEMPO-mediated) oxidation of nanocellulose has arisen as an alternative NC isolation route to replace the conventional acid hydrolysis method due to its environmentally friendly and facile synthesis nature. The synthesis starts by using TEMPO/NaBr/NaClO or TEMPO/NaClO$_2$/NaClO as a reagent. TEMPO (stable nitroxyl radical) forms as the catalyst for NC synthesis, which further transforms into N-oxoammonium salt ($R_1R_2N^+=O$) under certain conditions while the NaClO acts as a primary oxidant [46]. Both NaClO and NaBr can reversibly transform the N-oxoammonium salt into TEMPO form. The hydroxymethyl groups of NC (primary hydroxyl groups located on C6) are selectively transformed into carboxylated groups while the secondary hydroxyl groups remain unchanged (secondary hydroxy groups located on C2 and C3) [66]. The incorporated carboxyl groups imparted negative surface charges from the change in the environment pH, which leads to improved colloidal stability.

As reported by Montanari et al. [154], TEMPO-mediated oxidation with the degree of oxidation 0.24 has imparted negative charges on the crystalline regions of nanocellulose, which provide dispersibility and individualization improvement time decreasing the crystallite size [154]. Meanwhile, Habibi et al. [150] underlined that the TEMPO-mediated oxidation did not affect the morphological and crystallinity of NCCs. Furthermore, they highlighted that the ratio of primary oxidizing agents affected the negative charge of NCCs [150].

A novel oxidation system of TEMPO/laccase/O_2 has been utilized to modify NC. The TEMPO/laccase/O_2 system with sufficient catalytic amounts of laccase and TEMPO reagent produced reactive TEMPO$^{+,}$ which subsequently transformed primary hydroxyl moieties into aldehyde moieties through oxidation. After the oxidation, the reactive TEMPO$^+$ was reduced into N-hydroxyl TEMPO. However, no-cycle regeneration occurred between TEMPO$^+$ and N-hydroxyl TEMPO due to the breakdown of the primary hydroxyl groups of polysaccharides and laccase molecules. Furthermore, the N-hydroxyl-TEMPO was accumulated in the reaction environment due to the absence of active laccase in the system. Therefore, a large amount of TEMPO and laccase and prolonged reaction time are required to oxidize the primary hydroxyl groups, which are considered major disadvantages of this process [151].

TEMPO-mediated oxidation was mainly used to modify NFCs before mechanical defibrillation to promote the fiber's individualization. TEMPO-mediated oxidation leads to the breakage of the strong intra-fiber hydrogen coordination to facilitate the softening and impairing of its rigid structure, which is beneficial in converting TEMPO-oxidized cellulose fiber into highly crystalline individual nanofibers through mechanical treatment. The NaClO concentration and mechanical treatment strength were considered crucial factors in determining the polymerization degree, carboxylate group numbers, and CNFs yield.

Carlsson et al. [155] emphasized the influence of surface charges in nanocellulose formulation as a drug carrier by introducing TEMPO-mediated oxidation in mesoporous claodophora cellulose for aspirin degradation. The surface charge negativity (carboxylate content 0.44 ± 0.01 mmol g^{-1}) significantly accelerated the degradation of aspirin compared to the native source of CNFs, which had a deficient surface charge (0.06 ± 0.01 mmol g^{-1}). This phenomenon is caused by the strong interaction of opposite charge entities between aspirin and TEMPO-oxidized cellulose nanofibers (TOCNFs), leading to increased partial amorphization ability inside the mesoporous TOCNFs [155].

Without a chlorine-containing oxidant, 1.1 mmol g^{-1} of carboxyl groups were incorporated onto wood cellulose. High in carboxylate content, wood cellulose underwent tremendous depolymerization during oxidation. In addition, a long reaction duration of up to 15 h was required to achieve 0.6 mmol g^{-1} carboxylate content, while 1.1 mmol g^{-1} was achieved by increasing the reaction time up to 20 h. Prolonged reaction time is considered the major disadvantage of this process. This method has been utilized for nanocellulose modification in drug delivery applications [156]. The sequential periodate-chlorine oxidation selectively and simultaneously incorporates two carboxyl groups through the oxidative transformation of two vicinal secondary hydroxyl groups (located in C2 and C3 instead of C6 position), enabling higher surface charge density introduction. The increase of surface charge density is essential in retaining the colloidal stability of drug carrier and improving the electrostatic interaction between drug and carrier, which increase the loading uptake of drugs.

Plappert et al. [152] investigated the pretreatment effect of sequential chlorite periodic oxidation on open-porous anisotropic CNF hydrogel membrane assembly. Hydrogel membranes were used for transdermal drug delivery systems for nonsteroidal anti-inflammatory drugs (NSAIDs) and piroxicam (PRX). By tuning the surface charge density and the amount of carboxylated groups (0.74–2.00 mmol g^{-1}) by varying the reagent concentration, the drug carrier uptake capacity can be increased to within the range of 30–60 mg g^{-1} with the surface charge -66 mV to -128 mV. The electrostatic interaction between the cationic drug (PCX) and the anionic characterized surface of CNF membranes is the main driving factor behind the loading of drugs in the membrane [152].

5.3. Functionalization through Post Chemical Modification via Covalent and Physical Bonding Strategy

Maintaining the structural integrity of nanocellulose to prevent the polymorphic transformation and maintaining the crystalline area while modifying its surface are considered the main challenges. Therefore, several post-chemical modifications have been studied for surface modification and functionalization of nanocellulose surfaces before the drug upload. Sulfonation treatment is the most common strategy to introduce sulfate groups into hydroxyl moieties of nanocellulose, which produces a highly negatively charged surface. Nevertheless, the degree of sulfonation was highly determined by several factors such as temperature, acid concentration, and hydrolysis time. Treatment of NC with sulfuric acid or sulfonation followed by acid hydrolysis [137,157] can improve the characteristics of NCs. However, these improvements may lead to lowering the colloidal stability of NC due to the reduction in the sulfonate degree. Since the primary goal of the drug delivery system is to achieve higher colloidal stability and strong electronegativity for further electrostatic drug adsorption or modification, straight H_2SO_4 hydrolysis remains the primary treatment for NC modification.

On account of the simple and straightforward treatment, modification of hydroxyl groups at the NC surface by Fischer esterification is another common approach. Several reactants have been used to acetylate the surface of nanocellulose, such as acetic, citric, malonic, and malic acid with the combination of HCl or H_2SO_4. The utilization of H_3PO_4 provides NC modification with higher thermal stability than sulfonated NC. Camarero Espinosa et al. [135] suggested that only one hydroxyl group was incorporated by one ester bond of phosphoric groups. Another study by Kokol et al. [138] revealed the possibility of phosphate-modified nanocellulose (P-NC) originating from two structural isomers, either of which can behave as monobasic acid or dibasic groups. Acetylation of hydroxyl groups of NC can also be performed using enzymatic modification. In an environmentally friendly approach, enzymatic modification serves as a favorable modification route without the need for any addition of chemicals and has low energy requirements, improving biocompatibility and lowering the cytotoxicity of NC for drug delivery.

The acid hydrolysis and oxidation treatments are mainly considered as a primary synthesis of nanocellulose. Indeed, during acid-based hydrolysis or TEMPO-oxidation, hydroxyl groups of nanocellulose grafted by anionic sulfate ester groups ($-OSO_3^-$) and carboxylate groups (-COOH) produce the negative electrostatic layer of nanocellulose. Consequently, high stability of nanocellulose occurs in the aqueous solution resulting in electrostatic repulsion between individual particles. Maintaining the structural integrity of nanocellulose to prevent the polymorphic transformation and maintaining the crystalline area while modifying its surface are considered the main challenges. Several post-chemical modifications have been studied for surface modification and functionalization of nanocellulose surfaces before the drug upload.

Silylation is another approach to modify the surface nanocellulose by conjugating small molecules. A series of alkyl dimethyl-dimethylchlorosilane (alkyl-DMSiCl) with various alkyl groups such as isopropyl, N-butyl, N-octyl, and N-dodecyl can be grafted on the surface of NCC in the presence of toluene. However, the high price and high toxicity of the reagents limit the progress of silylation modification in the drug delivery field. Recently, Li et al. [158] developed an NC template for mesoporous hollow silica material (R-nCHMSNs) for ibuprofen and lysozyme drug delivery. The presence of NC as a template increases the content of geminal silanols on the R-nCHMSNs surface. Nanoparticles with high content of geminal silanols present outstanding delivery characteristics for various drugs [158].

The amine derivatives can covalently bond the surface of NC through a carbodiimide amidation reaction. The majority of amidation-mediated couplings were incorporated on the carboxylic groups of pre-oxidized NC without re-molding the morphology and crystalline native structure. N-ethyl-N-(3-dimethylaminopropyl) carbodiimide hydrochloride (EDAC) has been widely used for the amidation among carbodiimide derivatives. The addition of n-hydroxysuccinimide (NHS) is required to avoid unstable intermediate O-acyl urea formation and to achieve the direct formation of the stable N-acyl urea. The amidation approach was presented by Akhlagi et al. [159] to create a drug delivery system based on chitosan oligosaccharides (CSOS) and TEMPO-oxidized NCC. The carboxylic moieties on the oxidized NCC were coordinated into the primary alcohol and amino moieties of CSOS. Several limiting factors such as medium reaction, time reaction, pH, and the molar ratio of reagent and cross-linker reaction can be altered, translating to the modified grafting behavior and degree of substitution of CSOS into oxidized NCC. Electrostatic interactions were performed to achieve 21.5% of binding efficiency loading and 14% w/w of procaine hydrochloride (PrHy) loading. The rapid release profile observed in this study is suitable for local drug delivery by the oral system [159].

Direct covalent drug attachment towards the NC crystal backbone via a novel spacer arm through amine-mediated couplings is another potential strategy [160]. Tortorella et al. [160] modified NCC via periodate-oxidation-generated NCC-DAC (dialdehyde cellulose) and inserted them into molecules of g-aminobutyric acid (GABA) via the Schiff base condensation reaction. Subsequently, the nucleophilic substitution of 4-hydroxy benzyl alcohol (HBA) occurred and was followed by an acylation reaction with 4-nitrophenylchloroformiat that exerted a carbonate group for nucleophilic substitution of amino contained doxorubicin as model drug nucleophilic. Carbamate linkage adjacent to the linker presents highly stable conditions in an aqueous environment with harsh conditions, either basic or acidic. The drug release of active drugs was achieved only by hydrolysis in cells utilizing suitable enzymes to cleave a carbamate linkage (Figure 7).

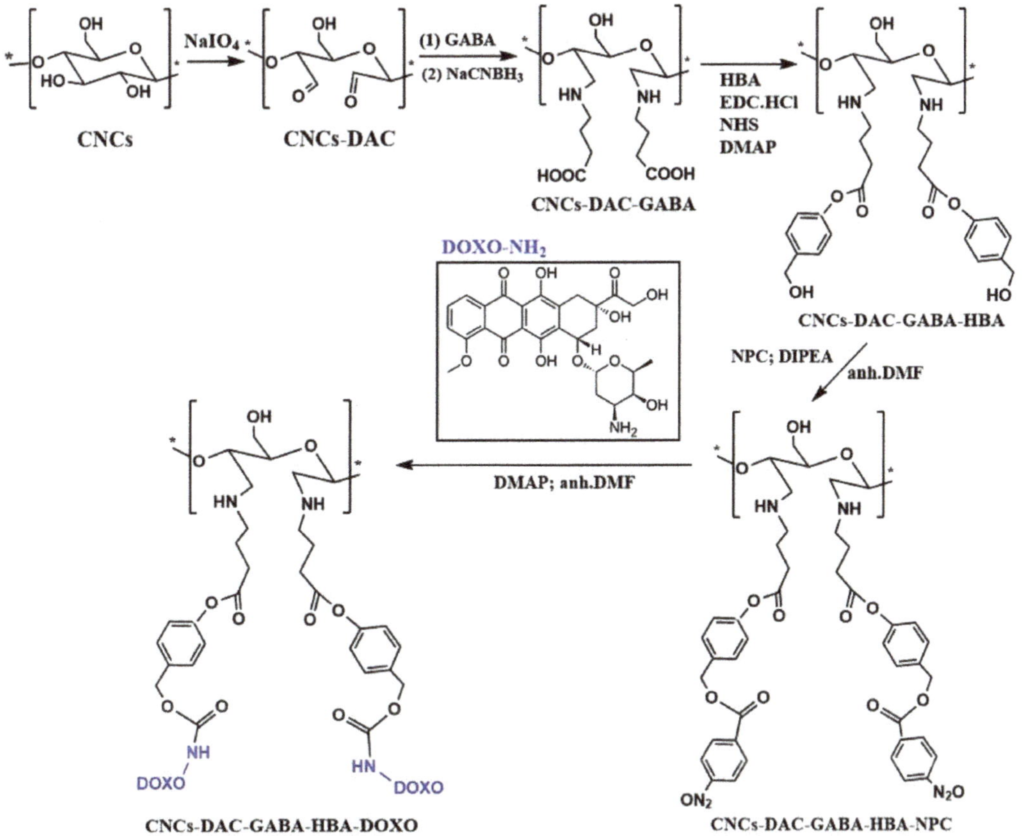

Figure 7. Illustrative representation of conjugated doxorubicin onto NCCs through chemical bonding (this picture is re-drawn from Tortorella et al. [160]. Copyright © Springer Fachmedien Wiesbaden GmbH). Abbreviations: NCCs = cellulose nanocrystals, DAC = cellulose dialdehyde, GABA = c-amino butyric acid, HBA = 4-hydroxy benzyl alcohol, EDC HCl = 1-ethyl-3-(3-dimethylaminopropyl) carbodiimide hydrochloride, NHS = N-hydroxy succinimide, DMAP = 4-dimethylamino pyridine, NPC = 4-nitrophenyl chloroformate, DIPEA = N,N-diisopropyl-N-ethylamine, anh. DMF = anhydrous dimethylformamide, DOXONH2 = doxorubicin, * is repetitive monomer molecules.

5.4. Polymer Grafting Modified Nanocellulose

Polymer-grafted NC has been introduced as the sought-after functionalization strategy to refine the drug delivery performance. Different techniques have been developed to introduce functional groups onto NC covalently, i.e., (i) Thiolene reaction; (ii) Oxime reaction; (iii) Michael addition; and (iv) imine and hydrazone synthesis. These reactions have been well-developed for polymer functionalization for drug delivery systems.

Integrating polymer onto the NC surface can be performed by the 'grafting onto' or 'grafting from' strategy. The 'grafting onto' technique requires pre-synthesized polymer attachment that can bear the reactive end groups onto either modified or non-modified hydroxyl groups of the NC surface. The adherence of polymer onto the NC surface's specific moieties can be performed via physical or chemical attachment. The 'grafting onto' approach offers the possibility of characterizing polymer before grafting and modulating the resultant carriers' characteristics.

Strong electrostatic interaction can be used to initiate the polymer grafting onto NC. There is a possibility of incorporating polydopamine (PDA) into the NCC surface to fortify the PDA material and develop NC's colloidal stability. The presence of functional groups in PDA, such as amine, imine, and catechol groups, can serve as the anchors for NC and the drug through the Van der Waals interaction, the π–π interaction, and hydrogel bonding [161].

Wang et al. [162] assembled poly(ethyl ethylene phosphate) (PEEP) that bears propargyl functionality onto azide modified nanocrystalline cellulose (NCC) Cu via Cu(I)-catalyzed azide-alkyne cycloaddition (CuAAC) "click" chemistry. In parallel, azide-modified NCC was constructed by two steps, i.e., (i) partial desulfation treatment of NCC followed by tosylation (NCC-Cl); and (ii) conversion of NCC-Cl into azido-NCC through nucleophilic substitution using sodium azide. Propargyl-PEEP was grafted onto azide modified NCC (NCC-g-PEEP) (Figure 8a). The as-synthesized suspension with negative charge can be utilized for doxorubicin (DOX) confinement through electrostatic interaction, exhibiting pH-responsive delivery in the tumor cell environment [162].

Kumar et al. [163] explored Diels-alder "click" chemistry by attachment of the metronidazole drug onto the CNFs. Initially, the TEMPO-oxidized CNFs underwent amidation with furfuryl amine. Subsequently, esterification occurred between metronidazole as a drug model and maleimide-hexanoic acid to introduce the ester function between the drug and the maleimide ring. Finally, the Diels–Alder reaction occurs between the furan functionalized CNF-t (CNF-fur) and metronidazole containing maleimide. Thus, the novel system of carrier provides the ester function on the linking chain for innovative drug carrier formulation, which induces the release in the presence of esterases enzyme [163].

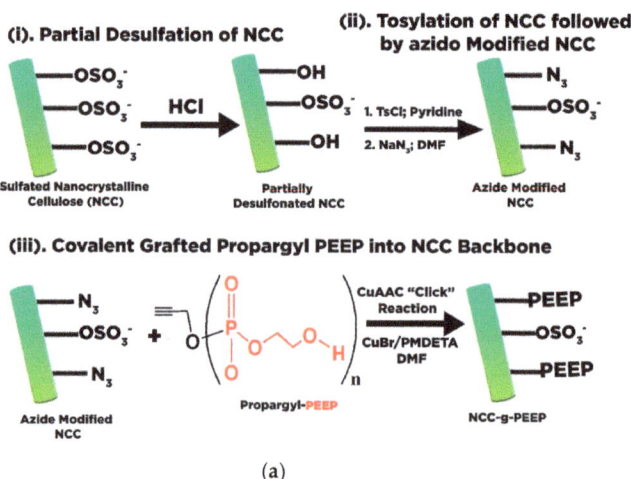

(a)

Figure 8. Cont.

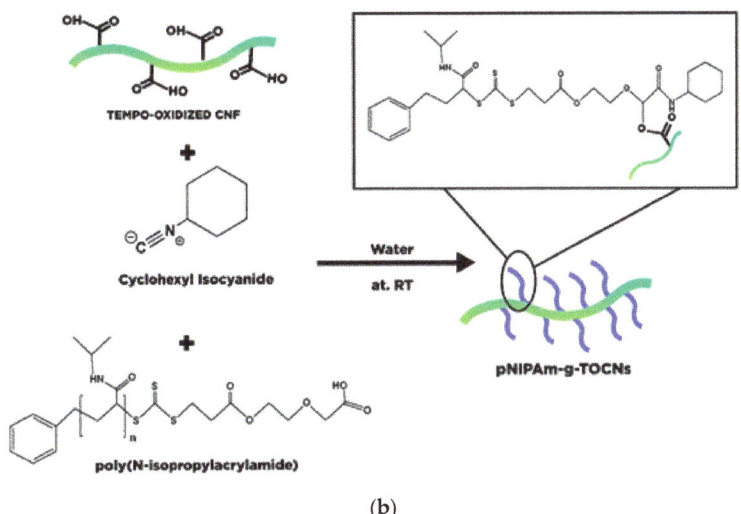

Figure 8. Schematic representation of the polymer grafting technique (**a**). CuAAC "click" reaction for NCC-gPEEP synthesis (this picture is redrawn from Wang et al. [162], Copyright 2010 Royal Society of Chemistry); (**b**) polymer-grafted cellulose fibrils (pNIPAm-g-TOCNs) via Passerini one-pot reaction (this figure is redrawn from Khine et al. [164]. Copyright © 2018 American Chemical Society).

A versatile grafting strategy for numerous functional groups is the Passerini reaction. This reaction is a multicomponent reaction (MCR) that comprises three substances, i.e., a carboxylic acid, an isocyanide, and aldehyde/a ketone, in one pot of reaction. For example, Khine et al. [164] modified poly(N-isopropylacrylamide) pNIPAm carrying aldehyde end groups via the Reversible Addition−Fragmentation Chain Transfer (RAFT) polymerization technique. Subsequently, the polymer with aldehyde functionality was further chemically grafted into TEMPO-oxidized CNFs. As a result, these materials exhibit thermal responsiveness, which is promising for use in stimuli-responsive carriers (Figure 8b) [164].

Another way of modifying NC with polymer in an aqueous solution is the NICAL reaction. For example, Khine et al. [132] demonstrated photo-induced "click" chemistry for (TEMPO)-oxidized CNF bearing carboxylic acid moieties (TOCNs) modified with the nitrile imine-mediated tetrazole under ultraviolet (UV) irradiation. The presence of fluorescence characteristics allowed for direct monitoring of NC throughout the cancer cells' incubation. In addition, doxorubicin as a drug model can be attached via electrostatic interaction to introduce excess negative charge onto carboxyl groups in the polymeric-grafted NC [132].

Undesirable reduction in surface grafting density is nonetheless observed as the major limitation. The steric barrier can hinder the optimum grafting throughout the reaction because the layer of attached polymer covered the available active sites. Therefore, an alternative strategy has been proposed, called 'grafting from'. Using this method, the polymer chains can be grown in situ on the surface hydroxyl groups of NC via ring-opening polymerization (ROP) with the presence of stannous octoate (Sn(Oct)$_2$) as an ROP agent. Another approach is atom transfer radical polymerization (ATRP) with 2-bromoisobutyrylbromide (BIBB) as the ATRP agent. These standard approaches for drug delivery have been well-reviewed elsewhere [132].

5.5. Surfactant Modified Nanocellulose

The adsorption of surfactants represents a promising alternative for the chemical modification of NC. Surfactants are classified into cationic, anionic, zwitterionic, and non-ionic. The distinct properties of the surfactant manifest through its micelle formulation in the aqueous solution, which is highly beneficial in the drug delivery system. The lack of a strong covalent bond is considered the significant drawback towards enabling molecule release. Therefore, it is necessary to study several factors affecting the interaction of surfactant and NC and their impact on drug uptake and release. Tardy et al. [165] reviewed several factors that influence the affinity of NC and the surfactant. This study provides some additional information on the affinity of NC and the surfactant on the drug delivery system.

The opposite charge between the NCC surface and CTAB drove the electrostatic interaction and physical adsorption for the NCC surface modified with the surfactant. NCC's negative charge creates a non-covalent interaction towards the cationic charge of CTAB, resulting in a strong electrostatic interaction. Zainuddin et al. [166] pointed out several factors that mainly involve the interaction between NCC and the surfactant, i.e., pH and ionic strength, the CTAB concentration, and the ratio of CTAB to NCC. They highlighted that the CTAB concentration and mass ratio of NCC: CTAB affects the interaction of surfactant-modified NCC with curcumin as a hydrophobic drug model. Increasing the CTAB concentration intensifies the hydrophobic character of the carrier, which is intensely coordinated with curcumin. However, at a high concentration of CTAB, the amount of curcumin attached tends to decrease [166].

Low surfactant concentration favors the electrostatic interaction between the monomer CTAB head with the negative charge of NCC surface, giving hydrophobic properties. While the CTAB concentration increases progressively, the adsorbed monomer of the surfactant tends to restructure and initiate surfactant cluster formation induced by hydrophobic coordination between surfactant alkyl chains. The CTAB cluster molecules can be absorbed through the NCC surface by hydrophobic interaction. However, the hydrophobic coordination of the surfactant and NCC manifested as a weak electrostatic interaction, which easily releases CTAB from NCC surfaces through the washing. Moreover, an excessive amount of CTAB concentration over the boundary of the surfactant critical micelles concentration (0.93 mM CTAB) might provoke the surfactant micelles formation on the NCC surface, which degrades the hydrophobic characters. Only ionic interaction between the cationic head of CTAB and anionic sulfate ester groups remains unaffected, which acts as available active sites for hydrophobic drug loading (Figure 9).

Raghav and Sharma [167] reported the coordination of the hydrophobic tail of CTAB in phosphate NCC. They also observed that the surfactant types (CTAB and TBAB) influence the capability of modified NCC to bind and release the drug. By observing the structure configuration, stearic near the central nitrogen in TBAB-NCC causes the insufficiency of drug binding, which exacerbates the coordination and controlled release of the carrier [168].

Putro et al. [25] modified the NCC with various types of surfactants such as cationic (CTAB), anionic (sodium dodecyl sulfate), and non-ionic surfactant (Tween 20). Different types of surfactants exhibit distinct interactions towards NCC, which influenced the electronegativity of modified NCC itself and the drug adsorption–desorption behavior. The presence of salt in the system had a significant influence on the uptake of paclitaxel. Different behavior of surfactants due to the salt effect significantly influences the interaction of NCC and drugs. They concluded that (1) electrostatic and Van der Waals interactions are the primary mechanism of paclitaxel adsorption towards surfactant-modified NCC, which can be enhanced through salt addition; and (2) pH played a significant role in the drug adsorption and release of paclitaxel by altering the surface charge of surfactant-modified NCC and the electrostatic interaction of hydroxyl ions and paclitaxel in solution.

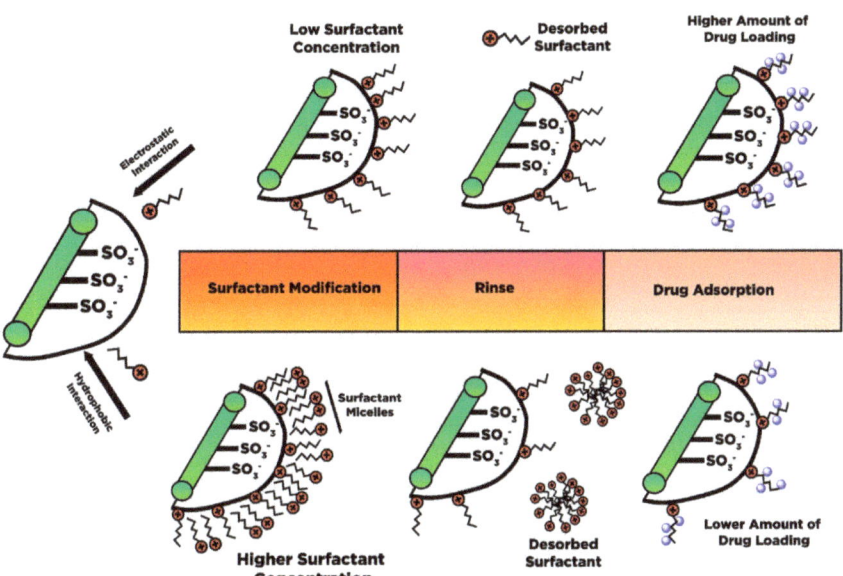

Figure 9. Schematic representation of the surfactant and nanocrystalline cellulose mechanism and its effect on drug adsorption (this figure is re-drawn from Bundjaja et al. [26]. Copyright © 2020 Elsevier B.V.).

Surfactants have also been widely used to modify cellulose nanofiber for poorly soluble drug adsorption performance. The surfactant attachment on the CNF surface is vital to overcome various limitations of CNF's chemical and physical characteristics during modification for drug conveyance systems. The physical interaction of the surfactant and CNF may overcome the aggregation tendency of CNF in an organic solvent, thus increasing the solvent's ability to assist CNF modification and adsorption of hydrophobic drugs. The presence of the surfactant strengthens the cationic and hydrophobic characters of CNF. A carrier's physical and chemical characteristics can be refined by adding a surfactant (CNF film and foams-based CNF).

The synthetic surfactant can induce membrane cell lysis based on biocompatibility, which is considered a toxic material for cells. Therefore, the naturally available surfactants have been considered to replace synthetic surfactants given their low toxicity. For instance, Bundjaja et al. [26] utilized natural surfactant (rarasaponin) extracted from *Sapindus rarak* DC fruits to modify nanocellulose via hydrophobic interaction. The results of their study indicated that the *rarasaponin*-modified NCC exhibits a lower adsorption capability of tetracycline relative to synthetic surfactant-modified NCC (CTAB, Tween20, and SDS). The utilization of natural surfactants for the modification of nanocellulose materials remains a challenge. Other bioactive compounds that were attached to the surface of NCC may cause limited interaction for tetracycline molecules.

5.6. Polyelectrolytes-Based Nanocellulose

Polyelectrolytes are charged polymers in which their repeating units contain the electrolyte group. In polar solutions such as water, these polymers dissociate into cations or anions. The most common approach to make a functional polyelectrolyte carrier is to create a multilayer carrier through electrostatically assembling layer-by-layer (LbL) the nanocellulose (either negative or positive surface) with an oppositely charged polyelectrolyte. Currently, the development of drug delivery carriers through LbL assembly has drawn considerable attention due to their unique properties. Various physical interactions such as hydrogen bonding, hydrophobic interaction, and Van der Waals interaction are present in functional polyelectrolyte carriers. Those interactions act as the driving force in drug binding and maintain the stability of the multilayer [169]. LbL hybridization assembly of nanocellulose with other organic and inorganic materials usually instigates an outstanding performance improvement for the entire LbL system to stimuli-responsive and localized drug delivery. Early development of the LbL approaches was demonstrated on the flat substrates and is currently extended to spherical particles.

Coating LbL film on a spherical sacrificial template becomes another layer-by-layer assembly approach for hollow polyelectrolytes capsule formation to encapsulate and release the drug. Melamine formaldehyde (MF) is a popular template for microcapsules preparation via LbL assembly due to narrow-sized distribution and optimized dissolution conditions [170]. The physicochemical characteristics of templates such as size, shape, porosity, colloidal stability, and template solubility modulate the characteristic of as-synthesized hollow capsules. For instance, the capsule size can be adjusted depending on the size of the template, which is common in the range of 150 nm to a few micrometers [171]. Nanocellulose has been used to construct the interior of multilayer thin film and hollow microcapsules for various types of therapeutic molecules loading such as DNA, RNA, protein, and drugs.

Several aspects should be considered to assemble suitable polyelectrolyte complexes through the LbL system, i.e., charge stoichiometry, charge density, molecular weight, polyelectrolytes concentration, pH, ionic strength, order of addition, mixing ratio, and mode of mixing the polyelectrolyte solution. These factors greatly influence the drug carrier thickness, the surface charge, and the morphological structure, such as the size, shape, and porous structure of the drug delivery system. Reviews on some crucial aspects that influence the stability of polyelectrolyte complexes for drug delivery systems are available elsewhere [172].

Mohanta et al. [173] produced an NCC multilayer thin film with counterionic polyelectrolytes (chitosan) on a quartz crystal microbalance (QCM) plate through LbL growth assembly. They also developed hollow microcapsules using MF as a template. By varying the concentration of the polyelectrolyte (either NCC and chitosan) and the number of depositions, a homogeneous multilayer thickness with a porous structure can be obtained. The thin film and microcapsule were utilized as carriers for hydrophilic drugs (doxorubicin) and hydrophobic drugs (curcumin). The protonation of amine groups in acidic conditions becomes the driving force for doxorubicin release, while the concentration difference between the medium and carrier is considered the primary factor affecting curcumin release. The stimulus-responsive pH in LbL system-based nanocellulose may apply to local drug transport and tumor therapy [173].

Other types of layer-by-layer assembly approaches were also used to construct PEC-based nanocellulose by incorporating various types of polyelectrolytes. For instance, Li et al. [174] proposed the buildup technique of LbL for opposite-charge building blocks (e.g., cellulose nanocrystal (NCC), polyethyleneimine (PEI), cis-aconityl-doxorubicin (CAD), and building blocks of folate (FA)). The highly negative charge of NCC serves as an anchor to carry the positive-charge PEI through electrostatic interaction as an intermediary layer. The coordination of NCC-PEI resulted in positive-charge material for electrostatic adsorption of the negative charge of FA and CAD to construct the outermost layer, which took place sequentially (denoted as FA/CAD@PEI@NCC). The presence of FA on the surface carrier increased the active targeting ability towards folate receptors in the tumor cell. Cis-aconityl amide linkage in doxorubicin (CAD) can specifically release DOX at the lysosomal pH due to the pH labile characteristic and hydrolysis cis-aconityl amide linkage by β-carboxylic acid under low pH. The integration of each layer can increase the uptake to 20 times larger than its counterpart due to the strong electrostatic charge. Besides the surface chemistry carrier, the carrier's morphological structure also helps the carrier delivery reach tumor cells [174]. Another potential form of polyelectrolytes, including hydrogel, aerogel, lightweight porous materials, and integrated inorganic–organic composites, are thoroughly discussed in the following sections.

6. Hydrogel Based Nanocellulose for Drug Delivery

Hydrogels are three-dimensional (3D) cross-linked polymeric networks that carry absorbed water and store a large quantity of water in the swelling state. The hydrogels can be cross-linked through physical (non-covalent interaction), chemical (covalent coordination), or an integration of both physical and chemical cross-links [175]. Given its biocompatibility and stimulus-responsive swelling behavior, the hydrogel has gained attention for drug delivery application. As a drug carrier the physically cross-linked hydrogel is preferable to the chemically cross-linked hydrogel. The covalently cross-linked hydrogel generates a permanent structure that limits the swelling ability, and therefore, most chemically cross-linked hydrogels are used as implantables. Furthermore, the incorporation of the drug via adsorption towards chemically cross-linked hydrogel restrains the loading efficacy. Although the cross-linked reaction may perform drug conjugation on the hydrogel, it sacrifices the chemical integrity of the drugs. Therefore, it is more desirable to construct a hydrogel delivery system where simultaneous gel formation and drug adsorption can occur in an aqueous environment without covalent cross-linking.

Due to the presence of sol-gel transition characteristics (such as swelling behavior, mechanical strength, and network structure), which are affected by the external stimulus such as pH, thermal, light wavelength, ultrasonic waves, pressure, magnetic field, and electrical field; the smart hydrogel-based nanocellulose has been well-developed for various drug delivery formulation. Diverse types of polyelectrolytes can modify the substantial charge of nanocellulose (either positive and negative) to form a variety of intelligent hydrogels such as injectable hydrogel [161], stimuli-responsive hydrogel [176], double-membrane hydrogel [177], supramolecular hydrogel [178], microsphere hydrogels, bacteria cellulose hydrogel [179], shape memory-based bacteria cellulose [180], and aerogel/cryogel [174]. All those hydrogels have desirable physical and chemical characteristics to be adapted to various drug delivery systems. Liu et al. [161] reviewed the current development of nanocellulose-based hydrogel and its modification for drug delivery systems. However, double-membrane hydrogel and supramolecular hydrogel are excluded from their review [161].

Different types of hydrogels have diverse morphological structures, network coordination, and functional groups, affecting the drug's diffusional path during adsorption and release. Double-membrane hydrogel was developed by Lin et al. [177], consisting of an external membrane composed of alginate and consolidation of cationic NCC (CNCC). Two different drugs were introduced on different layers of the membrane with contrasting types of release behavior. The outer hydrogel releases the drug rapidly, while sustained drug release occurs in the inner membrane hydrogel. This phenomenon occurred due to the 'nano-obstruction effect' and 'nano-locking effect' induced by CNCC components in the hydrogels. The 'nano-obstruction effect' offers sustained drug release throughout fragmentary disintegration, and the 'Nano-locking effect' is responsible for restricting the burst of drug release through progressive hydrogel disintegration (Figure 10). The different compositions and properties of external and internal hydrogels affect the drug's behavior and diffusional path [177].

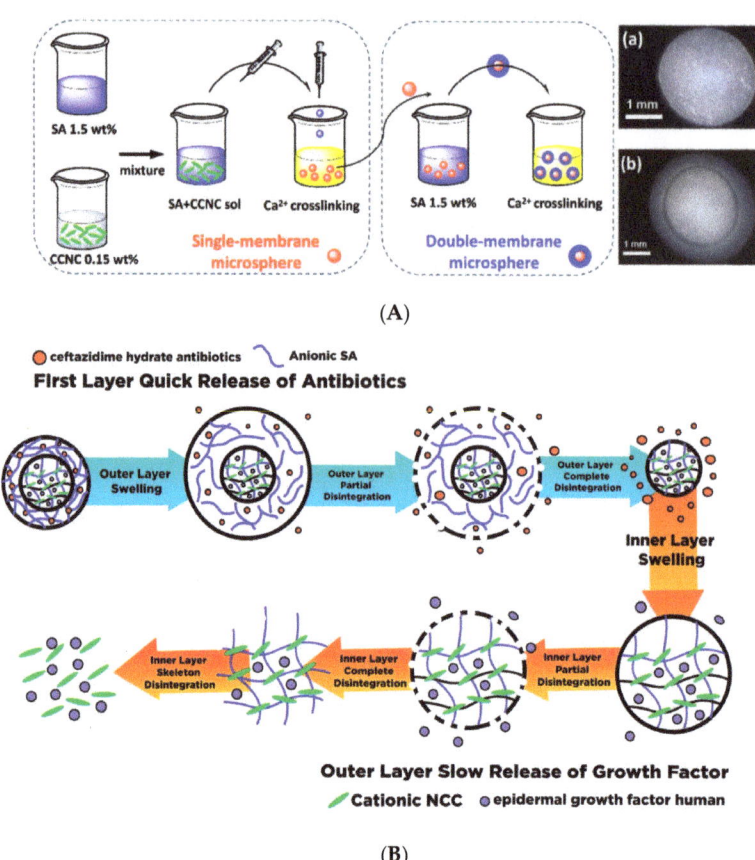

Figure 10. (**A**) The route fabrication of single membrane and double-membrane microsphere hydrogel with its optical microscope of single membrane SA/CNCC microsphere hydrogel and SA/CNCC double-membrane microsphere hydrogel (this figure is reprinted with permission from ref. [177]. Copyright © 2016 American Chemical Society); (**B**) schematic illustration of dual drug release mechanism from a double-layer membrane hydrogel constructed from Cationic NCC and Alginate (this figure is redrawn from Lin et al. [177]. Copyright © 2016 American Chemical Society).

Supramolecular hydrogels have been characterized as the 3-D solid network hydrogel organized by non-covalent interactions such as hydrogen linkage, hydrophobic coordination, and cation-π and π-π interactions. In contrast to the chemically cross-linked hydrogels, gel morphology is equilibrated through covalent coordination; the supramolecular hydrogel morphology is stabilized by a non-covalent interaction. Supramolecular hydrogel has been synthesized through extensive, diverse supramolecular configurations, including host–guest complexation, biomimetic interaction, hydrogen bonding, stereo-complex formation, and ionic and metal-ligand. Hydrogel-based supramolecular self-assembly through host–guest complexation is the most widely explored method for supramolecular hydrogel formation. Specifically, supramolecular hydrogels constructed by host–guest inclusion between polymer and cyclodextrin demonstrated the thixotropic reversibility, which is advantageous for syringe drug delivery.

Lin and Dufresne et al. [178] produced supramolecular hydrogel DDS by self-assembly of a covalently grafted α-cyclodextrin (α-CD) NCC surface with epichlorohydrin as a coupling agent through a one-step process. Furthermore, pluronic composed of triblock copolymers with different molecular weights (Pluronic F68 or F108), both bearing hydrophobic poly-(propylene glycol) (PPG) and hydrophilic poly(ethylene glycol) (PEG) segments (PEG-b-PPG-b-PEG), were immobilized on NCCs via the inclusion interaction between the hydrophobic segment of polymer and cyclodextrin (Figure 11). The supramolecular hydrogel-based NCC was utilized as a drug carrier for anti-cancer in vitro release of doxorubicin, which exhibited sustained drug release behavior (6.5 days). The kinetic release mechanism follows the 'obstruction' and 'locking' effects. They found that supramolecular hydrogels, upon being modified with NCC, induce a physical obstruction effect. Moreover, the adequate loading of NCC gave strong interaction (e.g., hydrogen bonding) inside supramolecular hydrogels and enabled the polymers to associate in the tridimensional percolating network, which provides a "locking effect" to delay the diffusion of doxorubicin molecules (Figure 12). The sustained release depends on the a-CD content, the chain length of the pluronic polymer, and the amount of NCC loaded in supramolecular hydrogels [178].

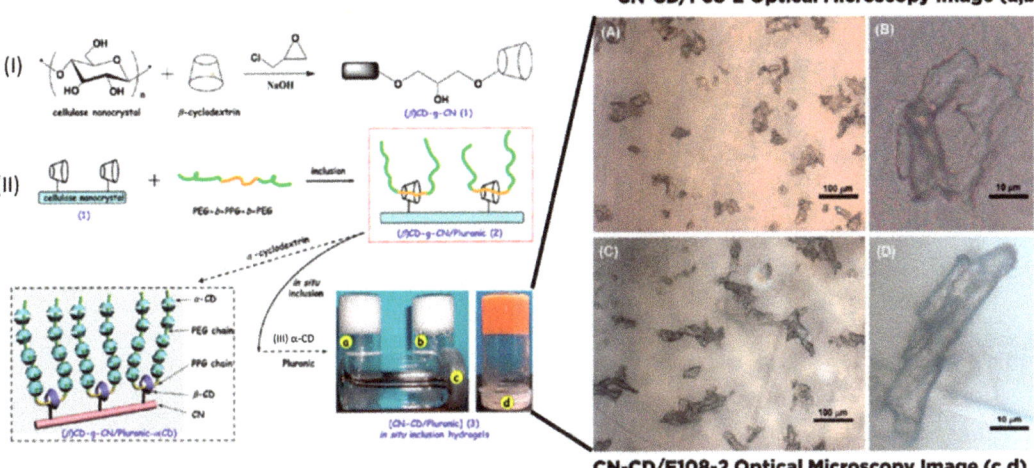

Figure 11. Construction pathway of (**I**) cellulose nanocrystal (β)CD-g-CN grafted β-cyclodextrin; (**II**) complex inclusion between Pluronic polymers and (β)CD-g-CN; (**III**) supramolecular hydrogels comprising an in situ inclusion between (β)CD-g-CN/Pluronic and α-CD (**a**) hydrogel CN-CD/F68-2 and its and its morphological evidence, (**b**) hydrogel CN-CD/F108-2 and its morphological evidence, (**c**) water, (**d**) drug-loaded hydrogel CN-CD/F108-2-Dox. This figure is reprinted with permission from [178]. Copyright © 2013 American Chemical Society.

Kopac et al. [181] pointed out that the main parameter for controlling the drug delivery rate in an anionic hydrogel-based nanocellulose is the average pore size (mesh size), controlled by selecting cross-linked and biopolymer concentration along with the adjustment of pH and temperature. The changes in the ionic strength and hydrogen bonding of functional groups in the internal hydrogel structure are responsible for altering the polymeric hydrogel network, which affected the average pore size of hydrogel (Figure 13). Due to the smaller hydrodynamic size of the drug relative to the mesh size, the drug can rapidly diffuse through the hydrogel network and vice versa without a steric barrier. However, both drugs can have a similar drug release rate by modulating the mesh size through cross-linking density and biopolymer ratio variation [181].

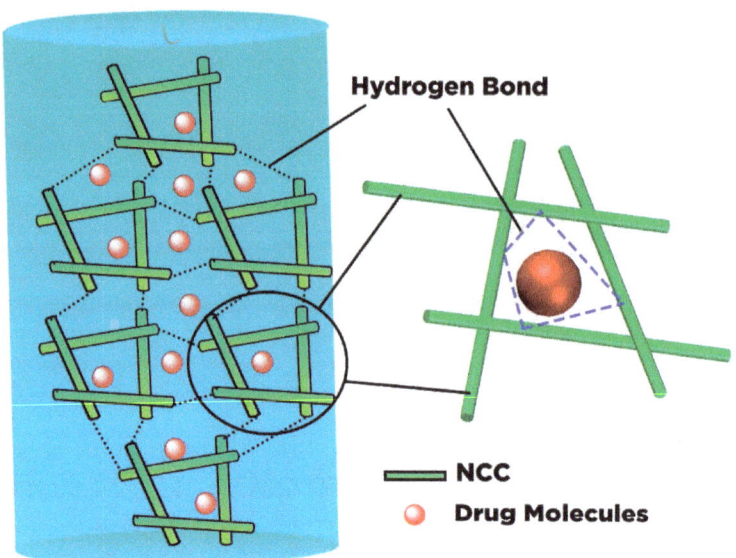

Figure 12. Schematic illustration of possible locking effect of the drug via host–guest inclusion in supramolecular hydrogel constructed from cyclodextrin and chemically modified nanocrystalline cellulose (this picture is redrawn from [178]. Copyright © 2013 American Chemical Society.)

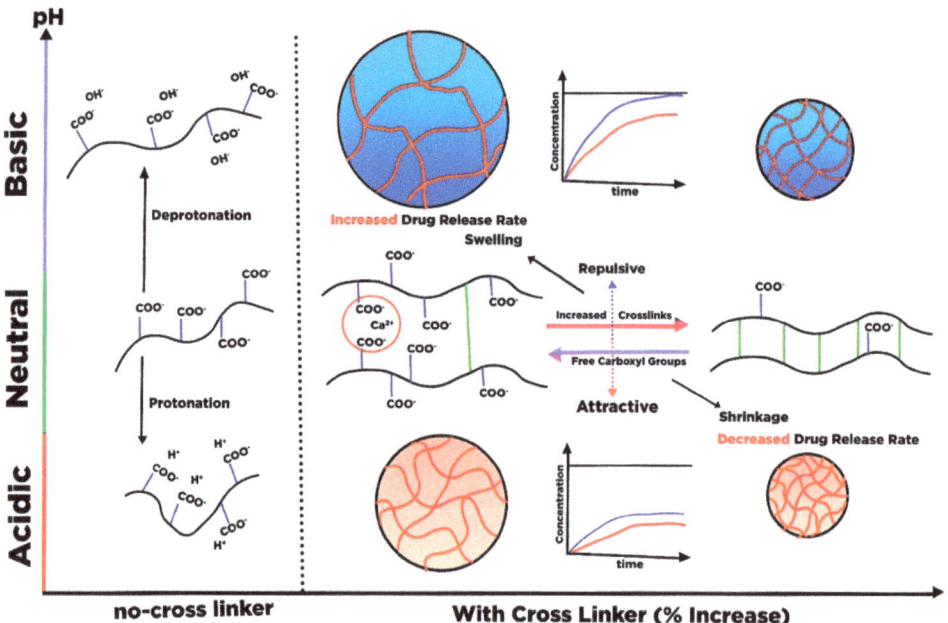

Figure 13. Schematic illustration of the swelling mechanism of hydrogel fabricated from TEMPO-mediated CNFs and alginate towards drug release (this figure is redrawn from [181]. Copyright © 2020 Elsevier B.V.).

7. Lightweight Porous Based Nanocellulose for Drug Delivery

Lightweight porous materials have been classified as a 3-D solid class of material with several features such as high specific surface area, very low density (<50%), and diverse pore structure with various pore sizes ranging from nanometer to micron. Sponge, foam, and aerogels are the three major categories of lightweight porous materials. The sponge is constructed by gas dispersion in the solid matrix, commonly present as an open cell structure of low density porous elastic polymer. The sponge has a macroporous structure full of gaps and channels, permitting easy access to water or molecules flow [182]. Similarly, foam can be made through the steady gas dispersion into a hydrogel or solid matrix and even liquid. Foam is commonly characterized as having a bubble diameter (pore diameter) greater than 50 nm [183]. Aerogel is a three-dimensional (3D) porous material constructed by self-assembly of the colloidal component or polymeric chains, creating nano-porous networks that can be filled up with a gaseous dispersion medium. Aerogel is prepared through the wet-gel drying process by removing the liquid component in the hydrogel, which is replaced by a gas constituent while still preserving the gel network [184]. The specific surface area of aerogel can reach up to 1000 $m^2\ g^{-1}$ with a porosity range between 80 and 99.8%. On the other hand, other aerogels, namely xerogel and cryogel, have been prepared by evaporation and freeze-drying. Detailed preparation of light-weight porous material-based nanocellulose has been reviewed elsewhere [185].

For the drug delivery field, carrier morphology, especially the porosity structure, controls the drug adsorption and release since the drug will pass through the internal pore to be retained inside and release outside regardless of the chemistry interaction. Sun et al. [186] underlined that the critical factor in controlling and modulating the pore structure of ultralightweight porous materials is selecting a drying method [186]. Initially, freeze-drying, supercritical drying, and evaporation drying have been utilized in fabricating ultralightweight porous materials. Evaporation drying has emerged as a conventional technique of synthesizing nanocellulose-based porous materials. However, there are several major drawbacks, such as internal network structure collapse due to the capillary forces of the solid matrix and the difficulty to prevent the shrinkage. Therefore, freeze-drying and supercritical drying have been used as drying methods to overcome these drawbacks. Freeze drying can retain the porous structure through the sublimation of liquid into gas. It is also possible to cross the solid–gas interface bypassing the liquid critical point through adjusting the temperature and pressure (supercritical drying). Both methods effectively retain the pore structure and refine the porosity and specific surface area of nanocellulose-based porous materials.

Aerogel-, xerogel-, and cryogel-based nanocelluloses are promising materials as the vehicle for a drug delivery system. Before the drying process, physical and chemical cross-linking are vital in controlling the 3D network formation and porous material performance. Physical cross-linking is commonly established by weaker interactions such as Van der Waals, hydrogen bonding, and electrostatic interaction. In contrast, covalent cross-linking can create a 3-D robust mechanical framework through the action of covalent coordination and polymerization. Chemical cross-linking exhibits better mechanical stiffness and structural stability compared to physical cross-linking.

Muller et al. [180] synthesized water-responsive xerogel to retain its original shape by submerging it in water through moisture utilization as the stimulus. The post-modification of BNC with the different supplementary hydrophilic substances was performed to achieve the re-swelling behavior. Rapid re-swelling behavior can be acquired by supplementary magnesium chloride, glucose, sucrose, and sorbitol with up to 88% maximum rehydration. Their findings of re-swelling modified BNC showed the possibility of developing a carrier with controlled release properties for hydrophilic drug model azorubine in the drug delivery system.

Li et al. [174] synthesized two types of nanocellulose/gelatin composite cryogels through hydrogen bonding and chemical cross-linking with dialdehyde starch (DAS) for controlled drug delivery of 5-fluorouracil (5-FU). DAS subsequently reacted with both gelatin and CNF to form the chemically cross-linked network. The reaction of aldehyde groups with the hydroxyl groups of CNFs led to the formation of hemiacetal/acetal types of structures. Furthermore, the aldehyde group presence effectively integrates with the ε-amino groups of gelatin to generate a Schiff base coordination. They found that the chemical cross-linking of Schiff bases and hemiacetal/acetate is crucial to regulate the structural porosity of cryogel composite. Since the porosity and cross-linking degree mainly control drug loading, selecting the chemical cross-linking method is crucial.

Moreover, the presence of gelatin hydration capability and reversible hydrolysis characteristic of hemiacetal/acetate, along with its morphological structure, is also responsible for achieving controllable and sustained release of 5-FU in a simulated intestinal environment. In addition, the cross-linking degree and the porosity can be tuned by the composition and ratio of CNF, gelatin, and dialdehyde starch. The addition of CNF increases the drug loading and the cross-linking degree [174]. Figure 14A shows that the improvement of surface roughness and cross-section morphology reduces the pore size of cryogel, leading to an increase in the cryogel resistance against ice crystal growth during freeze-drying, resulting in the smaller pore size, higher specific surface area, and lower density. The smaller pore size leads to better drug loading and releases efficiency since the smaller pore structure limits the drug looseness (Figure 14B).

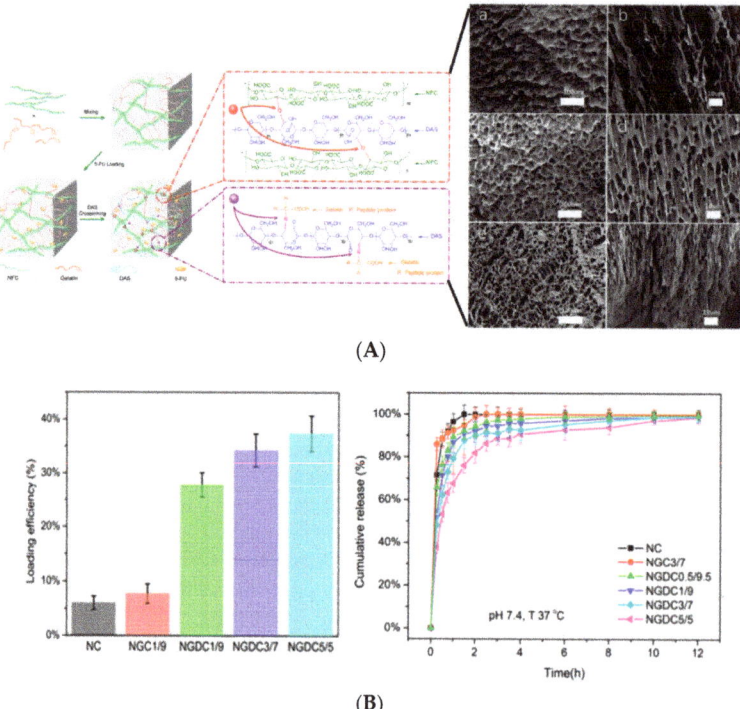

Figure 14. (A) Synthesis pathway and morphological structure of different ratio of NFC/Gelatin $R_{NFC/Gelatin}$: (**a,b**) NGDC1/9; (**c,d**) NGDC3/7; (**e,f**) NGDC5/5. Surface (**a–c**); cross-section (**b,d,f**); (B) the influence of morphological structure of NFC/Gelatin Cryogel towards drug loading (left side) and release efficiency (right side) (This figure is reprinted with permission from [174]. Copyright © 2019 American Chemical Society).

Zhao et al. [187] prepared polyethyleneimine (PEI) grafted to amine-modified CNF and cross-linked using glutaraldehyde to form an aerogel (CNFs-PEI). The success of the aerogel formation depends on the polymerization of methyl methacrylate (MMA) on the surface of CNFs, which induced the formation of the network between PEI and CNF. Polyethyleneimine (PEI) carries some primary and secondary amine functional groups, which increase the loading of sodium salicylate (NaSA) to 20 times higher than its counterpart (CNFs-based aerogel). The sustained and controlled release was achieved by the CNFs-PEI aerogel, which is highly responsive to pH because of the protonation and deprotonation of amine groups in PEI [187].

Chemically cross-linked PEI with TEMPO-mediated BC CNF (abbreviated as PEI-BC) for aspirin, gentamicin, and bovine serum albumin (BSA) carrier has been studied by Chen et al. [156]. The PEI cross-linking induced the morphological changes of BC by increasing the density of interconnected structures and thickening the pore walls, which provide the CNF interpenetrated network with improved mechanical strength [156]. Liang et al. [188] proposed a well-balanced dual responsive polymer (temperature and pH) by modifying branched PEI with N-isopropyl acrylamide (NIPAM), which was further grafted onto CNF through the condensation reaction (abbreviated as CNF-PEI-NIPAM). Remarkably, the pH and temperature of the carrier can alter the hydrophobic and hydrophilic characteristics of CNF-PEI-NIPAM [188].

CNF has been combined with the non-edible surfactant to make air bubble confinement by the Pickering technique, generating stable air bubbles encapsulated in wet-stable foams. Using the unique drying technique, the dry-foams with closed holes (cellular solid material-CSM) were made. Although the three-dimensional closed-hole structure presents a fascinating drug delivery system for the prolonged release of the drug because of confined stable air in the internal foam's structure, such structure may induce an elongated diffusional path of medicine to modify the characteristic drug release. CNFs foam as a drug carrier with the positive buoyancy characteristic was synthesized by Svagan et al. [189]. Positive buoyancy characteristics resulting from the presence of air are retained in the closed cells. These primary characteristics highlight the practicability of CNFs foam as a floating agent for gastro retentive drug delivery systems for site-specific drug release such as intestinal and stomach systems.

CNF foams were synthesized by combining the cationic suspension of CNF with the consumable surfactant (lauric acid sodium salt) as a foaming reagent. Subsequently, hydrophilic drug riboflavin was confined in the wet-stable CNFs foam structure and was further dried to acquire dry foam with a close hole structure with up to 50% drug loading (Figure 15A). The CNFs foam offers structural flexibility with different porosity and tortuosity, which can be modified in terms of shape and thickness and can be sliced into different pieces. An increase in the foam thickness leads to a decrease in the riboflavin release rate. In addition, the morphological foam structure showed a long and tortuous diffusion path, prolonging drug diffusion (Figure 15B). Therefore, the diffusion coefficient of the drug through the porous foam structure was lower than the diffusivity of the drug in the film structure [189].

The addition of surfactant is required to synthesize stable dry-foam-based cellulose nanofibers. Lobmann et al. [190] proposed an innovative way to synthesize stable foams by combining cationic CNF and hydrophobic drug indomethacin. Hydrophobic drugs provided a positive molecular interaction by partially covering the hydrophobic side of CNFs, which further changes the surface energy of CNFs. However, the indomethacin loading in the foams was limited to up to 21% of the loaded drug. An excessive amount of drug loading would destabilize and collapse the foam's structure since a higher fraction of free indomethacin and solvent in the solution was present in the air–water interface, which limited the surface-modified CNF aggregation [190].

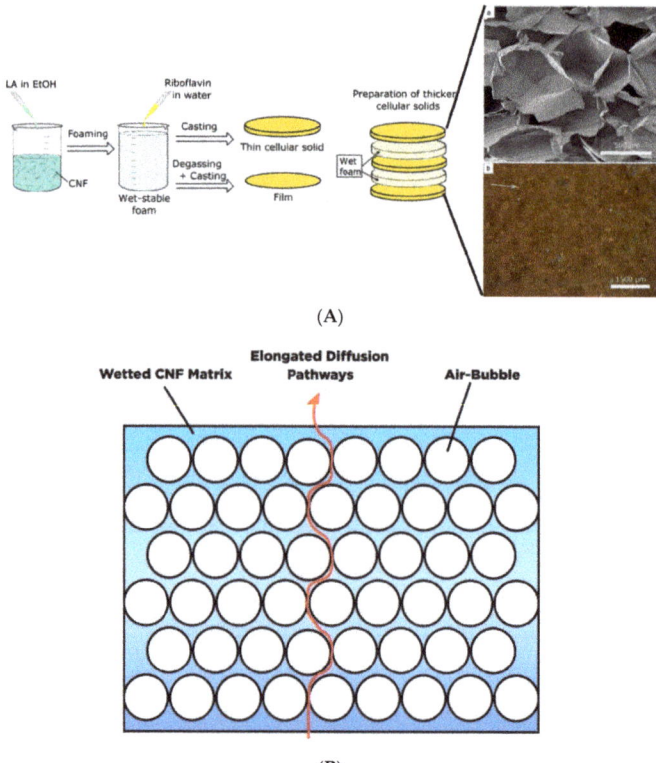

Figure 15. (**A**). Preparation route of CNF-based foams and its morphological, structural characteristic: (**a**): CNF-based foams cross-section morphological image; (**b**) cell structure image of CNF/LA loaded with riboflavin (the arrow points to riboflavin). (**B**) elongated diffusional pathways of the drug in foam-based CNFs (this figure is redrawn from Svagan et al. [189]. Copyright © 2016 Elsevier B.V.).

Svagan et al. [191] performed similar assembling of controlled-release CNFs foam with buoyance characteristics utilizing the poorly soluble drug furosemide as a foaming agent. They highlighted several factors such as the amount of drug loading, the foam piece dimension, and the solid-state of the incorporated drug that influenced the kinetic release of the drugs. Regarding the solid-state of the drug within the closed cell of foam, at 21% furosemide loading in foam, furosemide mainly exists in an amorphous state of furosemide salt, which leads to rapid release with the increase of the drug loading. In addition, the mass of incorporated drugs inside the foam structure can provide different foam dimensions, which alter the drug release kinetics. Bannow et al. [192] investigated the influence of processing parameters on the foaming characteristic and structure of nanofoam CNF/indomethacin. They found that the nanofoam density and the number of entrapped air bubbles depend on the pH, the mass of confined drugs, and the preparation route (pre- or post-adjustment of pH) [192].

The development of sponge-based nanocellulose for the drug delivery system by adding citric acid (CA) as a co-cross-linker between branched polyethyleneimine (bPEI) and TOCNFs was conducted by Fiorati and coworkers [193]. CA was added as an auxiliary carboxyl moieties source to improve the cross-linking process to bPEI. They investigated the as-synthesized sponge capability as a drug vehicle for amoxicillin and ibuprofen. The confined drug in the sponge structure with non-contained citric acid moieties exhibited a higher drug release percentage than that with the cross-linker. The presence of citric acid progressively increased the ibuprofen adsorption, while no significant effect was observed for amoxicillin adsorption. The presence of citric acid provided an additional carboxylic group, which was actively involved in the particular interaction with the ibuprofen molecules. In addition, the existence of CA also refined the mechanical strength and chemical stability of the material through the occurrence of amide bond formation between the primary amines of bPEI and with carboxylic groups of TOCNFs and CA (Figure 16).

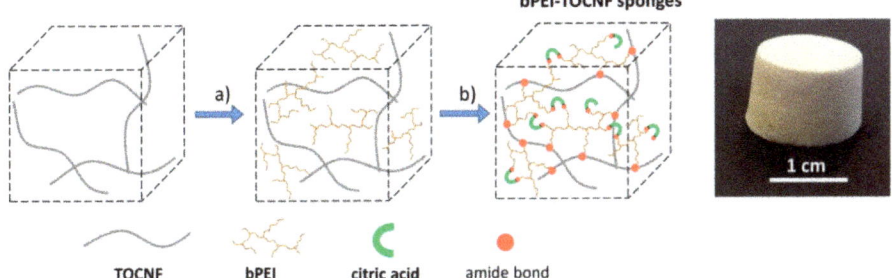

Figure 16. Preparation route of sponge-based TOCNFs via crosslinking of bPEI and TOCNFs with CA as a crosslinker; arrow (**a**) is cross-linking process, and arrow (**b**) is auxiliary carboxyl addition (this figure is reprinted with permission from ref. [193]. Copyright © 2017 Wiley-VCH Verlag GmbH & Co. KGaA, Weinheim.).

The progress of nanocellulose based sponges is still quite limited in the drug delivery field. Nevertheless, several types of sponge-based nanocellulose have been used in other biomedical applications. For instance, Xiao et al. [194] developed sponge-based CNFs via multiple cross-linking of CNFs by cellulose acetoacetate (CAA) and aminopropyl (triethoxy) silane (APTES), which further covalently bonded with surface-modified gentamicin through enamine coordination. The sponges' composite exhibits outstanding antibacterial characteristics towards *S. aureus* and *E. coli*, allowing 99.9% sterilization capability [194].

8. Integrated Inorganic/Organic-Based Nanocellulose for Drug Delivery

Recently, magnetic nanocomposites in drug delivery, particularly in cancer therapy, have drawn considerable attention. The targeted delivery of antitumor agents towards cancerous tissues can be carried through the advanced hybrid material with stimuli or specific recognition characteristics to pass through the targeted sites selectively. Nanohybrids with the stimuli effect respond to the external stimulus (e.g., pH, temperature, magnetic, and ultrasound) and further alter their physiological characteristic to release the therapeutic agent with a specific concentration towards the affected tissues. Therefore, the treatment system and the drug specificity can be improved, contributing to lessening systemic toxicity. Nonetheless, the drug carrier biocompatibility, immunogenicity, toxicity, responsiveness to magnetic gradients, and proper drug transportability still need much improvement.

NCC may also be utilized as a nanoparticle coating for colloidal stability improvement, biodegradability, biocompatibility, and chemical functionalization. Rahimi et al. [195] functionalized NCC with tris(2-aminoethyl)amine (AMFC) for Fe_3O_4 magnetic nanoparticles coating (AMFC-NPs). Initially, the nanocellulose underwent tosyl chloride treatment for tris(2-aminoethyl)amine functionalization (AMFC was chosen to be assigned to the amino moieties and cationic characteristics). The presence of amino groups in AMFC-NPs was linked to the methotrexate (MTX, an anticancer immunosuppressive drug) carboxyl groups. This method was employed to surpass the MTX limitation by keeping down the off-target side-impact towards healthy cells while optimizing the efficacy of anticancer drug delivery. The drug confinement efficacy reached 91.2% with 30.4% efficiency of drug loading in the AMFC-NPs. The MTX-AMFC-MNPs system exhibited pH responsivity in which, at an acidic condition (pH of 5.4), up to 79% of the drug was released, while over five days, it exhibited up to 29% drug release by the protonation behavior of MTX carboxylic groups. In addition, the nanoparticles containing MTX exhibit a higher uptake of cellular compared to the AMFC-MNPs, which were contributed to by the chemical similarity of MTX with folic acid (FA), which assists the internalization of receptor-mediated cellulose. This enhances the potential of MTX for cancer cell targeting (Figure 17) [195].

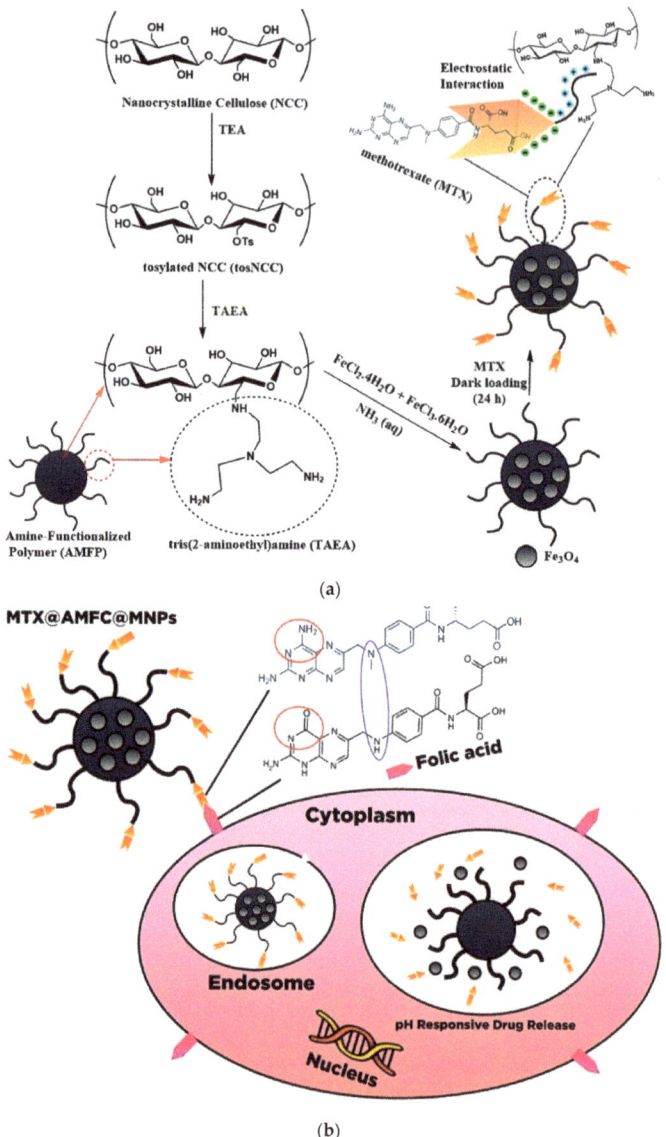

Figure 17. (a) Magnetic NCC-based nanocarrier with pH-responsive capability construction. The nanocellulose was undergoing tosylation, which reacted with tosyl chloride for tris(2-aminoethyl)amine (AMFC) functionalization, incorporating amino moieties for electrostatic interaction improvement, which connected into the methotrexate (MTX, anticancer drug) carboxyl groups (MTX@AMFC@MNPs); and (b) schematic illustration of pH-responsive and localization of cancer treatment that benefited from the structural similarity between folic acid and MTX, which assists the folate-receptor-mediated cell internalization (this figure is redrawn from [195]. Copyright 1987 Royal Society Of Chemistry).

Recently, Supramaniam et al. [196] introduced magnetic characteristics towards nanocellulos-based hydrogels, which were further utilized for controlling drug delivery. The co-precipitation with Fe (II) and Fe (III) ions was incorporated into the NCCs, followed by the insertion of the magnetic characteristic, and subsequently, morphologically modified into beads with sodium alginate. It was observed that the magnetic nanocellulose existence refines the physical and mechanical characteristics of hydrogel beads and swelling degree improvement and limits the drug release due to the formation of physical entanglement inside the hydrogel. Jeddi and Mahkam [197] developed magnetic hydrogel beads composite-based carboxymethyl nanocellulose to deliver dexamethasone. The composite can control the dexamethasone delivery up to 12 h.

Carbon nanotubes have outstanding characteristics such as high thermal stability, homogenous pore arrangement, high specific area, and excellent electrical features. This advanced material has also been employed as a vehicle in the drug delivery system in recent years. The combination of this material with the NC material provides some advantages. The incorporation of nanocellulose in the composite increased biocompatibility and biodegradability while the CNTs provided good stability, magnetic and electromagnetic behavior, and high cellular uptake [198]. Although the cytotoxicity of material still became an issue, CNTs were widely exploited in drug delivery systems, particularly cancer therapy applications [199].

Integration of nanocellulose into graphene-based materials through the layer-by-layer assembly as a drug carrier was carried out by Anirudhan et al. [200]. Chemically modified GO was used as a template for the layer-by-layer assembly of aminated nano-dextran (AND) and carboxylic acid functionalized nanocellulose (NCCs) to form a MGO-AND/NCCs nanocomposite. Curcumin can be loaded into the carrier through $\pi-\pi$ stacking and hydrogen bonding interactions due to the phenolic and aromatic rings of curcumin. Based on the release study, the acidic environment promotes COO- groups' protonation and amino in aminated in nano-dextran to form NH_3^+. This phenomenon decreased the static interaction between MGO-AND/NCC, resulting in the electrostatic repulsion of each component, consequently provoking the drug release. In addition, a cytotoxicity assay on HCT116 cells exhibited high efficacy of curcumin-loaded MGO-AND/NCC.

The electrochemical activity of the carbon nanotube was utilized to modulate the drug release. The release of ibuprofen from a novel hybrid hydrogel composed of sodium alginate (SA), bacterial cellulose (BC), and multi-walled carbon nanotubes (MWCNTs) was studied by Shi et al. [176]. The release of ibuprofen can be provoked by electrostatic repulsion. Thereby, the on–off release mechanism can be attained by introducing electrochemical potential [176].

9. Conclusions

Modified and functionalized nanocelluloses with low toxicity and high biocompatibility render them promising materials as advanced drug carriers. Various hydroxyl groups on the surface of the nanocellulose serve as attachment sites of drugs through covalent and/or physical interactions. In addition, nanocellulose modification results in a different morphological structure for the carrier, which contributes to an increase in the diffusion pathway of the drug within the carrier. Therefore, surface chemistry is a crucial factor that should be considered in the design of nanocellulose as a drug carrier for effective drug delivery. High-purity nanocelluloses are also required to obtain drug carriers with the well-constructed framework, thus facilitating drug adsorption and release control. Considering all these factors, carrier-based nanocellulose is a promising candidate for developing novel sustained drug delivery systems.

Author Contributions: Conceptualization V.B.L., F.E.S. and S.I.; redraw the figures V.B.L. and M.Y.; drafting the manuscript V.B.L., J.N.P., S.P.S. and S.I.; Supervision S.I. and F.E.S.; review and editing, J.S, and Y.H.J.; proof reading J.S. and Y.H.J. All authors have read and agreed to the published version of the manuscript.

Funding: This study was supported by the Directorate of Research and Community Service, Deputy for Strengthening Research and Development, Ministry of Research and Technology/National Research and Innovation Agency, Number: 150J/WM01.5/N/2021.

Institutional Review Board Statement: Not applicable.

Informed Consent Statement: Not applicable.

Conflicts of Interest: The authors declare no conflict of interest.

References

1. Jain, K.K. An overview of drug delivery systems. In *Drug Delivery Systems*; Springer: New York, NY, USA, 2020.
2. Sunasee, R.; Hemraz, U.D.; Ckless, K. Cellulose nanocrystals: A versatile nanoplatform for emerging biomedical applications. *Expert Opin. Drug Deliv.* **2016**, *13*, 1243–1256. [CrossRef] [PubMed]
3. Bamrungsap, S.; Zhao, Z.; Chen, T.; Wand, L.; Li, C.; Fu, T.; Tan, W. Nanotechnology in therapeutics: A focus on nanoparticles as a drug delivery system. *Nanomedicine* **2012**, *7*, 1253–1271. [CrossRef]
4. Tong, X.; Pan, W.; Su, T.; Zhang, M.; Dong, W.; Qi, X. Recent advances in natural polymer-based drug delivery systems. *React. Funct. Polym.* **2020**, *148*, 104501. [CrossRef]
5. Cavallaro, G.; Micciulla, S.; Chiappisi, L.; Lazzara, G. Chitosan-based smart hybrid materials: A physico-chemical perspective. *J. Mater. Chem. B* **2021**, *9*, 594–611. [CrossRef]
6. Bertolino, V.; Cavallaro, G.; Milioto, S.; Lazzara, G. Polysaccharides/Halloysite nanotubes for smart bionanocomposite materials. *Carbohydr. Polym.* **2020**, *245*, 116502. [CrossRef]
7. Ahmad, A.; Mubarak, N.; Jannat, F.T.; Ashfaq, T.; Santulli, C.; Rizwan, M.; Najda, A.; Bin-Jumah, M.; Abdel-Daim, M.M.; Hussain, S. A Critical Review on the Synthesis of Natural Sodium Alginate Based Composite Materials: An Innovative Biological Polymer for Biomedical Delivery Applications. *Processes* **2021**, *9*, 137. [CrossRef]
8. Chen, J.; Ouyang, J.; Chen, Q.; Deng, C.; Meng, F.; Zhang, J.; Cheng, R.; Lan, Q.; Zhong, Z. EGFR and CD44 dual-targeted multifunctional hyaluronic acid nanogels boost protein delivery to ovarian and breast cancers in vitro and in vivo. *ACS Appl. Mater. Interfaces* **2017**, *9*, 24140–24147. [CrossRef] [PubMed]
9. Qi, X.; Wei, W.; Shen, J.; Dong, W. Salecan polysaccharide-based hydrogels and their applications: A review. *J. Mater. Chem. B* **2019**, *7*, 2577–2587. [CrossRef] [PubMed]
10. Dragan, E.S.; Dinu, M.V. Polysaccharides constructed hydrogels as vehicles for proteins and peptides. A review. *Carbohydr. Polym.* **2019**, *225*, 115210. [CrossRef] [PubMed]
11. Jorfi, M.; Foster, E.J. Recent advances in nanocellulose for biomedical applications. *J. Appl. Polym. Sci.* **2015**, *132*, 41719. [CrossRef]
12. Xue, Y.; Mou, Z.; Xiao, H. Nanocellulose as a sustainable biomass material: Structure, properties, present status and future prospects in biomedical applications. *Nanoscale* **2017**, *9*, 14758–14781. [CrossRef]
13. Habibi, Y.; Lucia, L.A.; Rojas, O.J. Cellulose nanocrystals: Chemistry, self-assembly, and applications. *Chem. Rev.* **2010**, *110*, 3479–3500. [CrossRef]
14. Lin, N.; Dufresne, A. Nanocellulose in biomedicine: Current status and future prospect. *Eur. Polym. J.* **2014**, *59*, 302–325. [CrossRef]
15. Plackett, D.; Letchford, K.; Jackson, J.; Burt, H. A review of nanocellulose as a novel vehicle for drug delivery. *Nord. Pulp Pap. Res. J.* **2014**, *29*, 105–118. [CrossRef]
16. Jawaid, M.; Mohammad, F. *Nanocellulose and Nanohydrogel Matrices: Biotechnological and Biomedical Applications*; John Wiley & Sons: Hoboken, NJ, USA, 2017.
17. Hasan, N.; Rahman, L.; Kim, S.-H.; Cao, J.; Arjuna, A.; Lallo, S.; Jhun, B.H.; Yoo, J.-W. Recent advances of nanocellulose in drug delivery systems. *J. Pharm. Investig.* **2020**, *50*, 553–572. [CrossRef]
18. Grishkewich, N.; Mohammed, N.; Tang, J.; Tam, K.C. Recent advances in the application of cellulose nanocrystals. *Curr. Opin. Colloid Interface Sci.* **2017**, *29*, 32–45. [CrossRef]
19. Yan, G.; Chen, B.; Zeng, X.; Sun, Y.; Tan, X.; Lin, L. Recent advances on sustainable cellulosic materials for pharmaceutical carrier applications. *Carbohydr. Polym.* **2020**, *244*, 116492. [CrossRef] [PubMed]
20. Rajinipriya, M.; Nagalakshmaiah, M.; Robert, M.; Elkoun, S. Importance of agricultural and industrial waste in the field of nanocellulose and recent industrial developments of wood based nanocellulose: A review. *ACS Sustain. Chem. Eng.* **2018**, *6*, 2807–2828. [CrossRef]
21. Karimian, A.; Parsian, H.; Majidinia, M.; Rahimi, M.; Mir, S.M.; Kafil, H.S.; Shafiei-Irannejad, V.; Kheyrollah, M.; Ostadi, H.; Yousefi, B. Nanocrystalline cellulose: Preparation, physicochemical properties, and applications in drug delivery systems. *Int. J. Biol. Macromol.* **2019**, *133*, 850–859. [CrossRef] [PubMed]
22. Salimi, S.; Sotudeh-Gharebagh, R.; Zarghami, R.; Chan, S.Y.; Yuen, K.H. Production of nanocellulose and its applications in drug delivery: A critical review. *ACS Sustain. Chem. Eng.* **2019**, *7*, 15800–15827. [CrossRef]
23. Seabra, A.B.; Bernandes, J.S.; Favaro, W.J.; Paula, A.J.; Duran, N. Cellulose nanocrystals as carriers in medicine and their toxicities: A review. *Carbohydr. Polym.* **2018**, *181*, 514–527. [CrossRef] [PubMed]

24. Sheikhi, A.; Hayashi, J.; Eichenbaum, J.; Gutin, M.; Kuntjoro, N.; Khorsandi, D.; Khademhosseini, A. Recent advances in nanoengineering cellulose for cargo delivery. *J. Control. Release* **2019**, *294*, 53–76. [CrossRef] [PubMed]
25. Putro, J.N.; Ismadji, S.; Gunarto, C.; Yuliana, M.; Santoso, S.P.; Soetaredjo, F.E.; Ju, Y.H. The effect of surfactants modification on nanocrystalline cellulose for paclitaxel loading and release study. *J. Mol. Liq.* **2019**, *282*, 407–414. [CrossRef]
26. Bundjaja, V.; Sari, T.M.; Soetaredjo, F.E.; Yuliana, M.; Angkawijaya, A.E.; Ismadji, S.; Cheng, K.-C.; Santoso, S.P. Aqueous sorption of tetracycline using rarasaponin-modified nanocrystalline cellulose. *J. Mol. Liq.* **2020**, *301*, 112433. [CrossRef]
27. Huang, Y.; Zhu, C.; Yang, J.; Nie, Y.; Chen, C.; Sun, D. Recent advances in bacterial cellulose. *Cellulose* **2014**, *21*, 1–30. [CrossRef]
28. Foo, M.L.; Tan, C.R.; Lim, P.D.; Ooi, C.W.; Tan, K.W.; Chew, I.M.L. Surface-modified nanocrystalline cellulose from oil palm empty fruit bunch for effective binding of curcumin. *Int. J. Biol. Macromol.* **2019**, *138*, 1064–1071. [CrossRef]
29. Wijaya, C.J.; Saputra, S.N.; Soetaredjo, F.E.; Putro, J.N.; Lin, C.X.; Kurniawan, A.; Ju, Y.-H.; Ismadji, S. Cellulose nanocrystals from passion fruit peels waste as antibiotic drug carrier. *Carbohydr. Polym.* **2017**, *175*, 370–376. [CrossRef] [PubMed]
30. Dufresne, A. Cellulose nanomaterial reinforced polymer nanocomposites. *Curr. Opin. Colloid Interface Sci.* **2017**, *29*, 1–8. [CrossRef]
31. Gopinath, V.; Saravanan, S.; Al-Maleki, A.; Ramesh, M.; Vadivelu, J. A review of natural polysaccharides for drug delivery applications: Special focus on cellulose, starch and glycogen. *Biomed. Pharmacother.* **2018**, *107*, 96–108. [CrossRef]
32. Wertz, J.-L.; Bedue, O.; Mercier, J.P. *Cellulose Science and Technology*; EPFL Press: Lausanne, Switzerland, 2010.
33. George, J.; Sabapathi, S. Cellulose nanocrystals: Synthesis, functional properties, and applications. *Nanotechnol. Sci. Appl.* **2015**, *8*, 45. [CrossRef]
34. Moon, R.J.; Martini, A.; Nairn, J.; Simonsen, J.; Youngblood, J. Cellulose nanomaterials review: Structure, properties and nanocomposites. *Chem. Soc. Rev.* **2011**, *40*, 3941–3994. [CrossRef]
35. Medronho, B.; Lindman, B. Brief overview on cellulose dissolution/regeneration interactions and mechanisms. *Adv. Colloid Interface Sci.* **2015**, *222*, 502–508. [CrossRef]
36. Kamel, R.; El-Wakil, N.A.; Dufresne, A.; Elkasabgy, N.A. Nanocellulose: From an agricultural waste to a valuable pharmaceutical ingredient. *Int. J. Biol. Macromol.* **2020**, *163*, 1579–1590. [CrossRef] [PubMed]
37. Miao, C.; Hamad, W.Y. Cellulose reinforced polymer composites and nanocomposites: A critical review. *Cellulose* **2013**, *20*, 2221–2262. [CrossRef]
38. Li, N.; Lu, W.; Yu, J.; Xiao, Y.; Liu, S.; Gan, L.; Huang, J. Rod-like cellulose nanocrystal/cis-aconityl-doxorubicin prodrug: A fluorescence-visible drug delivery system with enhanced cellular uptake and intracellular drug controlled release. *Mater. Sci. Eng. C* **2018**, *91*, 179–189. [CrossRef] [PubMed]
39. Jia, B.; Li, Y.; Yang, B.; Xiao, D.; Zhang, S.; Rajulu, A.V.; Kondo, T.; Zhang, L.; Zhou, J. Effect of microcrystal cellulose and cellulose whisker on biocompatibility of cellulose-based electrospun scaffolds. *Cellulose* **2013**, *20*, 1911–1923. [CrossRef]
40. Dash, R.; Ragauskas, A.J. Synthesis of a novel cellulose nanowhisker-based drug delivery system. *RSC Adv.* **2012**, *2*, 3403–3409. [CrossRef]
41. Biswas, S.K.; Sano, H.; Yang, X.; Tanphichai, S.; Shams, M.I.; Yano, H. Highly thermal-resilient AgNW transparent electrode and optical device on thermomechanically superstable cellulose nanorod-reinforced nanocomposites. *Adv. Opt. Mater.* **2019**, *7*, 1900532. [CrossRef]
42. Wang, N.; Ding, E.; Cheng, R. Thermal degradation behaviors of spherical cellulose nanocrystals with sulfate groups. *Polymer* **2007**, *48*, 3486–3493. [CrossRef]
43. Ram, B.; Chauhan, G.S.; Mehta, A.; Gupta, R.; Chauhan, K. Spherical nanocellulose-based highly efficient and rapid multifunctional naked-eye Cr (VI) ion chemosensor and adsorbent with mild antimicrobial properties. *Chem. Eng. J.* **2018**, *349*, 146–155. [CrossRef]
44. Mariano, M.; El Kissi, N.; Dufresne, A. Cellulose nanocrystals and related nanocomposites: Review of some properties and challenges. *J. Polym. Sci. B Polym. Phys.* **2014**, *52*, 791–806. [CrossRef]
45. Brinchi, L.; Cotana, F.; Fortunati, E.; Kenny, J. Production of nanocrystalline cellulose from lignocellulosic biomass: Technology and applications. *Carbohydr. Polym.* **2013**, *94*, 154–169. [CrossRef]
46. Phanthong, P.; Reybroycharoen, P.; Hao, X.; Xu, G.; Abudula, A.; Guan, G. Nanocellulose: Extraction and application. *Carbon Resour. Convers.* **2018**, *1*, 32–43. [CrossRef]
47. Lam, E.; Male, K.B.; Chong, J.H.; Leung, A.C.; Luong, J.H. Applications of functionalized and nanoparticle-modified nanocrystalline cellulose. *Trends Biotechnol.* **2012**, *30*, 283–290. [CrossRef] [PubMed]
48. Prathapan, R.; Tabor, R.F.; Garnier, G.; Hu, J. Recent progress in cellulose nanocrystal alignment and its applications. *ACS Appl. Bio Mater.* **2020**, *3*, 1828–1844. [CrossRef]
49. Abitbol, T.; Rivkin, A.; Cao, Y.; Nevo, Y.; Abraham, E.; Ben-Shalom, T.; Lapidot, S.; Shoseyov, O. Nanocellulose, a tiny fiber with huge applications. *Curr. Opin. Biotechnol.* **2016**, *39*, 76–88. [CrossRef] [PubMed]
50. Henriksson, M.; Henriksson, G.; Berglund, L.; Lindström, T. An environmentally friendly method for enzyme-assisted preparation of microfibrillated cellulose (MFC) nanofibers. *Eur. Polym. J.* **2007**, *43*, 3434–3441. [CrossRef]
51. Kim, C.-W.; Kim, D.-S.; Kang, S.-Y.; Marquez, M.; Joo, Y.L. Structural studies of electrospun cellulose nanofibers. *Polymer* **2006**, *47*, 5097–5107. [CrossRef]
52. Dufresne, A. Nanocellulose: A new ageless bionanomaterial. *Mater. Today* **2013**, *16*, 220–227. [CrossRef]
53. Lavoine, N.; Desloges, I.; Dufresne, A.; Bras, J. Microfibrillated cellulose–Its barrier properties and applications in cellulosic materials: A review. *Carbohydr. Polym.* **2012**, *90*, 735–764. [CrossRef]

54. Blanco Parte, F.G.; Santoso, S.P.; Chou, C.-C.; Verma, V.; Wang, H.-T.; Ismadji, S.; Cheng, K.-C. Current progress on the production, modification, and applications of bacterial cellulose. *Crit. Rev. Biotechnol.* **2020**, *40*, 397–414. [CrossRef] [PubMed]
55. Siro, I.; Plackett, D. Microfibrillated cellulose and new nanocomposite materials: A review. *Cellulose* **2010**, *17*, 459–494. [CrossRef]
56. Huang, J.; Dufresne, A.; Lin, N. *Nanocellulose: From Fundamentals to Advanced Materials*; John Wiley & Sons: Hoboken, NJ, USA, 2019.
57. Jozala, A.F.; De Lencastre-Novaes, L.C.; Lopes, A.M.; De Carvalho Santos-Ebinuma, V.; Mazzola, P.G.; Pessoa-JR, A.; Grotto, D.; Gerenutti, M.; Chaud, M.V. Bacterial nanocellulose production and application: A 10-year overview. *Appl. Microbiol. Biotechnol.* **2016**, *100*, 2063–2072. [CrossRef] [PubMed]
58. Wei, B.; Yang, G.; Hong, F. Preparation and evaluation of a kind of bacterial cellulose dry films with antibacterial properties. *Carbohydr. Polym.* **2011**, *84*, 533–538. [CrossRef]
59. Wan, Y.; Wang, J.; Gama, M.; Guo, R.; Zhang, Q.; Zhang, P.; Yao, F.; Luo, H. Biofabrication of a novel bacteria/bacterial cellulose composite for improved adsorption capacity. *Compos. Part A Appl. Sci. Manuf.* **2019**, *125*, 105560. [CrossRef]
60. Gorgieva, S.; Trcek, J. Bacterial cellulose: Production, modification and perspectives in biomedical applications. *Nanomaterials* **2019**, *9*, 1352. [CrossRef] [PubMed]
61. Kim, J.-H.; Shim, B.S.; Kim, H.S.; Lee, Y.-J.; Min, S.-K.; Jang, D.; Abas, Z.; Kim, J. Review of nanocellulose for sustainable future materials. *Int. J. Precis. Eng. Manuf. Green Technol.* **2015**, *2*, 197–213. [CrossRef]
62. Dufresne, A. Cellulose-based composites and nanocomposites. In *Monomers, Polymers, and Composites from Renewable Resources*; Elsevier: Amsterdam, The Netherlands, 2008.
63. Song, F.; Li, X.; Wang, Q.; Liao, L.; Zhang, C. Nanocomposite hydrogels and their applications in drug delivery and tissue engineering. *J. Biomed. Nanotechnol.* **2015**, *11*, 40–52. [CrossRef]
64. Ibrahim, M.M.; Fahmy, T.Y.; Salaheldin, E.I.; Mobarak, F.; Youssef, M.A.; Mabrook, M.R. Synthesis of tosylated and trimethylsilylated methyl cellulose as pH-sensitive carrier matrix. *Life Sci. J.* **2015**, *1*, 29–37.
65. Ximenes, F.A.; Gardner, W.D.; Kathuria, A. Proportion of above-ground biomass in commercial logs and residues following the harvest of five commercial forest species in Australia. *For. Ecol. Manag.* **2008**, *256*, 335–346. [CrossRef]
66. Kargarzadeh, H.; Ioelovich, M.; Ahmad, I.; Thomas, S.; Dufresne, A. Methods for extraction of nanocellulose from various sources. *Handb. Nanocellulose Cellul. Nanocomposites* **2017**, *2*, 1–49.
67. Kumar, A.K.; Sharma, S. Recent updates on different methods of pretreatment of lignocellulosic feedstocks: A review. *Bioresour. Bioproces.* **2017**, *4*, 1–19. [CrossRef]
68. Ravindran, R.; Jaiswal, S.; Abu-Ghannam, N.; Jaiswal, A.K. A comparative analysis of pretreatment strategies on the properties and hydrolysis of brewers' spent grain. *Bioresour. Technol.* **2018**, *248*, 272–279. [CrossRef] [PubMed]
69. Pires, J.R.; De Souza, V.G.L.; Fernando, A.L. Production of nanocellulose from lignocellulosic biomass wastes: Prospects and limitations. In Proceedings of the International Conference on Innovation, Engineering and Entrepreneurship, Guimaraes, Portugal, 27–29 June 2018; pp. 719–725.
70. Shirkavand, E.; Baroutian, S.; Gapes, D.J.; Young, B.R. Combination of fungal and physicochemical processes for lignocellulosic biomass pretreatment–A review. *Renew. Sust. Energy Rev.* **2016**, *54*, 217–234. [CrossRef]
71. Baramee, S.; Siriatcharanon, A.-K.; Ketbot, P.; Teeravivattanakit, T.; Waeonukul, R.; Pason, P.; Tachaapaikoon, C.; Ratanakhanokchai, K.; Phitsuwan, P. Biological pretreatment of rice straw with cellulase-free xylanolytic enzyme-producing Bacillus firmus K-1: Structural modification and biomass digestibility. *Renew. Energy* **2020**, *160*, 555–563. [CrossRef]
72. Khalil, H.A.; Davoudpour, Y.; Islam, M.N.; Mustapha, A.; Sudesh, K.; Dungani, R.; Jawaid, M. Production and modification of nanofibrillated cellulose using various mechanical processes: A review. *Carbohydr. Polym.* **2014**, *99*, 649–665. [CrossRef]
73. Mood, S.H.; Golfeshan, A.H.; Tabatabaei, M.; Jouzani, G.S.; Najafi, G.H.; Gholami, M.; Ardjmand, M. Lignocellulosic biomass to bioethanol, a comprehensive review with a focus on pretreatment. *Renew. Sust. Energy Rev.* **2013**, *27*, 77–93. [CrossRef]
74. Hassan, S.S.; Williams, G.A.; Jaiswal, A.K. Emerging technologies for the pretreatment of lignocellulosic biomass. *Bioresour. Technol.* **2018**, *262*, 310–318. [CrossRef]
75. De Carvalho Benini, K.C.C.; Voorwald, H.J.C.; Cioffi, M.O.H.; Rezende, M.C.; Arantes, V. Preparation of nanocellulose from Imperata brasiliensis grass using Taguchi method. *Carbohydr. Polym.* **2018**, *192*, 337–346. [CrossRef]
76. Maciel, M.M.Á.D.; De Carvalho Benini, K.C.C.; Voorwald, H.J.C.; Cioffi, M.O.H. Obtainment and characterization of nanocellulose from an unwoven industrial textile cotton waste: Effect of acid hydrolysis conditions. *Int. J. Biol. Macromol.* **2019**, *126*, 496–506. [CrossRef]
77. Liu, J.; Korpinen, R.; Mikkonen, K.S.; Willför, S.; Xu, C. Nanofibrillated cellulose originated from birch sawdust after sequential extractions: A promising polymeric material from waste to films. *Cellulose* **2014**, *21*, 2587–2598. [CrossRef]
78. Couret, L.; Irle, M.; Belloncle, C.; Cathala, B. Extraction and characterization of cellulose nanocrystals from post-consumer wood fiberboard waste. *Cellulose* **2017**, *24*, 2125–2137. [CrossRef]
79. Vallejos, M.E.; Felissia, F.E.; Area, M.C.; Ehman, N.V.; Tarrés, Q.; Mutjé, P. Nanofibrillated cellulose (CNF) from eucalyptus sawdust as a dry strength agent of unrefined eucalyptus handsheets. *Carbohydr. Polym.* **2016**, *139*, 99–105. [CrossRef] [PubMed]
80. Rambabu, N.; Panthapulakkal, S.; Sain, M.; Dalai, A. 2016. Production of nanocellulose fibers from pinecone biomass: Evaluation and optimization of chemical and mechanical treatment conditions on mechanical properties of nanocellulose films. *Ind. Crops Prod.* **2016**, *83*, 746–754. [CrossRef]

81. Moriana, R.; Vilaplana, F.; Ek, M. Cellulose nanocrystals from forest residues as reinforcing agents for composites: A study from macro-to nano-dimensions. *Carbohydr. Polym.* **2016**, *139*, 139–149. [CrossRef]
82. Lu, H.; Zhang, L.; Liu, C.; He, Z.; Zhou, X.; Ni, Y. A novel method to prepare lignocellulose nanofibrils directly from bamboo chips. *Cellulose* **2018**, *25*, 7043–7051. [CrossRef]
83. Hua, K.; Strømme, M.; Mihranyan, A.; Ferraz, N. Nanocellulose from green algae modulates the in vitro inflammatory response of monocytes/macrophages. *Cellulose* **2015**, *22*, 3673–3688. [CrossRef]
84. Son, H.N.; Seo, Y.B. Physical and bio-composite properties of nanocrystalline cellulose from wood, cotton liners, cattail, and red algae. *Cellulose* **2015**, *22*, 1789–1798.
85. Rathod, M.; Haldar, S.; Basha, S. Nanocrystalline cellulose for removal of tetracycline hydrochloride from water via biosorption: Equilibrium, kinetic and thermodynamic studies. *Ecol. Eng.* **2015**, *84*, 240–249. [CrossRef]
86. Liu, Z.; Li, X.; Xie, W.; Deng, H. Extraction, isolation and characterization of nanocrystalline cellulose from industrial kelp (*Laminaria japonica*) waste. *Carbohydr. Polym.* **2017**, *173*, 353–359. [CrossRef] [PubMed]
87. Feng, X.; Meng, X.; Zhao, J.; Miao, M.; Shi, L.; Zhang, S.; Fang, J. Extraction and preparation of cellulose nanocrystals from dealginate kelp residue: Structures and morphological characterization. *Cellulose* **2015**, *22*, 1763–1772. [CrossRef]
88. Bhutiya, P.L.; Misra, N.; Rasheed, M.A.; Hasan, S.Z. Nested seaweed cellulose fiber deposited with cuprous oxide nanorods for antimicrobial activity. *Int. J. Biol. Macromol.* **2018**, *117*, 435–444. [CrossRef]
89. De Oliveira, J.P.; Bruni, G.P.; Fabra, M.J.; Da Rosa Zavareze, E.; López-Rubio, A.; Martínez-Sanz, M. Development of food packaging bioactive aerogels through the valorization of *Gelidium sesquipedale* seaweed. *Food Hydrocoll.* **2019**, *89*, 337–350. [CrossRef]
90. Chen, Y.W.; Lee, H.V.; Juan, J.C.; Phang, S.-M. Production of new cellulose nanomaterial from red algae marine biomass *Gelidium elegans*. *Carbohydr. Polym.* **2016**, *151*, 1210–1219. [CrossRef] [PubMed]
91. Mandal, A.; Chakrabarty, D. Isolation of nanocellulose from waste sugarcane bagasse (SCB) and its characterization. *Carbohydr. Polym.* **2011**, *86*, 1291–1299. [CrossRef]
92. Thomas, M.G.; Abraham, E.; Jyotishkumar, P.; Maria, H.J.; Pothen, L.A.; Thomas, S. Nanocelluloses from jute fibers and their nanocomposites with natural rubber: Preparation and characterization. *Int. J. Biol. Macromol.* **2015**, *81*, 768–777. [CrossRef] [PubMed]
93. Wu, J.; Du, X.; Yin, Z.; Xu, S.; Xu, S.; Zhang, Y. Preparation and characterization of cellulose nanofibrils from coconut coir fibers and their reinforcements in biodegradable composite films. *Carbohydr. Polym.* **2019**, *211*, 49–56. [CrossRef] [PubMed]
94. Mariño, M.; Lopes Da Silva, L.; Durán, N.; Tasic, L. Enhanced materials from nature: Nanocellulose from citrus waste. *Molecules* **2015**, *20*, 5908–5923. [CrossRef]
95. Dilamian, M.; Noroozi, B. A combined homogenization-high intensity ultrasonication process for individualizaion of cellulose micro-nano fibers from rice straw. *Cellulose* **2019**, *26*, 5831–5849. [CrossRef]
96. Kang, X.; Sun, P.; Kuga, S.; Wang, C.; Zhao, Y.; Wu, M.; Huang, Y. Thin cellulose nanofiber from corncob cellulose and its performance in transparent nanopaper. *ACS Sust. Chem. Eng.* **2017**, *5*, 2529–2534. [CrossRef]
97. Karimi, S.; Tahir, P.M.; Karimi, A.; Dufresne, A.; Abdulkhani, A. Kenaf bast cellulosic fibers hierarchy: A comprehensive approach from micro to nano. *Carbohydr. Polym.* **2014**, *101*, 878–885. [CrossRef]
98. Jodeh, S.; Hamed, O.; Melhem, A.; Salghi, R.; Jodeh, D.; Azzaoui, K.; Benmassaoud, Y.; Murtada, K. Magnetic nanocellulose from olive industry solid waste for the effective removal of methylene blue from wastewater. *Environ. Sci. Poll. Res.* **2018**, *25*, 22060–22074. [CrossRef]
99. Jongaroontaprangsee, S.; Chiewchan, N.; Devahastin, S. Production of nanofibrillated cellulose with superior water redispersibility from lime residues via a chemical-free process. *Carbohydr. Polym.* **2018**, *193*, 249–258. [CrossRef] [PubMed]
100. Diop, C.I.K.; Lavoie, J.-M. Isolation of nanocrystalline cellulose: A technological route for valorizing recycled tetra pak aseptic multilayered food packaging wastes. *Waste Biomass Valor.* **2017**, *8*, 41–56. [CrossRef]
101. Putro, J.N.; Santoso, S.P.; Soetaredjo, F.E.; Ismadji, S.; Ju, Y.-H. Nanocrystalline cellulose from waste paper: Adsorbent for azo dyes removal. *Environ. Nanotech. Monit. Manag.* **2019**, *12*, 100260. [CrossRef]
102. Ogundare, S.A.; Moodley, V.; Van Zyl, W.E. Nanocrystalline cellulose isolated from discarded cigarette filters. *Carbohydr. Polym.* **2017**, *175*, 273–281. [CrossRef] [PubMed]
103. Peretz, R.; Sterenzon, E.; Gerchman, Y.; Vadivel, V.K.; Luxbacher, T.; Mamane, H. Nanocellulose production from recycled paper mill sludge using ozonation pretreatment followed by recyclable maleic acid hydrolysis. *Carbohydr. Polym.* **2019**, *216*, 343–351. [CrossRef] [PubMed]
104. Cypriano, D.Z.; Da Silva, L.L.; Tasic, L. High value-added products from the orange juice industry waste. *Waste Manag.* **2018**, *79*, 71–78. [CrossRef] [PubMed]
105. Dubey, S.; Singh, J.; Singh, R. Biotransformation of sweet lime pulp waste into high-quality nanocellulose with an excellent productivity using Komagataeibacter europaeus SGP37 under static intermittent fed-batch cultivation. *Bioresour. Technol.* **2018**, *247*, 73–80. [CrossRef] [PubMed]
106. Mondal, S. Preparation, properties and applications of nanocellulosic materials. *Carbohydr. Polym.* **2017**, *163*, 301–316. [CrossRef]
107. Turbak, A.F.; Snyder, F.W.; Sandberg, K.R. Microfibrillated cellulose, a new cellulose product: Properties, uses, and commercial potential. *J. Appl. Polym. Sci. Appl. Polym. Sym.* **1983**, *37*, 815–827.

108. Herrick, F.W.; Casebier, R.L.; Hamilton, J.K.; Sandberg, K.R. Microfibrillated cellulose: Morphology and accessibility. *J. Appl. Polym. Sci. Appl. Polym. Sym.* **1983**, *37*, 797–813.
109. Kawee, N.; Lam, N.T.; Sukyai, P. Homogenous isolation of individualized bacterial nanofibrillated cellulose by high pressure homogenization. *Carbohydr. Polym.* **2018**, *179*, 394–401. [CrossRef]
110. Yusra, A.I.; Juahir, H.; Firdaus, N.N.A.; Bhat, A.; Endut, A.; Khalil, H.A.; Adiana, G. Controlling of green nanocellulose fiber properties produced by chemo-mechanical treatment process via SEM, TEM, AFM and image analyzer characterization. *J. Fund. Appl. Sci.* **2018**, *10*, 1–17.
111. Kalia, S.; Boufi, S.; Celli, A.; Kango, S. Nanofibrillated cellulose: Surface modification and potential applications. *Colloid Polym. Sci.* **2014**, *292*, 5–31. [CrossRef]
112. Niu, F.; Li, M.; Huang, Q.; Zhang, X.; Pan, W.; Yang, J.; Li, J. The characteristic and dispersion stability of nanocellulose produced by mixed acid hydrolysis and ultrasonic assistance. *Carbohydr. Polym.* **2017**, *165*, 197–204. [CrossRef] [PubMed]
113. Trache, D.; Hussin, M.H.; Haafiz, M.M.; Thakur, V.K. Recent progress in cellulose nanocrystals: Sources and production. *Nanoscale* **2017**, *9*, 1763–1786. [CrossRef]
114. Wang, J.; Tavakoli, J.; Tang, Y. Bacterial cellulose production, properties and applications with different culture methods–A review. *Carbohydr. Polym.* **2019**, *219*, 63–76. [CrossRef]
115. Czaja, W.; Romanovicz, D.; Malcolm, B.R. Structural investigations of microbial cellulose produced in stationary and agitated culture. *Cellulose* **2004**, *11*, 403–411. [CrossRef]
116. Matsutani, M.; Ito, K.; Azuma, Y.; Ogino, H.; Shirai, M.; Yakushi, T.; Matsushita, K. Adaptive mutation related to cellulose producibility in *Komagataeibacter medellinensis* (*Gluconacetobacter xylinus*) NBRC 3288. *Appl. Microbiol. Biotechnol.* **2015**, *99*, 7229–7240. [CrossRef] [PubMed]
117. Liu, M.; Zhong, C.; Wu, X.-Y.; Wei, Y.-Q.; Bo, T.; Han, P.-P.; Jia, S.-R. Metabolomic profiling coupled with metabolic network reveals differences in *Gluconacetobacter xylinus* from static and agitated cultures. *Biochem. Eng. J.* **2015**, *101*, 85–98. [CrossRef]
118. Gullo, M.; La China, S.; Petroni, G.; Di Gregorio, S.; Giudici, P. Exploring K2G30 genome: A high bacterial cellulose producing strain in glucose and mannitol based media. *Front. Microbiol.* **2019**, *10*, 58. [CrossRef] [PubMed]
119. Thorat, M.N.; Dastager, S.G. High yield production of cellulose by a *Komagataeibacter rhaeticus* PG2 strain isolated from pomegranate as a new host. *RSC Adv.* **2018**, *8*, 29797–29805. [CrossRef]
120. Volova, T.G.; Prudnikova, S.V.; Sukovatyi, A.G.; Shishatskaya, E.I. Production and properties of bacterial cellulose by the strain *Komagataeibacter xylinus* B-12068. *App. Microbiol. Biotechnol.* **2018**, *102*, 7417–7428. [CrossRef] [PubMed]
121. Molina-Ramírez, C.; Castro, M.; Osorio, M.; Torres-Taborda, M.; Gómez, B.; Zuluaga, R.; Gómez, C.; Gañán, P.; Rojas, O.J.; Castro, C. Effect of different carbon sources on bacterial nanocellulose production and structure using the low pH resistant strain *Komagataeibacter medellinensis*. *Materials* **2017**, *10*, 639. [CrossRef]
122. Mohammadkazemi, F.; Azin, M.; Ashori, A. Production of bacterial cellulose using different carbon sources and culture media. *Carbohydr. Polym.* **2015**, *117*, 518–523. [CrossRef] [PubMed]
123. Mitchell, M.J.; Billingsley, M.M.; Haley, R.M.; Wechsler, M.E.; Peppas, N.A.; Langer, R. Engineering precision nanoparticles for drug delivery. *Nat. Rev. Drug Discov.* **2020**, *20*, 101–124. [CrossRef] [PubMed]
124. Hare, J.I.; Lammers, T.; Ashford, M.B.; Puri, S.; Strom, G.; Barry, S.T. Challenges and strategies in anti-cancer nanomedicine development: An industry perspective. *Adv. Drug Deliv. Rev.* **2017**, *108*, 25–38. [CrossRef] [PubMed]
125. Araki, J. Electrostatic or steric?–preparations and characterizations of well-dispersed systems containing rod-like nanowhiskers of crystalline polysaccharides. *Soft Mater.* **2013**, *9*, 4125–4141. [CrossRef]
126. Kargarzadeh, H.; Mariano, M.; Gopakumar, D.; Ahmad, I.; Thomas, S.; Dufresne, A.; Huang, J.; Lin, N. Advances in cellulose nanomaterials. *Cellulose* **2018**, *25*, 2151–2189. [CrossRef]
127. Hebeish, A., Guthrie, T. *The Chemistry and Technology of Cellulosic Copolymers*; Springer Science & Business Media: Belin, Germany, 2012.
128. De La Motte, H.; Hasani, M.; Brelid, H.; Westman, G. Molecular characterization of hydrolyzed cationized nanocrystalline cellulose, cotton cellulose and softwood kraft pulp using high resolution 1D and 2D NMR. *Carbohydr. Polym.* **2011**, *85*, 738–746. [CrossRef]
129. Habibi, Y. Key advances in the chemical modification of nanocelluloses. *Chem. Soc. Rev.* **2014**, *43*, 1519–1542. [CrossRef]
130. Pakharenko, V.; Pervaiz, M.; Pande, H.; Sain, M.; Sain, M. Chemical and physical techniques for surface modification of nanocellulose reinforcements. In *Interface/Interphase in Polymer Nanocomposites*; John Wiley & Sons: Hoboken, NJ, USA, 2016; pp. 283–310.
131. Meneguin, A.B.; da Silva Barud, H.; Sabio, R.M.; de Sousa, P.Z.; Manieri, K.F.; de Freitas, L.A.P.; Pacheco, G.; Alonso, J.D.; Chorilli, M. Spray-dried bacterial cellulose nanofibers: A new generation of pharmaceutical excipient intended for intestinal drug delivery. *Carbohydr. Polym.* **2020**, *249*, 116838. [CrossRef]
132. Khine, Y.Y.; Stenzel, M.H. Surface modified cellulose nanomaterials: A source of non-spherical nanoparticles for drug delivery. *Mater. Horiz.* **2020**, *7*, 1727–1758. [CrossRef]
133. Kusano, Y.; Madsen, B.; Berglund, L.; Oksman, K. Modification of cellulose nanofibre surfaces by He/NH3 plasma at atmospheric pressure. *Cellulose* **2019**, *26*, 7185–7194. [CrossRef]
134. Taheri, A.; Mohammadi, M. The use of cellulose nanocrystals for potential application in topical delivery of hydroquinone. *Chem. Biol. Drug Design* **2015**, *86*, 102–106. [CrossRef] [PubMed]

135. Camarero Espinosa, S.; Kuhnt, T.; Foster, E.J.; Weder, C. Isolation of thermally stable cellulose nanocrystals by phosphoric acid hydrolysis. *Biomacromol.* **2013**, *14*, 1223–1230. [CrossRef]
136. Wang, N.; Ding, E.; Cheng, R. Preparation and liquid crystalline properties of spherical cellulose nanocrystals. *Langmuir* **2008**, *24*, 5–8. [CrossRef] [PubMed]
137. Lin, N.; Dufresne, A. Surface chemistry, morphological analysis and properties of cellulose nanocrystals with gradiented sulfation degrees. *Nanoscale* **2014**, *6*, 5384–5393. [CrossRef]
138. Kokol, V.; Božič, M.; Vogrinčič, R.; Mathew, A.P. Characterisation and properties of homo- and heterogeneously phosphorylated nanocellulose. *Carbohydr. Polym.* **2015**, *125*, 301–313. [CrossRef]
139. Wang, H.; Xie, H.; Du, H.; Wang, X.; Liu, W.; Duan, Y.; Zhang, X.; Sun, L.; Zhang, X.; Si, C. Highly efficient preparation of functional and thermostable cellulose nanocrystals via H2SO4 intensified acetic acid hydrolysis. *Carbohydr. Polym.* **2020**, *239*, 116233. [CrossRef]
140. Braun, B.; Dorgan, J.R. Single-step method for the isolation and surface functionalization of cellulosic nanowhiskers. *Biomacromolecules* **2009**, *10*, 334–341. [CrossRef] [PubMed]
141. Chen, G.-Y.; Yu, H.-Y.; Zhang, C.-H.; Zhou, Y.; Yao, J.-M. A universal route for the simultaneous extraction and functionalization of cellulose nanocrystals from industrial and agricultural celluloses. *J. Nanopart. Res.* **2016**, *18*, 48. [CrossRef]
142. Du, H.; Liu, C.; Mu, X.; Gong, W.; Lv, D.; Hong, Y.; Si, C.; Li, B. Preparation and characterization of thermally stable cellulose nanocrystals via a sustainable approach of FeCl 3-catalyzed formic acid hydrolysis. *Cellulose* **2016**, *23*, 2389–2407. [CrossRef]
143. Spinella, S.; Re, G.L.; Liu, B.; Dorgan, J.; Habibi, Y.; Leclere, P.; Raquez, J.-M.; Dubois, P.; Gross, R.A. Polylactide/cellulose nanocrystal nanocomposites: Efficient routes for nanofiber modification and effects of nanofiber chemistry on PLA reinforcement. *Polymer* **2015**, *65*, 9–17. [CrossRef]
144. Chen, L.; Zhu, J.Y.; Baez, C.; Kitin, P.; Elder, T. Highly thermal-stable and functional cellulose nanocrystals and nanofibrils produced using fully recyclable organic acids. *Green Chem.* **2016**, *18*, 3835–3843. [CrossRef]
145. Bian, H.; Chen, L.; Dai, H.; Zhu, J. Integrated production of lignin containing cellulose nanocrystals (LCNC) and nanofibrils (LCNF) using an easily recyclable di-carboxylic acid. *Carbohydr. Polym.* **2017**, *167*, 167–176. [CrossRef]
146. Jia, W.; Liu, Y. Two characteristic cellulose nanocrystals (CNCs) obtained from oxalic acid and sulfuric acid processing. *Cellulose* **2019**, *26*, 8351–8365. [CrossRef]
147. Luo, J.; Huang, K.; Xu, Y.; Fan, Y. A comparative study of lignocellulosic nanofibrils isolated from celery using oxalic acid hydrolysis followed by sonication and mechanical fibrillation. *Cellulose* **2019**, *26*, 5237–5246. [CrossRef]
148. Spinella, S.; Maiorana, A.; Qian, Q.; Dawson, N.J.; Hepworth, V.; Mccallum, S.A.; Ganesh, M.; Singer, K.D.; Gross, R.A. 2016. Concurrent cellulose hydrolysis and esterification to prepare a surface-modified cellulose nanocrystal decorated with carboxylic acid moieties. *ACS Sust. Chem. Eng.* **2016**, *4*, 1538–1550. [CrossRef]
149. Ji, H.; Xiang, Z.; Qi, H.; Han, T.; Pranovich, A.; Song, T. Strategy towards one-step preparation of carboxylic cellulose nanocrystals and nanofibrils with high yield, carboxylation and highly stable dispersibility using innocuous citric acid. *Green Chem.* **2019**, *21*, 1956–1964. [CrossRef]
150. Habibi, Y.; Chanzy, H.; Vignon, M.R. TEMPO-mediated surface oxidation of cellulose whiskers. *Cellulose* **2006**, *13*, 679–687. [CrossRef]
151. Jiang, J.; Ye, W.; Liu, L.; Wang, Z.; Fan, Y.; Saito, T.; Isogai, A. Cellulose nanofibers prepared using the TEMPO/laccase/O2 system. *Biomacromolecules* **2017**, *18*, 288–294. [CrossRef] [PubMed]
152. Plappert, S.F.; Liebner, F.W.; Konnerth, J.; Nedelec, J.-M. Anisotropic nanocellulose gel–membranes for drug delivery: Tailoring structure and interface by sequential periodate–chlorite oxidation. *Carbohydr. Polym.* **2019**, *226*, 115306. [CrossRef] [PubMed]
153. Mascheroni, E.; Rampazzo, R.; Ortenzi, M.A.; Piva, G.; Bonetti, S.; Piergiovanni, L. Comparison of cellulose nanocrystals obtained by sulfuric acid hydrolysis and ammonium persulfate, to be used as coating on flexible food-packaging materials. *Cellulose* **2016**, *23*, 779–793. [CrossRef]
154. Montanari, S.; Roumani, M.; Heux, L.; Vignon, M.R. Topochemistry of carboxylated cellulose nanocrystals resulting from TEMPO-mediated oxidation. *Macromolecules* **2005**, *38*, 1665–1671. [CrossRef]
155. Carlsson, D.O.; Hua, K.; Forsgren, J.; Mihranyan, A. Aspirin degradation in surface-charged TEMPO-oxidized mesoporous crystalline nanocellulose. *Int. J. Pharm.* **2014**, *461*, 74–81. [CrossRef]
156. Chen, X.; Xu, X.; Li, W.; Sun, B.; Yan, J.; Chen, C.; Liu, J.; Qian, J.; Sun, D. Effective drug carrier based on polyethylenimine-functionalized bacterial cellulose with controllable release properties. *ACS Appl. Bio Mater.* **2018**, *1*, 42–50. [CrossRef]
157. Singhsa, P.; Narain, R.; Manuspiya, H. Bacterial cellulose nanocrystals (BNCC) preparation and characterization from three bacterial cellulose sources and development of functionalized BNCCs as nucleic acid delivery systems. *ACS Appl. Nano Mater.* **2017**, *1*, 209–221. [CrossRef]
158. Li, L.; Yu, C.; Yu, C.; Chen, Q.; Yu, S. Nanocellulose as template to prepare rough-hydroxy rich hollow silicon mesoporous nanospheres (R-nCHMSNs) for drug delivery. *Int. J. Biol. Macromol.* **2021**, *180*, 432–438. [CrossRef]
159. Akhlaghi, S.P.; Berry, R.C.; Tam, K.C. Surface modification of cellulose nanocrystal with chitosan oligosaccharide for drug delivery applications. *Cellulose* **2013**, *20*, 1747–1764. [CrossRef]
160. Tortorella, S.; Maturi, M.; Dapporto, F.; Spanu, C.; Sambri, L.; Franchini, M.C.; Chiariello, M.; Locatelli, E. Surface modification of nanocellulose through carbamate link for a selective release of chemotherapeutics. *Cellulose* **2020**, *27*, 8503–8511. [CrossRef]

161. Liu, Y.; Sui, Y.; Liu, C.; Liu, C.; Wu, M.; Li, B.; Li, Y. A physically crosslinked polydopamine/nanocellulose hydrogel as potential versatile vehicles for drug delivery and wound healing. *Carbohydr. Polym.* **2018**, *188*, 27–36. [CrossRef] [PubMed]
162. Wang, H.; He, J.; Zhang, M.; Tam, K.C.; Ni, P. A new pathway towards polymer modified cellulose nanocrystals via a "grafting onto" process for drug delivery. *Polym. Chem.* **2015**, *6*, 4206–4209. [CrossRef]
163. Kumar, A.; Durand, H.; Zeno, E.; Balsollier, C.; Watbled, B.; Sillard, C.; Fort, S.; Baussanne, I.; Belgacem, N.; Lee, D. The surface chemistry of a nanocellulose drug carrier unraveled by MAS-DNP. *Chem. Sci.* **2020**, *11*, 3868–3877. [CrossRef] [PubMed]
164. Khine, Y.Y.; Ganda, S.; Stenzel, M.H. Covalent tethering of temperature responsive pNIPAm onto TEMPO-oxidized cellulose nanofibrils via three-component passerini reaction. *ACS Macro Lett.* **2018**, *7*, 412–418. [CrossRef]
165. Tardy, B.L.; Yokota, S.; Ago, M.; Xiang, W.; Kondo, T.; Bordes, R.; Rojas, O.J. Nanocellulose–surfactant interactions. *Curr. Opi. Colloid Interface Sci.* **2017**, *29*, 57–67. [CrossRef]
166. Zainuddin, N.; Ahmad, I.; Kargarzadeh, H.; Ramli, S. Hydrophobic kenaf nanocrystalline cellulose for the binding of curcumin. *Carbohydr. Polym.* **2017**, *163*, 261–269. [CrossRef]
167. Raghav, N.; Sharma, M.R. Usage of nanocrystalline cellulose phosphate as novel sustained release system for anti-inflammatory drugs. *J. Mol. Struct.* **2021**, *1233*, 130108. [CrossRef]
168. Gupta, R.D.; Raghav, N. Differential effect of surfactants tetra-n-butyl ammonium bromide and N-Cetyl-N,N,N-trimethyl ammonium bromide bound to nano-cellulose on binding and sustained release of some non-steroidal anti-inflammatory drugs. *Int. J. Biol. Macromol.* **2020**, *164*, 2745–2752. [CrossRef]
169. Liu, X.Q.; Picart, C. Layer-by-layer assemblies for cancer treatment and diagnosis. *Adv. Mater.* **2016**, *28*, 1295–1301. [CrossRef] [PubMed]
170. De Koker, S.; Hoogenboom, R.; De Geest, B.G. Polymeric multilayer capsules for drug delivery. *Chem. Soc. Rev.* **2012**, *41*, 2867–2884. [CrossRef]
171. Eivazi, A.; Medronho, B.; Lindman, B.; Norgren, M. On the development of all-cellulose capsules by vesicle-templated layer-by-layer assembly. *Polymers* **2021**, *13*, 589. [CrossRef]
172. Kulkarni, A.D.; Vanjari, Y.H.; Sancheti, K.H.; Patel, H.M.; Belgamwar, V.S.; Surana, S.J.; Pardeshi, C.V. Polyelectrolyte complexes: Mechanisms, critical experimental aspects, and applications. *Artif. Cells Nanomed. Biotechnol.* **2016**, *44*, 1615–1625. [CrossRef]
173. Mohanta, V.; Madras, G.; Patil, S. Layer-by-layer assembled thin films and microcapsules of nanocrystalline cellulose for hydrophobic drug delivery. *ACS Appl. Mater. Interfaces* **2014**, *6*, 20093–20101. [CrossRef] [PubMed]
174. Li, J.; Wang, Y.; Zhang, L.; Xu, Z.; Dai, H.; Wu, W. Nanocellulose/gelatin composite cryogels for controlled drug release. *ACS Sust. Chem. Eng.* **2019**, *7*, 6381–6389. [CrossRef]
175. Hennink, W.E.; Van Nostrum, C.F. Novel crosslinking methods to design hydrogels. *Adv. Drug Deliv. Rev.* **2012**, *64*, 223–236. [CrossRef]
176. Shi, X.; Zheng, Y.; Wang, G.; Lin, Q.; Fan, J. pH-and electro-response characteristics of bacterial cellulose nanofiber/sodium alginate hybrid hydrogels for dual controlled drug delivery. *RSC Adv.* **2014**, *4*, 47056–47065. [CrossRef]
177. Lin, N.; Geze, A.; Wouessidjewe, D.; Huang, J.; Dufresne, A. Biocompatible double-membrane hydrogels from cationic cellulose nanocrystals and anionic alginate as complexing drugs codelivery. *ACS Appl. Mater. Interfaces* **2016**, *8*, 6880–6889. [CrossRef]
178. Lin, N.; Dufresne, A. Supramolecular hydrogels from in situ host–guest inclusion between chemically modified cellulose nanocrystals and cyclodextrin. *Biomacromolecules* **2013**, *14*, 871–880. [CrossRef]
179. Muller, A.; Ni, Z.; Hessler, N.; Wesarg, F.; Muller, F.A.; Kralisch, D.; Fischer, D. The biopolymer bacterial nanocellulose as drug delivery system: Investigation of drug loading and release using the model protein albumin. *J. Pharma. Sci.* **2013**, *102*, 579–592. [CrossRef]
180. Muller, A.; Zink, M.; Hessler, N.; Wesarg, F.; Muller, F.A.; Kralisch, D.; Fischer, D. Bacterial nanocellulose with a shape-memory effect as potential drug delivery system. *RSC Adv.* **2014**, *4*, 57173–57184. [CrossRef]
181. Kopac, T.; Krajnc, M.; Ručigaj, A. A mathematical model for pH-responsive ionically crosslinked TEMPO nanocellulose hydrogel design in drug delivery systems. *Int. J. Biol. Macromol.* **2021**, *168*, 695–707. [CrossRef] [PubMed]
182. Jiang, S.; Agarwal, S.; Greiner, A. Low-density open cellular sponges as functional materials. *Angew. Chem. Int. Ed.* **2017**, *56*, 15520–15538. [CrossRef] [PubMed]
183. Lavoine, N.; Bergstrom, L. Nanocellulose-based foams and aerogels: Processing, properties, and applications. *J. Mater. Chem. A* **2017**, *5*, 16105–16117. [CrossRef]
184. Pierre, A.C.; Pajonk, G.M. Chemistry of aerogels and their applications. *Chem. Rev.* **2002**, *102*, 4243–4266. [CrossRef]
185. Kargarzadeh, H.; Huang, J.; Lin, N.; Ahmad, I.; Mariano, M.; Dufresne, A.; Thomas, S.; Galeski, A. Recent developments in nanocellulose-based biodegradable polymers, thermoplastic polymers, and porous nanocomposites. *Prog. Polym. Sci.* **2018**, *87*, 197–227. [CrossRef]
186. Sun, Y.; Chu, Y.; Wu, W.; Xiao, H. Nanocellulose-based Lightweight Porous Materials: A Review. *Carbohydr. Polym.* **2021**, *255*, 117489. [CrossRef]
187. Zhao, J.; Lu, C.; He, X.; Zhang, X.; Zhang, W.; Zhang, X. Polyethylenimine-grafted cellulose nanofibril aerogels as versatile vehicles for drug delivery. *ACS Appl. Mater. Interfaces* **2015**, *7*, 2607–2615. [CrossRef] [PubMed]
188. Liang, Y.; Zhu, H.; Wang, L.; He, H.; Wang, S. Biocompatible smart cellulose nanofibres for sustained drug release via pH and temperature dual-responsive mechanism. *Carbohydr. Polym.* **2020**, *249*, 116876. [CrossRef] [PubMed]

189. Svagan, A.J.; Benjamins, J.-W.; Al-Ansari, Z.; Shalom, D.B.; Mullertz, A.; Wagberg, L.; Lobmann, K. Solid cellulose nanofiber based foams–towards facile design of sustained drug delivery systems. *J. Controlled Release* **2016**, *244*, 74–82. [CrossRef] [PubMed]
190. Lobmann, K.; Svagan, A.J. Cellulose nanofibers as excipient for the delivery of poorly soluble drugs. *Int. J. Pharm.* **2017**, *533*, 285–297. [CrossRef] [PubMed]
191. Svagan, A.J.; Mullertz, A.; Lobmann, K. Floating solid cellulose nanofibre nanofoams for sustained release of the poorly soluble model drug furosemide. *J. Pharm. Pharmacol.* **2017**, *69*, 1477–1484. [CrossRef]
192. Bannow, J.; Benjamins, J.-W.; Wohlert, J.; Lobmann, K.; Svagan, A.J. Solid nanofoams based on cellulose nanofibers and indomethacin—The effect of processing parameters and drug content on material structure. *Int. J. Pharm.* **2017**, *526*, 291–299. [CrossRef] [PubMed]
193. Fiorati, A.; Turco, G.; Travan, A.; Caneva, E.; Pastori, N.; Cametti, M.; Punta, C.; Melone, L. Mechanical and drug release properties of sponges from cross-linked cellulose nanofibers. *ChemPlusChem* **2017**, *82*, 848–858. [CrossRef] [PubMed]
194. Xiao, Y.; Rong, L.; Wang, B.; Mao, Z.; Xu, H.; Zhong, Y.; Zhang, L.; Sui, X. A light-weight and high-efficacy antibacterial nanocellulose-based sponge via covalent immobilization of gentamicin. *Carbohydr. Polym.* **2018**, *200*, 595–601. [CrossRef] [PubMed]
195. Rahimi, M.; Shojaei, S.; Safa, K.D.; Ghasemi, Z.; Salehi, R.; Yousefi, B.; Shafiei-Irannejad, V. Biocompatible magnetic tris (2-aminoethyl) amine functionalized nanocrystalline cellulose as a novel nanocarrier for anticancer drug delivery of methotrexate. *New J. Chem.* **2017**, *41*, 2160–2168. [CrossRef]
196. Supramaniam, J.; Adnan, R.; Kaus, N.H.M.; Bushra, R. Magnetic nanocellulose alginate hydrogel beads as potential drug delivery system. *Int. J. Biol. Macromol.* **2018**, *118*, 640–648. [CrossRef]
197. Jeddi, M.K.; Mahkam, M. Magnetic nano carboxymethyl cellulose-alginate/chitosan hydrogel beads as biodegradable devices for controlled drug delivery. *Int. J. Biol. Macromol.* **2019**, *135*, 829–838. [CrossRef]
198. Cirillo, G.; Hampel, S.; Spizzirri, U.G.; Parisi, O.I.; Picci, N.; Iemma, F. Carbon nanotubes hybrid hydrogels in drug delivery: A perspective review. *BioMed Res. Int.* **2014**, 825017. [CrossRef]
199. Zhang, W.; Zhang, Z.; Zhang, Y. The application of carbon nanotubes in target drug delivery systems for cancer therapies. *Nanoscale Res. Lett.* **2011**, *6*, 1–22. [CrossRef] [PubMed]
200. Anirudhan, T.; Shainy, F.; Thomas, J.P. Effect of dual stimuli-responsive dextran/nanocellulose polyelectrolyte complexes for chemo photothermal synergistic cancer therapy. *Int. J. Biol. Macromol.* **2019**, *135*, 776–789.

Article

Isolation and Characterization of Nanocellulose with a Novel Shape from Walnut (*Juglans Regia* L.) Shell Agricultural Waste

Dingyuan Zheng [1,2], Yangyang Zhang [1,2], Yunfeng Guo [1,2] and Jinquan Yue [1,2,*]

1. College of Material Science and Engineering, Northeast Forestry University, Harbin 150040, China
2. Key Laboratory of Bio-based Material Science and Technology of Ministry of Education, Northeast Forestry University, 26 Hexing Road, Harbin 150040, China
* Correspondence: yuejinq@163.com; Tel.: +86-135-0368-5163

Received: 3 June 2019; Accepted: 24 June 2019; Published: 3 July 2019

Abstract: Herein, walnut shell (WS) was utilized as the raw material for the production of purified cellulose. The production technique involves multiple treatments, including alkaline treatment and bleaching. Furthermore, two nanocellulose materials were derived from WS by 2,2,6,6-tetramethylpiperidine-1-oxyl radical (TEMPO) oxidation and sulfuric acid hydrolysis, demonstrating the broad applicability and value of walnuts. The micromorphologies, crystalline structures, chemical functional groups, and thermal stabilities of the nanocellulose obtained via TEMPO oxidation and sulfuric acid hydrolysis (TNC and SNC, respectively) were comprehensively characterized. The TNC exhibited an irregular block structure, whereas the SNC was rectangular in shape, with a length of 55–82 nm and a width of 49–81 nm. These observations are expected to provide insight into the potential of utilizing WSs as the raw material for preparing nanocellulose, which could address the problems of the low-valued utilization of walnuts and pollution because of unused WSs.

Keywords: Walnut shell; nanocellulose; TEMPO oxidation; sulfuric acid hydrolysis; ultrasonication

---

1. Introduction

With the depletion of fossil fuels and the increase in ecological and environmental problems caused by the usage of fossil fuels, the development and utilization of green biomass-based materials derived from renewable natural resources have been extensively studied around the world [1]. Cellulose, which can be observed in all the plant structures, is a common example of a renewable natural resource and offers the advantages of abundant availability, renewability, and biodegradability [2,3]. With the development of nanotechnology, nanocellulose, prepared from cellulose, has attracted significant attention from academic and industrial researchers because of its low cost, biocompatibility, biodegradability, nontoxicity, renewability, sustainability, strong surface reactivity, and desirable physical properties (it is lightweight and impermeable to gas and it also exhibits high stiffness, good optical transparency, and low thermal expansion) [4,5]. Compared with cellulose, nanocellulose exhibits large surface area, high crystallinity, high mechanical strength, high hydrophilicity and supramolecular structure [6]. These characteristics make nanocellulose promising for various applications such as polymer nanocomposites [7,8], packaging [9,10], electronics [11,12], and stimulus-responsive materials [13,14].

Nanocellulose can be extracted from several cellulose resources such as wood [15,16], cotton [17,18], ramie [19], bagasse [20,21], bamboo [22], sisal [23], corn straw [24], rice straw [18], and coconut shell [25]. It has been reported that the fundamental properties of the obtained nanocellulose, such as morphology, crystallinity, dimensions, and surface chemistry, vary highly depending on the raw material and the isolation process used to obtain it [26,27]. Hence, different categories of nanocellulose, such as cellulose nanofibrils, cellulose nanocrystals, and bacterial cellulose, which differ in terms of their dimensions and morphologies, exist [2,28]. These key properties are of critical importance for ensuring the end use of the isolated nanocellulose.

Walnut (*Juglans regia* L.), also known as a peach, is harvested from the seed of the walnut tree. Walnut is one of the world's four major "dried fruits" along with almonds, cashews, and hazelnuts. The global walnut production in 2017 exceeded 3.8 million tons. The walnut shell (WS) is the hard outer shell (endocarp) of walnut, which comprises cellulose, hemicellulose, lignin, and other small molecular substances and accounts for 67% of the total weight of walnut [29]. China, which is the top walnut producer in the world, produced 1.92 million tons of walnuts in 2017, which resulted in approximately 1.29 million tons of WSs (UN Food and Agriculture Organization, Corporate Statistical Databas, 2018). After the nut is removed from the WS, the shells are considered to be agricultural and forestry waste, which is often burned as fuel, seriously damaging the environment. Only a small portion of the produced WSs is used for preparing wood–plastic composite materials [30], carbon materials [31,32], or handcrafted products. Thus, the utilization of WSs is not much valued. The walnut contains desirable substances such as cellulose, hemicellulose, and lignin. It has been reported that the cellulose content of walnut shell is about 22% [33]. Hence, in this study, we investigate the preparation of nanocellulose using WS as the raw material for the first time.

Cellulose nanofibrils are mainly obtained from cellulosic fibers by mechanical treatments such as high-pressure homogenization, microfluidization, grinding, and ultrasonication [34]. However, such mechanical fibrillation methods are very energy intensive. Various pretreatment methods (such as enzymatic hydrolysis [35], 2,2,6,6-tetramethylpiperidine-1-oxyl radical (TEMPO)-mediated oxidation [36–38]) have been proposed to reduce the energy required for the mechanical deconstruction process by reducing the negative or positive charge on the fiber surfaces and by enhancing the colloidal stability of the final cellulose nanofibrils.

Unlike cellulose nanofibrils, cellulose nanocrystals have a rod-like morphology [39]. Cellulose nanocrystals are generally prepared by hydrolysis using a strong inorganic acid such as sulfuric acid [40,41], other inorganic strong acid hydrolysis are used for cellulose nanocrystals preparation as well, including hydrochloric acid, phosphoric acid, hydrobromic acid, and nitric acid [42]. Recent technological developments have resulted in a few sustainable and environment friendly methods that rely on recyclable chemicals. Examples include hydrolysis with solid acids (e.g., phosphor tungstic acid [43]) or treatment with ionic liquids [44] or deep eutectic solvents [45]. Among these, sulfuric acid is the most commonly used acid for producing sulfonated cellulose nanocrystals exhibiting good dispersibility in water [42]. During the hydrolysis process, the paracrystalline or disordered parts of cellulose are hydrolyzed and dissolved in the acid solution; however, the crystalline parts are chemically resistant to the acid and remain intact. Consequently, the cellulose fibrils are transversely cleaved, yielding short cellulose nanocrystals with a relatively high crystallinity [34].

In this study, we aim to isolate nanocellulose from a renewable, cheap, and currently underutilized raw material, i.e., WS. Herein, WS is pretreated using several mechanical and chemical processes, including grinding, extraction, alkali treatment, and bleaching. Nanocellulose was produced by TEMPO oxidation accompanied by ultrasonic treatment (yielding TNC) or sulfuric acid hydrolysis followed by ultrasonication (SNC). The resulting TNC and SNC and the intermediate products were characterized by transmission electron microscopy (TEM), X-ray diffraction (XRD), Fourier-transform infrared spectroscopy (FTIR), and thermogravimetric analysis (TGA). The proposed method of extracting nanocellulose from WSs is expected to reduce the pressure on natural resources. This will offer an alternative high-valued utilization for the currently wasted WSs and, furthermore, to serve as a reference for future studies to improve the utilization of the obtained nanocellulose for developing new biobased nanomaterials.

2. Materials and Methods

2.1. Materials

The WS collected from the Shanxi Province in China was used as the raw material. Ethanol, sodium hydroxide, sodium chlorite, sodium bromide, sodium hypochlorite, glacial acetic acid, sulfuric acid, and other chemicals were of analytical grade and used without any further purification. All of the chemicals were supplied by Tianjin Kemiou Chemical Reagent Co., Ltd. (Tianjin, China). Distilled water was used the whole process. TEMPO was purchased from Aladdin Reagent Co., Ltd. (Shanghai, China). Dialysis bags (MD44, viskase, Lombard, IL, USA) were provided by Beijing Biotopped Science & Technology CO., Ltd. (Beijing, China).

2.2. Preparation of Nanocellulose

The procedure used to prepare nanocellulose is outlined in Figure 1.

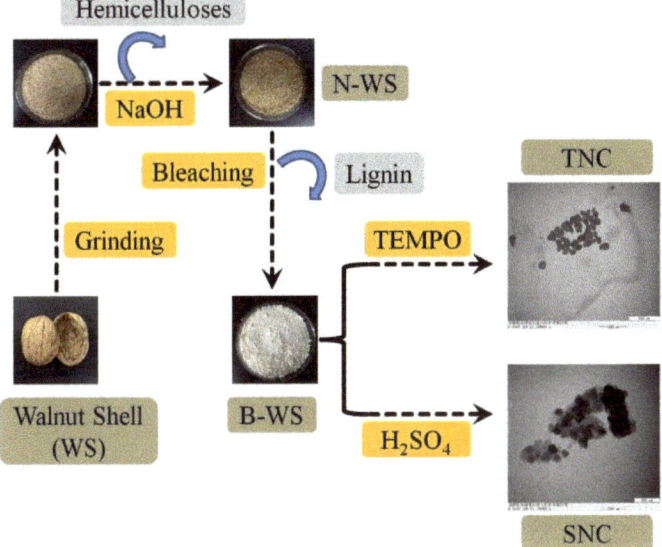

Figure 1. Procedure used to isolate nanocellulose from the walnut shell.

2.2.1. WS Pretreatment

In natural WSs, cellulose is embedded in a network structure in which lignin and hemicellulose are linked together, mainly by covalent bonds [46]. Thus, WS was initially pretreated to isolate the nanocellulose. Pretreatment of walnut shell was performed using the method described by de Rodriguez, Thielemans, and Dufresne [47]. The WS was ground and sieved through 60 mesh (0.25 mm), dried in an oven at 105 °C, and stored in a desiccator. To eliminate hemicellulose, the obtained WS powder was mixed with a 2 wt.% NaOH solution in a WS/NaOH solid-to-liquid ratio of 10 g/100 mL and stirred for 4 h at 100 °C. This NaOH treatment was repeated four times until no more discoloration occurred; the product obtained after the NaOH treatment can be referred to as N-WS. Further, N-WS was bleached in a solution containing equal amounts of acetate buffer and 1.7 wt.% $NaClO_2$ at a WS/solution solid-to-liquid ratio of 5 g/100 mL for 6 h at 80 °C (the reaction flask was shaken every 20 min to ensure the reaction occurs evenly). This bleaching process was repeated four times. The bleached products were then thoroughly washed using distilled water. The obtained product can be referred to as B-WS.

2.2.2. TNC Preparation

TNC was prepared using the method described by Kuramae, Saito, and Isogai [48], with few modifications. One gram of B-WS was dispersed in 100 mL of distilled water; further, 0.1 g of sodium bromide and 0.016 g of TEMPO were added, followed by 20 mmol of NaClO. During the TEMPO oxidation, the pH of the suspension was adjusted to 10 ± 0.5 using 0.1 M NaOH and 0.1 M HCl. The reaction was continued for 5 h at room temperature under magnetic stirring at 1000 rpm and was terminated by adding 10 mL of ethanol. The oxidized B-WS was washed and filtered; subsequently, it was stored with distilled water at 4 °C to avoid strong hydrogen bonding.

To convert the nonoxidized hydroxyl and aldehyde groups in the oxidized B-WS into carboxyl groups, the product obtained in the previous step was further processed. One gram of oxidized B-WS (dry weight) was dispersed in 65 mL of distilled water, and the pH value was adjusted to 4–5. Further, 0.6 g of $NaClO_2$ was added to the reaction system. The reaction was allowed to proceed for 1 h at 70 °C under magnetic stirring at 1000 rpm to yield carboxylated B-WS. The obtained carboxylated B-WS had a carboxylate content of 1.12 mmol/g as determined by conductivity titration [49]. A certain amount of carboxylated B-WS was weighed, and the mass fraction was adjusted to 0.5 wt.%. Finally, high-intensity ultrasonication was performed in an ultrasonic cell pulverizer (SCIENTZ-1200E, Ningbo Scientz Biotechnology Co., Ltd., Ningbo, China) at 600 w for 30 min in an ice/water bath to yield TNC.

2.2.3. SNC Preparation

SNC was prepared using the method described by Beck-Candanedo [41]. One gram of B-WS was added to 8.75 mL of the sulfuric acid solution (64 wt.%) under vigorous stirring, and the hydrolysis reaction was allowed to proceed for 1 h at 45 °C. The reaction was ended by adding distilled water in a volume that was ten times the reaction volume. The obtained suspension was washed by mixing with distilled water followed by centrifugation at 12,000 rpm for 15 min to eliminate the excess acid; the washing was repeated until the precipitate generation was terminated. The suspension was subsequently dialyzed against distilled water until the pH became 6.5–7. The dialyzed suspension was sonicated at 600 w for 2 min to yield SNC.

2.3. Characterization of TNC, SNC, and Intermediate Products

2.3.1. Analysis of the Chemical Components

The α-cellulose content was determined as follows [50]. Two grams of WS powder was weighed and transferred to a 250-mL Erlenmeyer flask, and 25 mL of nitric acid/ethanol solution (1:4 by volume) was added. The mixed solution was refluxed for 1 h, and the process was repeated several times until the sample turned white. The powder was repeatedly washed using distilled water and filtered until the pH became neutral; the obtained residue was dried at 105 °C. The α-cellulose content (X, %) was calculated as follows

$$X = \frac{G}{G_1(1-W)} \times 100 \tag{1}$$

where G denotes the weight of the obtained residue, G_1 denotes the weight of the WS, and W denotes the moisture content of the WS.

The remaining chemical components of WS (organic extracts, lignin, hemicellulose, ash, and holocellulose) were analyzed according to the Technical Association of Pulp and Paper Industry standards that have been previously described [51].

2.3.2. Scanning Electron Microscopy

The WS, N-WS, B-WS, TNC, and SNC microstructures were observed using a scanning electron microscope (SU8010, Hitachi, Japan) at an accelerating voltage of 5.0 kV. TNC and SNC were freeze-dried before observation. All the samples were coated with gold.

2.3.3. Transmission Electron Microscopy

TNC and SNC were imaged using a transmission electron microscope (H-7650 Hitachi, Japan) at a 100-kV acceleration voltage. The TNC and SNC suspensions were diluted to a concentration of 0.01% and deposited onto carbon-coated grids (230 mesh, Beijing Zhongjingkeyi Technology Co., Ltd., Beijing, China). After drying, the samples were negatively stained using a 1% phosphotungstic acid solution for 10 min. The TEM images were analyzed using Nanomeasurer 1.2 (Department of Chemistry, Fudan Univ., Shanghai, China) to determine the TNC and SNC size distributions.

2.3.4. Fourier-transform Infrared Spectroscopy

The FTIR spectra of WS, N-WS, B-WS, TNC, and SNC were recorded on a Fourier-transform infrared instrument (Nicolette 6700, Thermo Fisher Scientific Inc., Waltham, MA, USA) in 400–4000 cm^{-1} with a resolution of 4 cm^{-1}, and 20 scans for each sample were conducted. The FTIR spectra of all samples were collected using the attenuated total reflection technique (ATR, the ATR crystal material is zinc selenide (ZnSe)).

2.3.5. X-ray Diffraction Technique

XRD analysis was performed on WS, N-WS, B-WS, TNC, and SNC using a diffractometer (D/max 2200, Rigaku, Japan) equipped with Ni-filtered Cu Kα radiation (λ = 1.5406 Å) at 40 kV and 30 mA. The diffraction intensities were recorded in 2θ = 5°–60° with a scan rate of 5°/min.

The crystallinity index (CrI, %) was calculated according to the method reported by Segal [52] as follows

$$CrI = \frac{I_{200} - I_{am}}{I_{200}} \times 100 \quad (2)$$

where I_{200} is the maximum intensity of the diffraction at 200 peak (2θ = 22.6°) and I_{am} is the intensity of the diffraction at 2θ = 18°.

2.3.6. Thermogravimetric Analysis

The thermal stability of each sample was evaluated using a thermogravimetric analyzer (TG209F1, Netzsch Scientific Instruments Trading (Shanghai) Co., Ltd., Shanghai, China) from room temperature to 600 °C at a rate of 10 °C/min in a nitrogen atmosphere.

3. Results and Discussion

3.1. Chemical Components

Table 1 presents the WS, N-WS, and B-WS chemical compositions. The chemical compositions of the three samples were observed to be significantly different because of the applied chemical treatments.

Table 1. Chemical components of walnut shell WS, walnut shell treated by NaOH (N-WS), and walnut shell after bleaching (B-WS).

Sample	α-Cellulose (%)	Lignin (%)	Hemicelluloses (%)	Ash (%)	Benzene/Ethanol Extractives (%)
WS	27.4	36.31	31.3	3.6	1.57
N-WS	56.6	30.98	7.6	1.97	1.16
B-WS	87.9	0.17	1.8	1.64	0.41

It is found that WS has the lowest percentage of cellulose and highest percentage of noncellulosic components such as lignin and hemicelluloses. The chemical treatments aim to remove the noncellulosic components. When WS was subjected to NaOH treatment, the lignin and hemicelluloses contents of N-WS decreased to 30.98% and 7.6%, respectively, whereas the cellulose concentration of N-WS increased to 56.6%. After bleaching treatment, the cellulose content of B-WS increased to 87.9% due to the removal of remaining lignin and hemicelluloses, resulting in highly purified cellulose.

3.2. SEM Analysis

The WS, N-WS, B-WS, TNC, and SNC microstructures were observed using a scanning electron microscope. The SEM image of WS (Figure 2a) denotes that the WS had a rough surface. However, the surface of N-WS (as seen in Figure 2b) had an irregular porous structure due to the degradation of hemicellulose and the partial degradation of lignin by the repeated alkali treatment. Further, B-WS (shown in Figure 2c) exhibited a loose structure, which indicated the successful removal of residual lignin by the acetate buffer/$NaClO_2$ solution. These results are consistent with the changes in chemical composition denoted in Table 1.

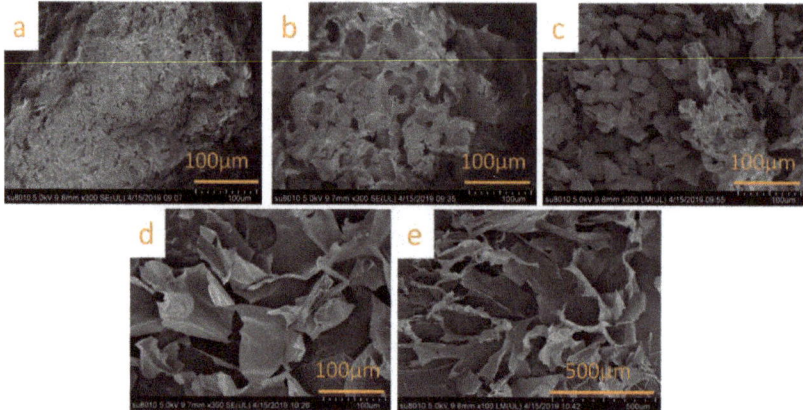

Figure 2. SEM micrographs of (**a**) WS, (**b**) N-WS, (**c**) B-WS, (**d**) nanocellulose obtained via TEMPO oxidation (TNC), and (**e**) nanocellulose obtained via sulfuric acid hydrolysis (SNC).

After the TNC was freeze-dried, the obtained aerogel was observed to be a porous network with a lamellar structure (Figure 2d). This structure is because the strong hydrogen bonding during freeze-drying caused the nanoparticles to self-assemble into the lamellar structures [53,54]. The aerogel formed by freeze-drying the SNC exhibited a similar porous structure (Figure 2e).

3.3. TEM Analysis

TEM was used to observe the morphology of the nanocellulose produced using different methods. The prepared TNC (Figure 3a) exhibited an irregular block structure, whereas the prepared SNC (Figure 3b) was rectangular with a length of 55–82 nm and a width of 49–81 nm. The TNC morphology observed in this study was significantly different from that of the nanocellulose produced using the same procedure in a previous study [36,54]; this may be attributed to the irregular morphological structure [30] of the WS powder used in this study, as depicted in Figure 2a. Furthermore, the SNC morphology obtained through sulfuric acid hydrolysis as observed by TEM differed from the typical rod- or needle-like morphologies of the typical cellulose nanocrystals [2]. As has been reported, the size and shape of nanocelluloses influence the properties (for example, optical characteristics, stability, and rheology) in aqueous media [55], which largely determines the application of nanocellulose. Nanocellulose with spherical or square structure makes them excellent candidates as stabilizer for Pickering emulsion [53] or drug delivery carrier for encapsulation [56].

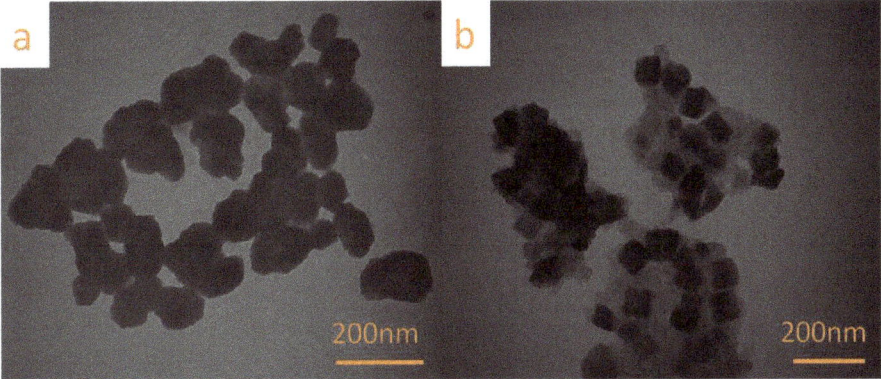

Figure 3. TEM images of (**a**) TNC and (**b**) SNC.

3.4. Chemical Structures

The changes in the chemical structures of the raw WS after various treatments were investigated using FTIR (Figure 4). Two main absorption regions appeared in all the curves among which one was in the high-wave-number region from 2800 to 3500 cm^{-1} and the other was in the low-wave-number region from 600 to 1750 cm^{-1} [57]. The peak at 2897 cm^{-1} was attributed to the stretching vibration of the C–H groups of cellulose, whereas the wide region around 3350 cm^{-1} was attributed to the O–H stretching vibration of the hydrogen-bonded hydroxyl groups in the cellulose molecules [51,58]. Further, the peak at 1738 cm^{-1} corresponds to the acetyl groups and ironic esters of the hemicellulose and the ester linkages of the carboxylic groups of the ferulic and p-coumaric acid in lignin and hemicellulose [59]. The absence of a peak at 1738 cm^{-1} in the FTIR spectrum of N-WS indicates that the hemicellulose in WS was effectively degraded by the NaOH treatment. The peaks at 1248 cm^{-1} and 1502 cm^{-1} in the FTIR spectrum of WS, corresponding to the aromatic skeletal vibrations of lignin [60], disappeared in the curve of B-WS, confirming that majority of the lignin was removed by the bleaching treatment. Furthermore, the absence of peaks at 1738 cm^{-1}, 1502 cm^{-1}, and 1248 cm^{-1} in the B-WS spectrum is consistent with the changes in chemical components during the alkaline and bleaching processes. The absorption peaks at 1162 cm^{-1} and 1034 cm^{-1} in all the curves correspond to the stretching vibration of the C–O–C bonds in the 1,4-glycosidic links linkages of the D-glucose units in cellulose, which were interpreted as typical for a cellulose structure [61]. The peaks at approximately 1645 cm^{-1} in the spectra of all the samples were attributed to the H–O–H stretching vibration of the adsorbed water due to the hydroxyl groups in cellulose [20], whereas those at 892 cm^{-1} represented the C_1–H deformation of cellulose [62]. There were no significant differences between the SNC and B-WS spectra, demonstrating that the characteristics of the cellulose molecular structure were maintained during sulfuric acid hydrolysis. The increase in the relative amount of cellulose in the sample due to the decrease in the amounts of other components upon hydrolysis may account for the slight increase in the intensity of the peak at 1034 cm^{-1} from B-WS to SNC. The peak at 1738 cm^{-1}, which is characteristic of the C=O stretching of carboxyl groups, reappeared in the FTIR spectrum of TNC when compared with that of B-WS due to the introduction of –COOH on the cellulose surface; this indicates that cellulose was successfully modified by the TEMPO oxidation [63].

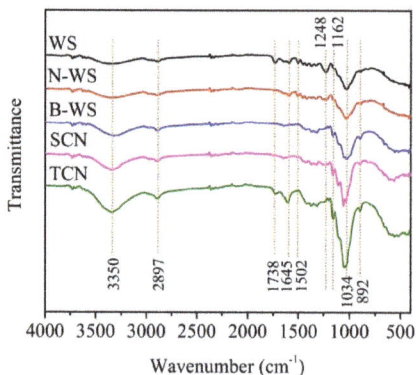

Figure 4. FTIR spectra of WS, N-WS, B-WS, TNC, and SNC.

3.5. Crystal Structures

The XRD patterns of WS, N-WS, B-WS, TNC, and SNC (Figure 5) were studied to further evaluate the influence of the processing treatment. All the samples exhibited diffraction peaks at approximately 16° (110) and 22.6° (200), whereas WS, N-WS, and B-WS also exhibited a diffraction peak at approximately 34° (004). These peaks indicate that the typical cellulose crystal structure was preserved in all the samples, indicating that the chemical and ultrasonic treatments did not change the integrity of the original cellulose crystal [64,65]. However, the crystallinity index changed with each step. The apparent crystallinity of WS was 29.5%. After NaOH treatment, the CrI of N-WS increased to 40.9% due to the dissolution and removal of lignin and majority of the hemicellulose (i.e., the amorphous hemicellulose) [57]. The CrI further increased to 42.9% as a result of the lignin removal by the acetate buffer/NaClO$_2$ solution. These results are consistent with the observed changes in the chemical composition and the FTIR analyses. The CrI of SNC (40.1%) decreased slightly after the sulfuric acid hydrolysis and ultrasonication, which may have been caused because the amorphous and crystalline regions were damaged by the strong acid hydrolysis [66]. There were no significant differences in the position of X-ray diffraction peaks between the TNC and B-WS XRD curves, indicating that the original crystal structure of the cellulose was unchanged after the oxidation (Figure 5). These results indicate that the carboxylate groups formed by the TEMPO-mediated oxidation are selectively introduced on the surfaces of the cellulose microfibril rather than the internal cellulose crystallites [38].

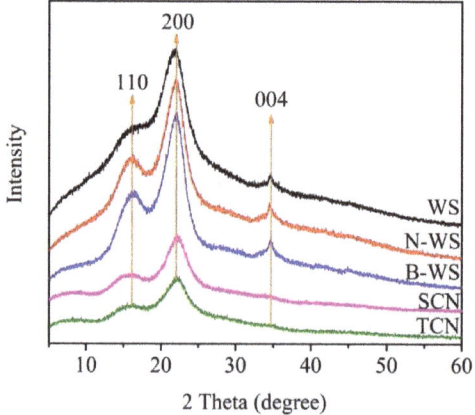

Figure 5. XRD patterns of WS, N-WS, B-WS, TNC, and SNC.

3.6. Thermal Stability

Figures 6 and 7 depicted the TGA and DTG curves, respectively, for WS, N-WS, B-WS, TNC, and SNC. All the TGA curves began with slight mass losses from room temperature to 105 °C, corresponding to the evaporation of slightly bound water from all the samples. As the temperature increased further, the WS degradation occurred in two phases. The first decomposition occurred between 214 °C and 300 °C, corresponding to the hemicellulose degradation and the beginning of lignin degradation [23,67]. After this change, the largest loss of mass in the material occurred between 300 °C and 380 °C, peaking at 342 °C, which corresponded to that of cellulose [68].

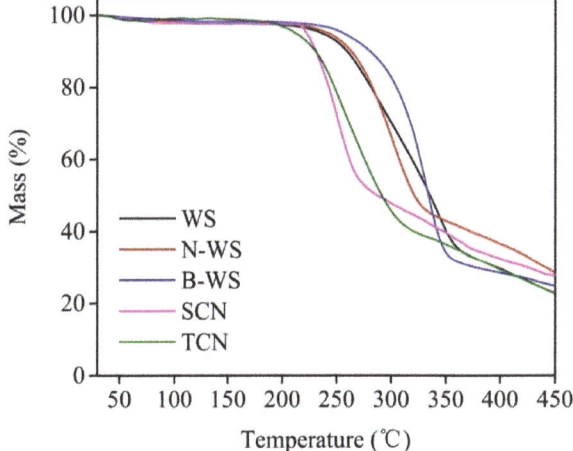

Figure 6. TGA curves of WS, N-WS, B-WS, TNC, and SNC.

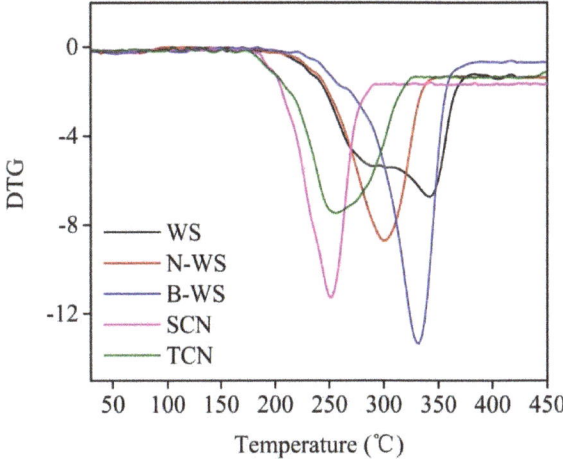

Figure 7. DTG curves of WS, N-WS, B-WS, TNC, and SNC.

In contrast, the decomposition of four other samples occurred in only one stage. N-WS began to degrade at 205 °C due to lignin degradation; however, the maximum mass loss occurred at 301 °C, which can be attributed to cellulose degradation. The absence of peaks related to the degradation of hemicellulose and the relatively small change associated with lignin degradation in this sample as compared with that associated with WS indicates that the alkaline treatment eliminated most of the hemicellulose and some of the lignin from the WS.

B-WS degraded from 222 to 380 °C, with the most significant mass loss occurring at 332 °C; this was mainly attributed to cellulose degradation, which was within the decomposition temperature range reported in previous studies [68,69] (315–400°C). The wide range, which was associated with lignin degradation, was not observed in this sample, indicating that lignin was eliminated after the bleaching reaction in the acetate buffer/$NaClO_2$ solution.

The thermal degradation of TNC (i.e., following TEMPO/NaClO/NaBr oxidation) began at 178 °C, which was approximately 50 °C lower than that in B-WS. It has been reported that the introduction of –COOH by the TEMPO-mediated oxidation of the C6 primary hydroxyl groups on the cellulose surface results in a significant decrease in thermal degradation [70].

SNC exhibited the worst thermal stability among the five tested samples, which can be attributed to the introduction of sulfate groups. It had been reported that the degradation of sulfate groups requires relatively low activation energy [71,72]. The thermal stability of TNC was likely to be lower than that of SNC because it exhibited fewer crystalline regions. This result is consistent with those in previous studies [73,74], which have shown that the crystalline regions of nanocellulose generally provide thermal stability. This finding is in agreement with the finding of XRD analyses, which showed that TNC had a high crystallinity index.

4. Conclusions

In this study, nanocellulose was produced using WS as the raw material; novel shapes were obtained by applying different production processes. The chemical composition analysis revealed that the hemicellulose and lignin in WS were effectively removed by alkali treatment and bleaching, increasing the cellulose content to 89%. The TEM images denoted that the produced TNC exhibited an irregular block structure, whereas SNC was rectangular with a length of 55–82 nm and a width of 49–81 nm. The results of the FTIR and XRD analyses of WS, N-WS, and B-WS were consistent with the observed chemical changes and confirmed that the typical cellulose structure remained unchanged through the TEMPO oxidation and sulfuric acid hydrolysis treatments. Our observations demonstrate that WSs, which are an abundant and sustainable agricultural waste, may be repurposed for the production of nanocellulose.

Author Contributions: Conceptualization, D.Z.; Formal analysis, D.Z.; Investigation, D.Z.; Methodology, D.Z.; Project administration, J.Y.; Resources, J.Y.; Supervision, J.Y.; Validation, Y.Z. and Y.G.; Writing—original draft, D.Z.; Writing—review & editing, D.Z. and J.Y.

Funding: This research received no external funding.

Acknowledgments: The authors express their appreciation to Key Laboratory of Biobased Material Science and Technology of Ministry of Education and College of Material Science and Engineering of Northeast Forestry University for full support of this research.

Conflicts of Interest: The authors declare no conflict of interest.

Abbreviations

WS	walnut shell
N-WS	walnut shell treated by NaOH
B-WS	walnut shell after bleaching
TNC	nanocellulose obtained via TEMPO oxidation
SNC	nanocellulose obtained via sulfuric acid hydrolysis

References

1. Harini, K.; Ramya, K.; Sukumar, M. Extraction of nano cellulose fibers from the banana peel and bract for production of acetyl and lauroyl cellulose. *Carbohydr. Polym.* **2018**, *201*, 329–339. [CrossRef] [PubMed]

2. Kontturi, E.; Laaksonen, P.; Linder, M.B.; Groschel, A.H.; Rojas, O.J.; Ikkala, O. Advanced Materials through Assembly of Nanocelluloses. *Adv. Mater.* **2018**, *30*, 39. [CrossRef] [PubMed]
3. Moreau, C.; Villares, A.; Capron, I.; Cathala, B. Tuning supramolecular interactions of cellulose nanocrystals to design innovative functional materials. *Ind. Crop. Prod.* **2016**, *93*, 96–107. [CrossRef]
4. Zhuo, X.; Liu, C.; Pan, R.T.; Dong, X.Y.; Li, Y.F. Nanocellulose Mechanically Isolated from Amorpha fruticosa Linn. *ACS Sustain. Chem. Eng.* **2017**, *5*, 4414–4420. [CrossRef]
5. Munch, A.S.; Wolk, M.; Malanin, M.; Eichhorn, K.J.; Simon, F.; Uhlmann, P. Smart functional polymer coatings for paper with anti-fouling properties. *J. Mat. Chem. B* **2018**, *6*, 830–843. [CrossRef]
6. Klemm, D.; Kramer, F.; Moritz, S.; Lindstrom, T.; Ankerfors, M.; Gray, D.; Dorris, A. Nanocelluloses: A New Family of Nature-Based Materials. *Angew. Chem. Int. Ed.* **2011**, *50*, 5438–5466. [CrossRef] [PubMed]
7. Kargarzadeh, H.; Mariano, M.; Huang, J.; Lin, N.; Ahmad, I.; Dufresne, A.; Thomas, S. Recent developments on nanocellulose reinforced polymer nanocomposites: A review. *Polymer* **2017**, *132*, 368–393. [CrossRef]
8. Fujisawa, S.; Togawa, E.; Kuroda, K. Facile Route to Transparent, Strong, and Thermally Stable Nanocellulose/Polymer Nanocomposites from an Aqueous Pickering Emulsion. *Biomacromolecules* **2017**, *18*, 266–271. [CrossRef] [PubMed]
9. Khan, A.; Huq, T.; Khan, R.A.; Riedl, B.; Lacroix, M. Nanocellulose-Based Composites and Bioactive Agents for Food Packaging. *Crit. Rev. Food Sci. Nutr.* **2014**, *54*, 163–174. [CrossRef] [PubMed]
10. El-Wakil, N.A.; Hassan, E.A.; Abou-Zeid, R.E.; Dufresne, A. Development of wheat gluten/nanocellulose/titanium dioxide nanocomposites for active food packaging. *Carbohydr. Polym.* **2015**, *124*, 337–346. [CrossRef] [PubMed]
11. Wang, Z.; Tammela, P.; Stromme, M.; Nyholm, L. Nanocellulose coupled flexible polypyrrole@graphene oxide composite paper electrodes with high volumetric capacitance. *Nanoscale* **2015**, *7*, 3418–3423. [CrossRef] [PubMed]
12. Chen, W.; Yu, H.; Lee, S.-Y.; Wei, T.; Li, J.; Fan, Z. Nanocellulose: A promising nanomaterial for advanced electrochemical energy storage. *Chem. Soc. Rev.* **2018**, *47*, 2837–2872. [CrossRef] [PubMed]
13. Liu, Y.; Li, Y.; Yang, G.; Zheng, X.T.; Zhou, S.B. Multi-Stimulus-Responsive Shape-Memory Polymer Nanocomposite Network Cross-Linked by Cellulose Nanocrystals. *ACS Appl. Mater. Interfaces* **2015**, *7*, 4118–4126. [CrossRef] [PubMed]
14. Wang, Y.; Heim, L.-O.; Xu, Y.; Buntkowsky, G.; Zhang, K. Transparent, Stimuli-Responsive Films from Cellulose-Based Organogel Nanoparticles. *Adv. Funct. Mater.* **2015**, *25*, 1434–1441. [CrossRef]
15. Rajinipriya, M.; Nagalakshmaiah, M.; Robert, M.; Elkoun, S. Importance of Agricultural and Industrial Waste in the Field of Nanocellulose and Recent Industrial Developments of Wood Based Nanocellulose: A Review. *ACS Sustain. Chem. Eng.* **2018**, *6*, 2807–2828. [CrossRef]
16. Inamochi, T.; Funahashi, R.; Nakamura, Y.; Saito, T.; Isogai, A. Effect of coexisting salt on TEMPO-mediated oxidation of wood cellulose for preparation of nanocellulose. *Cellulose* **2017**, *24*, 4097–4101. [CrossRef]
17. Saraiva Morais, J.P.; Rosa, M.D.F.; Moreira de Souza Filho, M.D.S.; Nascimento, L.D.; do Nascimento, D.M.; Cassales, A.R. Extraction and characterization of nanocellulose structures from raw cotton linter. *Carbohydr. Polym.* **2013**, *91*, 229–235. [CrossRef] [PubMed]
18. Hsieh, Y.-L. Cellulose nanocrystals and self-assembled nanostructures from cotton, rice straw and grape skin: A source perspective. *J. Mater. Sci.* **2013**, *48*, 7837–7846. [CrossRef]
19. Syafri, E.; Kasim, A.; Abral, H.; Sulungbudi, G.T.; Sanjay, M.R.; Sari, N.H. Synthesis and characterization of cellulose nanofibers (CNF) ramie reinforced cassava starch hybrid composites. *Int. J. Biol. Macromol.* **2018**, *120*, 578–586. [CrossRef]
20. Mandal, A.; Chakrabarty, D. Isolation of nanocellulose from waste sugarcane bagasse (SCB) and its characterization. *Carbohydr. Polym.* **2011**, *86*, 1291–1299. [CrossRef]
21. Li, J.; Wei, X.; Wang, Q.; Chen, J.; Chang, G.; Kong, L.; Su, J.; Liu, Y. Homogeneous isolation of nanocellulose from sugarcane bagasse by high pressure homogenization. *Carbohydr. Polym.* **2012**, *90*, 1609–1613. [CrossRef] [PubMed]
22. Brito, B.S.L.; Pereira, F.V.; Putaux, J.-L.; Jean, B. Preparation, morphology and structure of cellulose nanocrystals from bamboo fibers. *Cellulose* **2012**, *19*, 1527–1536. [CrossRef]
23. Moran, J.I.; Alvarez, V.A.; Cyras, V.P.; Vazquez, A. Extraction of cellulose and preparation of nanocellulose from sisal fibers. *Cellulose* **2008**, *15*, 149–159. [CrossRef]

24. Hernandez, C.C.; Ferreira, F.F.; Rosa, D.S. X-ray powder diffraction and other analyses of cellulose nanocrystals obtained from corn straw by chemical treatments. *Carbohydr. Polym.* **2018**, *193*, 39–44. [CrossRef] [PubMed]
25. Wan, C.; Lu, Y.; Jiao, Y.; Jin, C.; Sun, Q.; Li, J. Ultralight and hydrophobic nanofibrillated cellulose aerogels from coconut shell with ultrastrong adsorption properties. *J. Appl. Polym. Sci.* **2015**, *132*. [CrossRef]
26. Sacui, I.A.; Nieuwendaal, R.C.; Burnett, D.J.; Stranick, S.J.; Jorfi, M.; Weder, C.; Foster, E.J.; Olsson, R.T.; Gilman, J.W. Comparison of the Properties of Cellulose Nanocrystals and Cellulose Nanofibrils Isolated from Bacteria, Tunicate, and Wood Processed Using Acid, Enzymatic, Mechanical, and Oxidative Methods. *ACS Appl. Mater. Interfaces* **2014**, *6*, 6127–6138. [CrossRef] [PubMed]
27. Deepa, B.; Abraham, E.; Cordeiro, N.; Mozetic, M.; Mathew, A.P.; Oksman, K.; Faria, M.; Thomas, S.; Pothan, L.A. Utilization of various lignocellulosic biomass for the production of nanocellulose: A comparative study. *Cellulose* **2015**, *22*, 1075–1090. [CrossRef]
28. Garcia, A.; Gandini, A.; Labidi, J.; Belgacem, N.; Bras, J. Industrial and crop wastes: A new source for nanocellulose biorefinery. *Ind. Crop. Prod.* **2016**, *93*, 26–38. [CrossRef]
29. Pirayesh, H.; Khazaeian, A.; Tabarsa, T. The potential for using walnut (*Juglans regia* L.) shell as a raw material for wood-based particleboard manufacturing. *Compos. Part B Eng.* **2012**, *43*, 3276–3280. [CrossRef]
30. Ayrilmis, N.; Kaymakci, A.; Ozdemir, F. Physical, mechanical, and thermal properties of polypropylene composites filled with walnut shell flour. *J. Ind. Eng. Chem.* **2013**, *19*, 908–914. [CrossRef]
31. Yang, J.; Qiu, K. Preparation of activated carbons from walnut shells via vacuum chemical activation and their application for methylene blue removal. *Chem. Eng. J.* **2010**, *165*, 209–217. [CrossRef]
32. Martinez, M.L.; Torres, M.M.; Guzman, C.A.; Maestri, D.M. Preparation and characteristics of activated carbon from olive stones and walnut shells. *Ind. Crop. Prod.* **2006**, *23*, 23–28. [CrossRef]
33. Demirbas, A. Relationships between lignin contents and fixed carbon contents of biomass samples. *Energy Conv. Manag.* **2003**, *44*, 1481–1486. [CrossRef]
34. Kim, J.-H.; Lee, D.; Lee, Y.-H.; Chen, W.; Lee, S.-Y. Nanocellulose for Energy Storage Systems: Beyond the Limits of Synthetic Materials. *Adv. Mater.* **2019**, *31*, e1804826. [CrossRef]
35. Chen, X.-Q.; Deng, X.-Y.; Shen, W.-H.; Jia, M.-Y. Preparation and characterization of the spherical nanosized cellulose by the enzymatic hydrolysis of pulp fibers. *Carbohydr. Polym.* **2018**, *181*, 879–884. [CrossRef]
36. Saito, T.; Kimura, S.; Nishiyama, Y.; Isogai, A. Cellulose nanofibers prepared by TEMPO-mediated oxidation of native cellulose. *Biomacromolecules* **2007**, *8*, 2485–2491. [CrossRef]
37. Okita, Y.; Saito, T.; Isogai, A. Entire Surface Oxidation of Various Cellulose Microfibrils by TEMPO-Mediated Oxidation. *Biomacromolecules* **2010**, *11*, 1696–1700. [CrossRef]
38. Isogai, A.; Saito, T.; Fukuzumi, H. TEMPO-oxidized cellulose nanofibers. *Nanoscale* **2011**, *3*, 71–85. [CrossRef]
39. Capron, I.; Cathala, B. Surfactant-Free High Internal Phase Emulsions Stabilized by Cellulose Nanocrystals. *Biomacromolecules* **2013**, *14*, 291–296. [CrossRef]
40. Bondeson, D.; Mathew, A.; Oksman, K. Optimization of the isolation of nanocrystals from microcrystalline cellulose by acid hydrolysis. *Cellulose* **2006**, *13*, 171–180. [CrossRef]
41. Beck-Candanedo, S.; Roman, M.; Gray, D.G. Effect of reaction conditions on the properties and behavior of wood cellulose nanocrystal suspensions. *Biomacromolecules* **2005**, *6*, 1048–1054. [CrossRef]
42. Habibi, Y.; Lucia, L.A.; Rojas, O.J. Cellulose Nanocrystals: Chemistry, Self-Assembly, and Applications. *Chem. Rev.* **2010**, *110*, 3479–3500. [CrossRef]
43. Li, B.; Xu, W.; Kronlund, D.; Maattanen, A.; Liu, J.; Smatt, J.-H.; Peltonen, J.; Willfor, S.; Mu, X.; Xu, C. Cellulose nanocrystals prepared via formic acid hydrolysis followed by TEMPO-mediated oxidation. *Carbohydr. Polym.* **2015**, *133*, 605–612. [CrossRef]
44. Phanthong, P.; Karnjanakom, S.; Reubroycharoen, P.; Hao, X.; Abudula, A.; Guan, G. A facile one-step way for extraction of nanocellulose with high yield by ball milling with ionic liquid. *Cellulose* **2017**, *24*, 2083–2093. [CrossRef]
45. Sirvio, J.A.; Visanko, M.; Liimatainen, H. Acidic Deep Eutectic Solvents As Hydrolytic Media for Cellulose Nanocrystal Production. *Biomacromolecules* **2016**, *17*, 3025–3032. [CrossRef]
46. Zhang, X.L.; Yang, W.H.; Blasiak, W. Modeling Study of Woody Biomass: Interactions of Cellulose, Hemicellulose, and Lignin. *Energy Fuels* **2011**, *25*, 4786–4795. [CrossRef]
47. De Rodriguez, N.L.G.; Thielemans, W.; Dufresne, A. Sisal cellulose whiskers reinforced polyvinyl acetate nanocomposites. *Cellulose* **2006**, *13*, 261–270. [CrossRef]

48. Kuramae, R.; Saito, T.; Isogai, A. TEMPO-oxidized cellulose nanofibrils prepared from various plant holocelluloses. *React. Funct. Polym.* **2014**, *85*, 126–133. [CrossRef]
49. Saito, T.; Isogai, A. TEMPO-mediated oxidation of native cellulose. The effect of oxidation conditions on chemical and crystal structures of the water-insoluble fractions. *Biomacromolecules* **2004**, *5*, 1983–1989. [CrossRef]
50. Li, W.; Zhao, X.; Huang, Z.; Liu, S. Nanocellulose fibrils isolated from BHKP using ultrasonication and their reinforcing properties in transparent poly (vinyl alcohol) films. *J. Polymer Res.* **2013**, *20*. [CrossRef]
51. Chen, W.S.; Yu, H.P.; Liu, Y.X. Preparation of millimeter-long cellulose I nanofibers with diameters of 30-80 nm from bamboo fibers. *Carbohydr. Polym.* **2011**, *86*, 453–461. [CrossRef]
52. Segal, L.; Creely, J.J.; Martin, A.E.; Conrad, C.M. An Empirical Method for Estimating the Degree of Crystallinity of Native Cellulose Using the X-Ray Diffractometer. *Text. Res. J.* **1959**, *29*, 786–794. [CrossRef]
53. Kasiri, N.; Fathi, M. Production of cellulose nanocrystals from pistachio shells and their application for stabilizing Pickering emulsions. *Int. J. Biol. Macromol.* **2018**, *106*, 1023–1031. [CrossRef]
54. Jiang, F.; Hsieh, Y.-L. Chemically and mechanically isolated nanocellulose and their self-assembled structures. *Carbohydr. Polym.* **2013**, *95*, 32–40. [CrossRef]
55. Salas, C.; Nypelo, T.; Rodriguez-Abreu, C.; Carrillo, C.; Rojas, O.J. Nanocellulose properties and applications in colloids and interfaces. *Curr. Opin. Colloid Interface Sci.* **2014**, *19*, 383–396. [CrossRef]
56. Qing, W.; Wang, Y.; Wang, Y.; Zhao, D.; Liu, X.; Zhu, J. The modified nanocrystalline cellulose for hydrophobic drug delivery. *Appl. Surf. Sci.* **2016**, *366*, 404–409. [CrossRef]
57. Chandra, J.C.S.; George, N.; Narayanankutty, S.K. Isolation and characterization of cellulose nanofibrils from arecanut husk fibre. *Carbohydr. Polym.* **2016**, *142*, 158–166. [CrossRef]
58. Liu, C.; Li, B.; Du, H.; Lv, D.; Zhang, Y.; Yu, G.; Mu, X.; Peng, H. Properties of nanocellulose isolated from corncob residue using sulfuric acid, formic acid, oxidative and mechanical methods. *Carbohydr. Polym.* **2016**, *151*, 716–724. [CrossRef]
59. Sain, M.; Panthapulakkal, S. Bioprocess preparation of wheat straw fibers and their characterization. *Ind. Crop. Prod.* **2006**, *23*, 1–8. [CrossRef]
60. Tripathi, A.; Ferrer, A.; Khan, S.A.; Rojas, O.J. Morphological and Thermochemical Changes upon Autohydrolysis and Microemulsion Treatments of Coir and Empty Fruit Bunch Residual Biomass to Isolate Lignin-Rich Micro- and Nanofibrillar Cellulose. *ACS Sustain. Chem. Eng.* **2017**, *5*, 2483–2492. [CrossRef]
61. Luzi, F.; Puglia, D.; Sarasini, F.; Tirillo, J.; Maffei, G.; Zuorro, A.; Lavecchia, R.; Kenny, J.M.; Torre, L. Valorization and extraction of cellulose nanocrystals from North African grass: Ampelodesmos mauritanicus (Diss). *Carbohydr. Polym.* **2019**, *209*, 328–337. [CrossRef] [PubMed]
62. Camargo, L.A.; Pereira, S.C.; Correa, A.C.; Farinas, C.S.; Marconcini, J.M.; Mattoso, L.H.C. Feasibility of Manufacturing Cellulose Nanocrystals from the Solid Residues of Second-Generation Ethanol Production from Sugarcane Bagasse. *Bioenergy Res.* **2016**, *9*, 894–906. [CrossRef]
63. Saito, T.; Nishiyama, Y.; Putaux, J.-L.; Vignon, M.; Isogai, A. Homogeneous suspensions of individualized microfibrils from TEMPO-catalyzed oxidation of native cellulose. *Biomacromolecules* **2006**, *7*, 1687–1691. [CrossRef] [PubMed]
64. French, A.D.; Cintron, M.S. Cellulose polymorphy, crystallite size, and the Segal Crystallinity Index. *Cellulose* **2013**, *20*, 583–588. [CrossRef]
65. Marett, J.; Aning, A.; Foster, E.J. The isolation of cellulose nanocrystals from pistachio shells via acid hydrolysis. *Ind. Crop. Prod.* **2017**, *109*, 869–874. [CrossRef]
66. Kargarzadeh, H.; Ahmad, I.; Abdullah, I.; Dufresne, A.; Zainudin, S.Y.; Sheltami, R.M. Effects of hydrolysis conditions on the morphology, crystallinity, and thermal stability of cellulose nanocrystals extracted from kenaf bast fibers. *Cellulose* **2012**, *19*, 855–866. [CrossRef]
67. Ilyas, R.A.; Sapuan, S.M.; Ishak, M.R. Isolation and characterization of nanocrystalline cellulose from sugar palm fibres (Arenga Pinnata). *Carbohydr. Polym.* **2018**, *181*, 1038–1051. [CrossRef]
68. Reddy, K.O.; Maheswari, C.U.; Reddy, D.J.P.; Rajulu, A.V. Thermal properties of Napier grass fibers. *Mater. Lett.* **2009**, *63*, 2390–2392. [CrossRef]
69. Yang, H.; Yan, R.; Chen, H.; Lee, D.H.; Zheng, C. Characteristics of hemicellulose, cellulose and lignin pyrolysis. *Fuel* **2007**, *86*, 1781–1788. [CrossRef]

70. Fukuzumi, H.; Saito, T.; Wata, T.; Kumamoto, Y.; Isogai, A. Transparent and High Gas Barrier Films of Cellulose Nanofibers Prepared by TEMPO-Mediated Oxidation. *Biomacromolecules* **2009**, *10*, 162–165. [CrossRef]
71. Wang, N.; Ding, E.; Cheng, R. Thermal degradation behaviors of spherical cellulose nanocrystals with sulfate groups. *Polymer* **2007**, *48*, 3486–3493. [CrossRef]
72. Wang, H.; Li, D.; Yano, H.; Abe, K. Preparation of tough cellulose II nanofibers with high thermal stability from wood. *Cellulose* **2014**, *21*, 1505–1515. [CrossRef]
73. Yu, H.; Qin, Z.; Liang, B.; Liu, N.; Zhou, Z.; Chen, L. Facile extraction of thermally stable cellulose nanocrystals with a high yield of 93% through hydrochloric acid hydrolysis under hydrothermal conditions. *J. Mater. Chem. A* **2013**, *1*, 3938–3944. [CrossRef]
74. Cherian, B.M.; Pothan, L.A.; Nguyen-Chung, T.; Mennig, G.; Kottaisamy, M.; Thomas, S. A novel method for the synthesis of cellulose nanofibril whiskers from banana fibers and characterization. *J. Agric. Food Chem.* **2008**, *56*, 5617–5627. [CrossRef] [PubMed]

© 2019 by the authors. Licensee MDPI, Basel, Switzerland. This article is an open access article distributed under the terms and conditions of the Creative Commons Attribution (CC BY) license (http://creativecommons.org/licenses/by/4.0/).

Article

Effect of Cellulose Nanocrystals from Different Lignocellulosic Residues to Chitosan/Glycerol Films

Marina Reis de Andrade [1], Tatiana Barreto Rocha Nery [2], Taynã Isis de Santana e Santana [1], Ingrid Lessa Leal [2], Letícia Alencar Pereira Rodrigues [2], João Henrique de Oliveira Reis [3], Janice Izabel Druzian [3] and Bruna Aparecida Souza Machado [2,4,*]

1. Department of Materials, University Center SENAI CIMATEC, National Service of Industrial Learning, Salvador 41650-010, Brazil; marina.andrade@fbter.org.br (M.R.d.A.); tayisis.santana@gmail.com (T.I.d.S.e.S.)
2. Department of Food and Beverages, National Service of Industrial Learning, Applied Research Laboratory of Biotechnology and Food, University Center SENAI CIMATEC; Salvador 41650-010, Brazil; tatianabr@fieb.org.br (T.B.R.N.); ingrid.leal@fieb.org.br (I.L.L.); leticiap@fieb.org.br (L.A.P.R.)
3. Biotechnology, Pharmacy Faculty, Federal University of Bahia; Salvador 40170-115, Brazil; jhonhyba47@hotmail.com (J.H.d.O.R.); druzian@ufba.br (J.I.D.)
4. University Center SENAI CIMATEC, National Service of Industrial Learning, Laboratory of Pharmaceutical's Formulations, Health Institute of Technologies (ITS CIMATEC), Salvador 41650-010, Brazil
* Correspondence: brunamachado17@hotmail.com; Tel.: +55-(71)-3879-5624

Received: 11 March 2019; Accepted: 8 April 2019; Published: 10 April 2019

Abstract: Interest in nanocellulose obtained from natural resources has grown, mainly due to the characteristics that these materials provide when incorporated in biodegradable films as an alternative for the improvement of the properties of nanocomposites. The main purpose of this work was to investigate the effect of the incorporation of nanocellulose obtained from different fibers (corncob, corn husk, coconut shell, and wheat bran) into the chitosan/glycerol films. The nanocellulose were obtained through acid hydrolysis. The properties of the different nanobiocomposites were comparatively evaluated, including their barrier and mechanical properties. The nanocrystals obtained for coconut shell (CS), corn husk (CH), and corncob (CC) presented a length/diameter ratio of 40.18, 40.86, and 32.19, respectively. Wheat bran (WB) was not considered an interesting source of nanocrystals, which may be justified due to the low percentage of cellulose. Significant differences were observed in the properties of the films studied. The water activity varied from 0.601 (WB Film) to 0.658 (CH Film) and the moisture content from 15.13 (CS Film) to 20.86 (WB Film). The highest values for tensile strength were presented for CC (11.43 MPa) and CS (11.38 MPa) films, and this propriety was significantly increased by nanocellulose addition. The results showed that the source of the nanocrystal determined the properties of the chitosan/glycerol films.

Keywords: films; nanocellulose; nanocrystals; biodegradable packaging

1. Introduction

With the advancement of nanotechnology and nanoscience, materials are modulated in their technologies, generating new technologies to incorporate to the needs of the current society [1–5]. In this context, cellulose nanocrystals are being used to improve the mechanical and barrier properties of chitosan films and make their commercialization viable [6–11]. Nanocellulose or cellulose nanocrystals are the crystalline domains of cellulosic sources, obtained through acid hydrolysis, having characteristics of high rigidity, high crystallinity, and nanometric size [6–10]. The cellulose nanocrystals have been the object of several studies, since they present great potential of application as reinforcement in polymer matrices [12,13]. Cellulose nanocrystal is known as the most appropriate and efficient reinforcement additive due to its renewability, excellent mechanical properties, and economic cost [14], and can be obtained from waste, further improving the cost benefit.

Among the nanoreinforced materials, the application of the vegetal fibers stands out, since these have ample availability in almost all the countries, being usually designated as lignocellulosic materials [15–17]. Some fibers are found spontaneously in nature, while others are derived from agricultural activity and from waste generated mainly by agroindustry. Studies evaluate the application of nanocrystals obtained by several fibers in polymeric systems, for example, barley straw and husk in poly(vinyl alcohol) (PVA) blended with natural chitosan (CH) nanocomposites [18], pine cones in a biodegradable poly(3-hydroxybutyrate)/poly(ε-caprolactone) (PHB/PCL) [19], and sunflower stalks on wheat gluten bionanocomposites [20]. The high availability of lignocellulosic fibers, coupled with the need for a renewable source for the production of polymers, represents a great opportunity for technological advances that add value to the products or residues of the agroindustry and, at the same time, act in the fixation of carbon in nature [21–24]. This implies helping to reduce the emission of CO_2 into the atmosphere during the production cycle, increasing the economic potential of agribusiness due to the possibility of trading carbon credits in the production chain [25–29].

In relation to the development of new materials, there is also the growth of technologies using polymers from renewable sources for diverse applications. These materials have been important for the advancement of the sciences, and have several advantages such as being easily obtainable, biocompatible, and biodegradable [30,31]. The choice of material to be used in the formulation of the films is very important, as they will depend on the interactions between the components of the material, which may interfere with the barrier properties, mechanical properties, and the physical aspects of the films [32].

Several biopolymers such as polysaccharides, proteins, and lipids have been used as polymer matrices for the development of biodegradable packaging due to their availability, renewability, low cost, respect for the environment, and biodegradability. Among these, chitosan is considered favorable for the development of biocomposites [33,34], and based on production volumes, is the second most abundant polymer after cellulose [35,36]. Chitosan is a natural polysaccharide derived from chitin, and although the most important sources are commercial shellfish carapaces, studies indicate that this element can be found in insects, mollusks, and fungal cell walls [37]. The bioactivity of this material has aroused interest in the application as a packaging film due to its ability to form flexible and resistant films with efficient oxygen barrier and antimicrobial activity [6,7,9,38,39]. In addition, the use of chitosan in the field of biomedicine has been reported for its various important pharmacological properties and its role in tissue engineering, regenerative medicine, scaffold, and drug delivery systems is also well documented [40].

Chitosan based films are biocompatible and biodegradable, with excellent mechanical strength and cost effectiveness [41]. Previous studies have shown the efficiency of the incorporation of nanocrystals obtained from different sources into chitosan films, contributing to increase the mechanical and barrier properties of these materials [42–44]. Azeredo et al. [45] evaluated the effect of different concentrations of nanocellulose on tensile properties and water vapor permeability of chitosan films. Pereda et al. [46] demonstrated that the combined use of cellulose nanoparticles and olive oil proved to be an efficient method to reduce the inherently high water vapor permeability of plasticized chitosan films, improving their tensile behavior at the same time.

The films obtained by natural polymers are poorly flexible and brittle, thus, it becomes necessary to add plasticizer to the polymer matrix to improve its flexibility characteristics. Plasticizers reduce the interactions between adjacent molecules, increasing film flexibility [47]. For application of a plasticizer, it is extremely important that it is suitably compatible with the polymer used and the definition of proportionality between the components in order to tailor the final composition to a given application [48]. Several plasticizers are used in the preparation of biodegradable films and coatings, including mono-, di- and oligosaccharides (glucose, sucrose); polyols (glycerol, sorbitol, derivatives of glycerol); and lipids (saturated fatty acids, monoglycerides and ester derivatives, phospholipids and surfactants) [49]. The use of plasticizers in these films allows a greater percentage of elongation and adaptation of the matrix in the structure [50]. Glycerol is currently one of the plasticizers most used in the development of biodegradable films, which has caused scientific impact since this polyalcohol is a byproduct generated from the biodiesel chain, which is expanding worldwide [51], and is thus a low-cost and high-availability material.

With this background, the objective of this work was the extraction of cellulose nanocrystals from four different lignocellulosic fibers, considered as byproducts of the agro-food industries (Corncob (CC), corn husk (CH), wheat bran (WB) and coconut shell (CS)) and the investigation of the influence of the incorporation of these nanoparticles on the physical, barrier, and mechanical properties in chitosan biofilms. The main focus of the study is to identify sources of residues for the production of cellulose nanocrystals and, consequently, to evaluate the behavior of the inclusion of these nanoparticles in films using chitosan as a polymer matrix. Chitosan is identified as a highly attractive biomaterial for film owing to its properties seen in previous reports. Chitosan can be easily incorporated into gels, membranes, beads, and scaffolds, and these forms provide a wide variety of biomedical applications and food packaging.

2. Materials and Methods

For extraction of the cellulose nanocrystals were used corncob (CC) and corn husk (CH) bought in local commerce in Salvador, Bahia, Brazil, wheat bran (WB) provided by a local wheat mill, and coconut shell (CS) donated by Frisbraztech (Conde, Bahia, Brazil). The films were produced with chitosan (Sigma-Aldrich, Saint Louis, MO, USA, Cas Number: 9012-76-4, with a degree of deacetylation ≥ 75%), and the glycerol purchased from Synth (São Paulo, Brazil).

2.1. Characterization of the Fibers

The natural fibers were characterized as moisture, water activity, ash content, and crude fiber content.

The moisture content was determined using an infrared-heated scale (Shimadzu, MOC-120H, Kyoto, Japan) with the intensity of the emitted radiation set so that the sample would reach 105 °C. Measurements of water activity were performed using a Decagon (Novasina®, Lab Master aw, Neuheimstrasse, Switzerland) at temperature of 25 °C. The ash content was determined using an muffle Fornitec, and crude fiber content (lignin, hemicellulose and cellulose) in Ankom A200 Fiber Analyzer (New York, NY, USA), by the FDA (Acid Detergent Fiber) and NDF (Neutral Detergent Fiber), according to the methodology proposed by Van-Soest, Robertson and Lewis [52].

2.2. Extraction of Cellulose from Fibers

The extraction of the cellulose pulp was performed based on the works of Samir et al. [53] and Machado et al. [12]. The selected materials (CC, CH, WB and CS) were previously dried at 60 °C for 3 h to remove excess moisture and ground in a blender to obtain a fine powder. The crushed fibers (30 g) were washed in 1200 mL of 2% NaOH solution and at 80 °C for a period of 4 h under constant stirring. The resulting solutions were filtered and washed with water to obtain a pulp. The washing process was repeated three times for complete removal of the water soluble agents and obtaining the cellulose pulp. After washing, the process of delignification and bleaching of pulps was carried out, using 300 mL of sodium hypochlorite (1.7%) and 300 mL of a buffer solution. The resulting solution was placed under constant stirring at a temperature of 80 °C for 6 h (TE-394/2, Tecnal, São Paulo, Brazil), filtered and oven dried (40 °C) to obtain the cellulose pulp of each fiber.

2.3. Preparation of Cellulose Nanocrystals

The cellulose nanocrystals were prepared by acid hydrolysis using 45% H_2SO_4 [12,38]. Briefly, 12 mL/g of cellulose was subjected to constant stirring for period of 1 h, and temperature between 50 and 55 °C. After acid hydrolysis, the dispersions were cooled to 30 °C and the volume was completed to 40 mL in Falcon tubes. The tubes were centrifuged for 10 minutes at 4400 rpm and a temperature of 10 °C (Sigma 2-16KHL, Osterode am Harz, Germany) in order to separate the crystals in suspension. Centrifugation was repeated 4 to 6 times for better separation results.

Then the suspensions were subjected to dialysis using cellulose membranes (D9777-100 FTO, 12.000 Da cut off, Sigma-Aldrich, Saint Louis, MO, USA), and after reaching the pH between 6 to 7, the samples were placed in an ultrasonic bath (RMS, Quimis, São Paulo, Brazil) with a power of 200 W, frequency of 60 kHz, and a temperature of 25°C for 5 min for the dispersion of the nanocrystals.

2.4. Production of Films (Nanobiocomposites)

The films with nanoparticles extracted from different fibers were produced by the casting method, according to the methodology proposed by Yassue–Cordeiro et al. [31] and de Souza [54]. Chitosan (1.5%, g/100g) dissolved in glacial acetic acid (1.0%, g/100g) and glycerol (0.15%, g/100g) was used as the plasticizing agent. To this mixture was added 5% (w/v) cellulose nanocrystals (a filmogenic solution for each fiber) under constant shaking in Shaker Incubator (MA420, Marconi, São Paulo, Brazil) at 40 °C for 24 h (Table 1). After homogenization, 40 g of the solution was transferred to Petri dishes and subjected to dehydration in an air circulation oven (35 ± 2 °C) (Q314M, Quimis, São Paulo, Brazil) for 20 h. Other film was produced without the addition of cellulose nanocrystals for use as control. The films obtained were packed in a vacuum desiccator containing saturated sodium chloride solution (TE-3950, Tecnal, São Paulo, Brazil).

Table 1. Formulation of films containing nanocrystals of different lignocellulosic sources and control.

Formulations	Chitosan (%, g/100 g)	Acetic Acid (%, g/100 g)	Glycerol (%, g/100 g)	Cellulose Nanocrystals (%, g/100 g)
Control	1.50	1.00	0.15	0.00
CS	1.50	1.00	0.15	5.00
CH	1.50	1.00	0.15	5.00
CC	1.50	1.00	0.15	5.00
WB	1.50	1.00	0.15	5.00

Coconut shell (CS), corn husk (CH), corncob (CC) and wheat bran (WB).

2.5. Characterization of Cellulose Nanocrystals

The concentration of nanocrystalline cellulose in the suspensions was determined by gravimetric analysis. An aliquot with a known volume was dried at 40 °C for 24 h in an air circulation oven (TE-394/2, Tecnal, São Paulo, Brazil). The birefringence of the nanocellulose suspensions was determined by the methodology proposed by Flauzino et al. [55], an aliquot of aqueous suspension of the nanocrystals (5×10^{-3} g.mL^{-1}) was placed in a glass test tube, this tube was placed in front of a polarized light source, and it was then photographed with a camera equipped with a filter of polarized light. Cellulose nanocrystalline dispersions for all fibers were analyzed by transmission electron microscopy (TEM), in order to determine the length of the fibers (L), width (D), and aspect ratio (L/D) and indicate the state of aggregation of the crystals. Measurements were made directly from the micrographs using Image Tool 6.3 (Media Cybernetics, Rockville, USA) with 30 measurements to determine mean and standard deviation values according to the methodology proposed by Machado et al. [12].

2.6. Characterization of Films—Determination of Thickness and Mechanical Properties

The thickness of the films and control were evaluated in random positions using digital micrometer (Digimess, São Paulo, Brazil) with resolution of 0.001 mm. The tensile tests were performed using a Brookfield (Braseq CT310K, Middleboro, MA, USA), with a maximum load of 10 KN, with a speed of 0.5 mm s^{-1}, a temperature of 25 °C, trigger load of 7 g, test probe tip of TA3/100 and TA / TPB device [13,56]. Tensile tests were performed on six specimens with dimensions of 80 mm in length and 25 mm wide for each sample.

2.7. Statistical Analysis

The results of this study were expressed as the mean ± standard deviation (sd) (n = 3). The statistical analysis of the results was performed using the Statistica® 6.0 software from StatSoft (Tulsa, Hamburg, Germany). The results were treated by the Tukey test to identify if the changes in the parameters evaluated were significant at the 95% level of significance.

3. Results

3.1. Characterization of Fibers

The composition and structure of lignocellulosic biomass have great influence on the nature and yields of the hydrolysis processes. The moisture content (%), the water activity (a_w), and the ash content (%) of the fibers obtained from corn cob (CC), corn husk (CH), wheat bran (WB), and coconut shell (CS) are shown in Table 2.

Table 2. Mean values of moisture content, water activity, and ash of lignocellulosic fibers (mean ± standard deviation).

Fibers	Moisture (%)	Activity Water	Ash Content (%)
CS	88.7 ± 0.07 [a]	0.970 ± 0.05 [a]	5.37 ± 0.10 [a]
CH	67.1 ± 5.40 [b]	0.940 ± 0.06 [b]	0.92 ± 0.12 [b]
CC	72.7 ± 5.80 [b]	0.770 ± 0.05 [c]	3.64 ± 0.26 [c]
WB	12.4 ± 0.60 [c]	0.640 ± 0.03 [d]	4.75 ± 0.27 [d]

Coconut shell (CS), corn husk (CH), corncob (CC), and wheat bran (WB). Mean ± standard deviation of samples. Values with the same letter in the same column did not present significant differences ($p < 0.05$) by Tukey's test at 95% confidence (a–d).

Table 3 shows the cellulose, hemicellulose and lignin contents of the fibers studied. Each component of the lignocellulosic fibers is responsible for different functions, so different levels of these components influence the properties of the nanocellulose fibers obtained [38,57,58].

Table 3. Cellulose, hemicellulose, and lignin contents in natural lignocellulosic fibers (mean ± standard deviation).

Fibers	Cellulose (%)	Hemicellulose (%)	Lignin (%)
CS	47.16 ± 1.24 [b]	20.71 ± 0.66 [b]	30.71 ± 0.21 [a]
CH	24.09 ± 1.13 [c]	12.99 ± 0.58 [c]	0.50 ± 0.13 [c]
CC	52.99 ± 1.79 [a]	29.72 ± 0.69 [a]	4.56 ± 1.84 [b]
WB	10.86 ± 1.25 [d]	28.88 ± 0.32 [a]	4.89 ± 0.84 [b]

Coconut shell (CS), corn husk (CH), Corncob (CC), and wheat bran (WB). Mean ± standard deviation of samples. Values with the same letter in the same column did not present significant differences ($p < 0.05$) by Tukey's test at 95% confidence (a-d).

3.2. Characterization of Cellulose Nanocrystals

Figure 1 shows the four steps of washing the fiber with NaOH for the process of extracting cellulose from the corncob, as an example, until the bleaching stage.

For each 30 g of fiber submitted to the washing and bleaching process (Figure 1), different yields of cellulose pulp were obtained, and these results are presented in Table 4. The concentration of nanocrystals (g.10mL^{-1}) is a parameter for the evaluation of the nanocellulose dispersion, indicating if it needs a higher concentration for use in the production of the films.

The existence of nanocrystals can be proven by birefringence analysis and microscopy. According to Pereira et al. [59], flow birefringence results from the alignment of nanoparticles and indicates the existence of isolated nanocrystals in the dispersion. As the cellulose nanocrystals are rigid rod-like particles, they have a strong tendency to align a vector director and increase in relation to particle size/diameter. As a result of the strong birefringence of the native cellulose, this rod alignment creates a macroscopic birefringence that can be directly observed though cross polarizers [60]. The suspensions (CH, CC, CS and WB) were analyzed from a polarizing lens. In the birefringence analysis, the CC nanocrystals dispersion was the one that presented the nematic phase liquid crystals (N) clearly to prove the existence of the crystals (Figure 2), CH and CS presented nematic phase with less clarity, and the WB did not present nematic phase. This information is in agreement with the results presented in Table 5.

Figure 1. Cellulose pulp obtained by corncob (CC): (**a**) first wash with NaOH; (**b**) second washing; (**c**) third washing; (**d**) fourth washing; (**e**) bleaching step.

Table 4. Yield and concentration of cellulose pulp and nanocrystals in the different lignocellulosic sources.

Lignocellulosic Source	CS	CH	CC	WB
Pulp Cellulose (%)	12.50	25.40	38.70	28.00
Nanocellulose (g.10mL^{-1})	0.660	0.050	0.072	NA

Coconut shell (CS), corn husk (CH), corncob (CC), and wheat bran (WB).

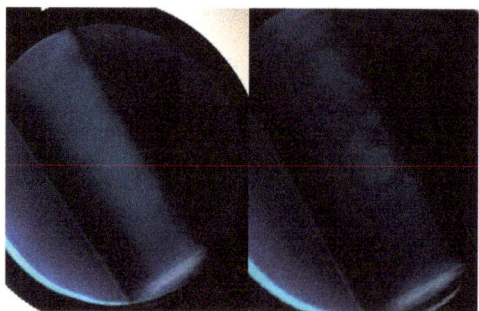

Figure 2. Phenomenon of birefringence observed through a polarized lens after dispersion of cellulose nanocrystals extracted from corncob (CC).

Table 5. Size of the nanocrystals of different lignocellulosic sources (mean ± standard deviation).

Nanocrystals	L ± sd (nm)	D ± sd (nm)	L/D
CS	254.0 ± 98	6.32 ± 1.02	40.18
CH	298.3 ± 97	7.30 ± 1.20	40.86
CC	302.0 ± 86	8.12 ± 0.96	32.19
WB	-	-	-

Coconut shell (CS), corn husk (CH), corncob (CC), and wheat bran (WB). L = length; D = width and L/D ratio.

Table 5 presents the mean values of the width (D), length (L), and the L/D ratio of the crystals.

CC, CH, and CS presented positive results for TEM analysis, where it was possible to visualize the aggregate crystals in a needle format (Figure 3). It was not possible to visualize crystal formation for wheat fiber, indicating that they are not present in a considerable amount.

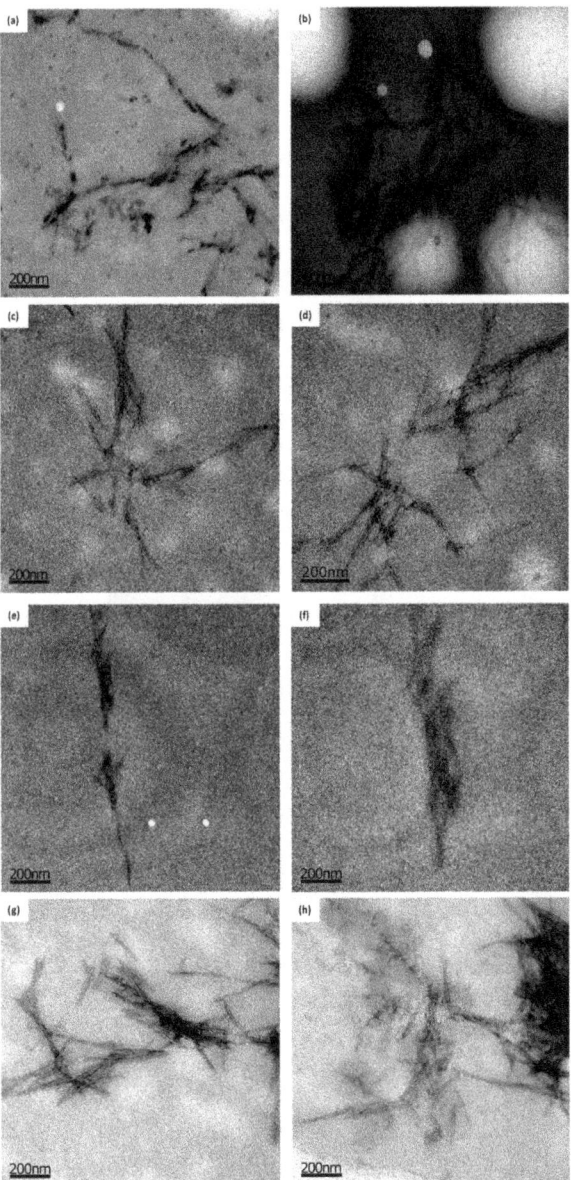

Figure 3. Cellulose nanocrystals obtained by Transmission Electron Microscopy (TEM) (PTA contrast and uranyl). (**a**) and (**b**) Corn husks; (**c**) and (**d**) Corncob; (**e**) and (**f**) Wheat bran; (**g**) and (**h**) Coconut shell (Scale: 200 nm).

The process conditions, whether concerning fiber preparation or hydrolysis for whisker isolation, affect the morphological characteristics of these nanomaterials. The acid used for hydrolysis may affect the characteristics of whisker dispersion in an aqueous system. The effect of reaction time and acid-wood pulp ratio on the properties and behavior of the whisker suspension, obtained by sulfuric acid hydrolysis, was observed that shorter whiskers, less variable in length, were obtained in longer reaction times [61]. Beck-Candanedo et al. [61] also found out that an increase in the acid-pulp ratio also leads to whiskers with reduced dimensions.

3.3. Production and Characterization of Films

The films produced from the formulations described in Table 1, using different types of nanocrystals, as well as the control film, were analyzed to determine their physical and barrier properties (water activity (a_w), moisture content (M), total solids (TS), and thickness (t)) and their mechanical properties (tensile strength (σ) and strain (ε)), described in Table 6. Figure 4 shows the physical appearance of CS film.

Table 6. Characterization of nanobiocomposites (mean ± standard deviation).

Film	a_w ± sd	M ± sd (%)	TS ± sd (%)	t ± sd (mm)	σ ± sd (MPa)	ε ± sd (%)
Control	0.610±0.01 [b]	20.75±0.78 [a]	78.92±0.78 [d]	0.049±0.02 [a]	4.08±1.87 [d]	115.9±4.36 [e]
CS Film	0.600±0.01 [b]	15.13±0.01 [c]	84.87±0.01 [a]	0.040±0.04 [ab]	11.38±3.53 [a]	274.2±1.35 [a]
CH Film	0.658±0.02 [a]	20.24±0.62 [a]	79.76±0.62 [c]	0.027±0.01 [bc]	6.99±4.56 [c]	155.2±5.13 [c]
CC Film	0.611±0.05 [b]	18.32±0.90 [b]	81.68±0.90 [b]	0.019±0.01 [c]	11.43±3.58 [a]	195.2±8.76 [b]
WB Film	0.601±0.01 [b]	20.86±0.06 [a]	79.14±0.06 [cd]	0.027±0.01 [bc]	4.03 ±1.67 [b]	141.0±9.09 [d]

Coconut shell (CS), corn husk (CH), corncob (CC), and wheat bran (WB). Values followed by same letter, in the same column, did not present significant differences ($p > 0.05$) by the Tukey test at 95% confidence (a-d).

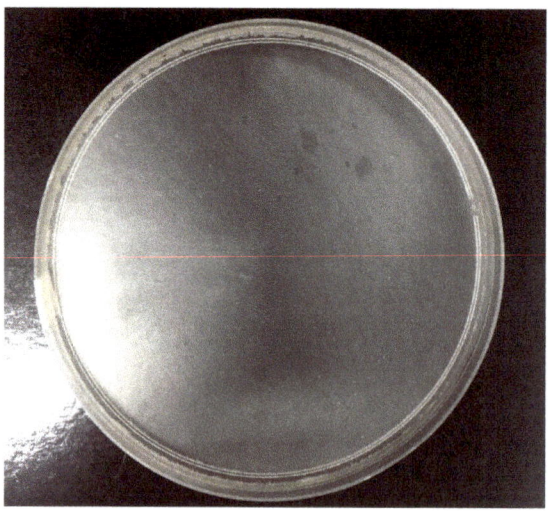

Figure 4. Physical appearance of CS film.

4. Discussion

Among the fibers, CS had the highest results to moisture (88.7% ± 0.07), water activity (0.97 ± 0.05), and ash (5.37% ± 0.10), the WB presented the smaller moisture content (12.4% ± 0.60), and water activity (0.640 ± 0.03) (Table 2). The moisture content of the fibers, as well as the storage conditions and the time, can interfere in the degree of crystallinity of the cellulose, and consequently, in obtaining the nanocrystals [15,29,54]. The CH and CC fibers analyzed in this study were collected from ears of green and not dry corn. For this reason, the values found for moisture and water activity were higher than those found in the literature. Corn husk presented higher moisture than that presented by Reference [62] (12.96%) and near ash content (1.52%), which can be explained mainly by the origin of the fiber and its level of maturation. Ziglio et al. [63] found moisture values of 8.9% for cob corn [63]. The results of moisture found for wheat bran are similar to those presented by De Lima Dantas et al. [64], 11.59%, however, the ash value was lower (0.62%). Differences can be justified by the wheat variety, the maturation stage of the samples, and the producer region.

The results in Table 3 show the differences in fiber composition of the different residues studied. The CC fiber showed the highest values of cellulose (52.99% ± 1.79) followed by the CS fiber (47.16% ± 1.24). The CC fiber presented the highest hemicellulose content while the CS fiber had the highest lignin content (30.71 ± 0.21). Souza et al. [65] found values of 37.6% cellulose, 34.5% hemicellulose, and 12.6% lignin for corn husk, and 31.7% cellulose, 34.7% hemicellulose, and 20.3% lignin for corn cob. In the literature, different values of cellulose (56.8%) and lignin (29.8%) were presented for coconut shell [66].

In their study, Merali et al. [67] found values of 18.5% cellulose, 54.8% hemicellulose, and 10.8 lignin in pretreated WB hydrothermally. These values may be associated to the exchange rate in the fiber, in the region where it was extracted and its botanical varieties. Mendes et al. [68] found values of 33–40% cellulose, 33–40% hemicellulose, and 2–16% lignin in CH residue samples. The values found by the authors show the large range of these constituents in the samples, which can be attributed to maize variety and harvest period.

The importance of determining cellulose at the fibers is that being a polysaccharide made of repeating beta 1,4-glycosidic bonds, it is characterized by having intercalated arrangements of highly ordered (crystalline) and amorphous (disordered) [69]. Thus, the isolation and obtaining of nanocrystals from lignocellulosic fibers depend directly on the proportion of these crystalline regions, as well as the lignin and hemicellulose content, since they interfere in the extraction process, the realization of pretreatment for removal of these components being necessary in many cases [28,31].

The visualization of the suspension of the cellulose nanocrystals obtained from the fibers studied using polarizers revealed a nematic phase, which was directly produced by light birefringence. This result was also important to confirm the presence of nanocrystals (except for WB) and is considered an important analysis to evaluate nanocrystals dispersion (Figure 2). Cerqueira et al. [13] and Alves et al. [17] similarly used crossed polarizers to visualize the birefringence phenomenon in suspension of cellulose nanocrystals obtained from coconut and eucalyptus, respectively. The nematic liquid crystal phase combines long-range orientational order with regular liquid-like short-range positional order. In a nematic CNC suspension, the nanocrystals align preferentially with their long axes along a common direction [70]. Perhaps there is a correlation among nematic phase (preferential orientation of the nanocrystals), refractive effects, and size of the CNC. In this study, a positive correlation was observed regarding the size of the nanocrystals obtained and the presence of birefringence. For example, the best nematic phase was presented by CC, which also has the highest L and D, and consequently, lower L/D ratio.

The morphological analysis is of great importance in determining the size and the state of agglomeration, considering that the source of cellulose as well as the technique used for the hydrolysis of the amorphous structure influence the size and the final properties of the nanoparticles [71,72]. Table 5 presents the mean values of the width (D), length (L), and the L/D ratio of the crystals. It is observed that the CC presented higher values of L and D and a lower L/D ratio. The values of (L) are in agreement with the literature [57,73] indicating a great potential of use of this fiber as reinforcement for bionanocomposites, as demonstrated in other studies. Similar results were reported by Machado et al. [74] (L = 98–430 nm, D = 6 nm e L/D = 38.9 ± 4.7), Rosa et al. [75] (L = 197 nm, D = 5.8 nm e L/D = 39), and Sarwar et al. [76] (240–280 nm).

Oliveira [77] notes that the resulting nanoparticle dimensions depend on the cellulose source and the hydrolysis process. Smaller nanoparticle diameters may be associated with higher amounts of hemicellulose present in the fiber structure, which would limit the organization of cellulose chains [78,79]. In this study, the CC fiber presented the highest concentration of cellulose and, consequently, presented the lowest L/D ratio for the nanocrystals obtained. The L/D ratio ranged from 32.19 to 40.86, being in agreement with other studies that presented values between 18.2 and 75.4 nm [13,79]. Some authors [11,80,81] found similar dimensions for the coconut shell. In this range, the crystals have great potential to be used as reinforcement in biodegradable films.

The film made with CH cellulose nanocrystals had a higher value of water activity (0.658), while the others presented approximate values of 0.600. The reduction of free water in packages for food products has the consequence of reducing the growth of microorganisms, avoiding undesirable chemical changes in the storage of the products [82–85]. Products that have water activity values lower than 0.600 are relatively protected against microbial contamination, whereas the proliferation of specific microorganisms can occur with water activity values above 0.600 [86–88]. Associated with the amount of free water of the product, the moisture of the film, which favors or inhibits proportionally the proliferation of microorganisms, is also evaluated [7,89].

The films produced with cellulose nanocrystals of CC and CS presented lower moisture content (18.32 ± 0.90 and 15.13 ± 0.01) in relation to films produced with WB and CH nanocrystals (20.86 and 20.24). In relation to the total solids, an increase is observed for the CC and CS, indicating that the amount of nonvolatile or water insoluble particles is slightly higher.

The thickness of the films (Table 6) was lower than the control film, maintaining unchanged barrier properties. Because it is a manual process, the standardization of the fluid distribution and the drying process are difficult. In a study, Hänninen et al. [90], observed that by adding a birch cellulose nanofiber, the thickness doubled compared to the control film containing only chitosan.

For this work, it was also observed that the incorporation of the nanocellulose dispersion, from the various fibers, to the film plasticized with glycerol, resulted in the improvement of the mechanical properties of the formulations studied. The tensile strength values (σ) varied between 4.08 MPa (control) to 11.38 MPa (CS Film) and 11.43 MPa (CC Film), presenting significant differences among the samples evaluated (except to CS and CC). The biofilms with the CS and CC nanocrystals dispersion presented a higher tensile strength at the rupture, indicating that a greater force was required for the rupture of the films, suggesting that there was an increase of the resistance. Regarding the deformation, the values found ranged from 115.9% (control) to 274.2% (CS film), and the WB film (110.0% ± 90.9) had a lower percentage of the evaluated fibers, with values close to the control. The lowest results in relation to the mechanical properties for the WB film may be justified due to the lowest concentration of cellulose in the fiber, and consequently the lowest efficiency to obtain the nanocrystals. This was also confirmed because of the impossibility of determining the size of the nanocrystals obtained by TEM (due to low concentration and formation of few clusters) as well as, no nematic phase formation. Benini (2011) [91] incorporated high impact polystyrene (HIPS) coconut fibers as a thermoplastic matrix, and the maximum stress and Young modulus were 23.7 MPa and 3.0 MPa for composites containing 30% of fibers.

The effect of surface and dispersion characteristics of the whiskers used as reinforcement material in a matrix with polypropylene was investigated by Ljungberg et al. [92]. These authors observed that the quality of nanocrystal dispersion is an important aspect, which affects the quality of the film, making the films more opaque and influencing film strength.

It is possible to visualize an increase in the mechanical resistance of the films with the incorporation of the nanocrystals dispersion, making them more rigid, but there was an increase in the percentage of deformation for all the films (Table 6). The stiffness in some materials can negatively interfere in the percentage of deformation of the same, the more rigid the material, the more easily it will break. The structure of the material and its composition interfere directly in this parameter. A study [93] showed that chitosan nanocomposites reinforced with chitin whiskers increases the tensile strength and significantly reduces the elongation at break. The work showed tensile strength of 52.23 MPa, and elongation at break of 21.32%.

Although microscopies of the films were not performed in this study, the incorporation of nanocellulose from CS, CH, and CC fibers into the films resulted in strong interactions between the plasticizer and the matrix modifying the mechanical profiles, which can demonstrate the compatibility between the phases. Similar results were identified by Marín–Silva et al. [42] when investigated chitosan nanocomposites with microcrystalline cellulose.

5. Conclusions

The results found in the study confirm that the nanocellulose crystals of coconut shell (CS), corn husk (CH), and corncob (CC) fibers incorporated into the chitosan/glycerol films are presented as promising materials for the development of biodegradable composites. The crystallization by acid hydrolysis was favorable, resulting in nanocrystals with great potential to be used as reinforcement due to their size. Although the CS did not present the highest amount of extractable cellulose (12.50%), the highest concentration of nanocellulose in the dispersion after the acid hydrolysis was obtained from this fiber. In this study, wheat bran (WB) was not considered an interesting source of nanocrystals, which may be justified due to the low percentage of cellulose present in this residue.

Significant differences were observed in the properties of the film formulations studied, showing that the source of the nanocrystals is an important parameter to be investigated. The water activity varied from 0.601 (WB Film) to 0.658 (CH Film) and the moisture content from 15.13 (CS Film) to 20.86 (WB Film). The highest values for tensile strength were presented for CC (11.43 MPa) and CS (11.38 MPa) films. Therefore, the films prepared with the nanocrystals dispersion from the CS and CC presented an improvement of the mechanical properties, and the analysis of TEM indicated that both presented good characteristics to be applied in the development of nanobiocomposites.

Author Contributions: Conceptualization, J.I.D. and B.A.S.M.; Data curation, M.R.d.A., T.B.R.N., I.L.L., L.A.P.R., J.H.d.O.R. and B.A.S.M.; Formal analysis, M.R.d.A., T.I.d.S.e.S., I.L.L., L.A.P.R. and J.H.d.O.R.; Investigation, M.R.d.A., T.I.d.S.e.S., J.I.D. and B.A.S.M.; Methodology, M.R.d.A., T.B.R.N., T.I.d.S.e.S., I.L.L., J.H.d.O.R. and B.A.S.M.; Project administration, J.I.D. and B.A.S.M.; Resources, B.A.S.M.; Software, I.L.L. and L.A.P.R.; Supervision, J.I.D. and B.A.S.M.; Validation, T.B.R.N., L.A.P.R. and B.A.S.M.; Visualization, T.B.R.N. and J.H.d.O.R.; Writing—original draft, T.B.R.N., I.L.L., J.H.d.O.R., J.I.D. and B.A.S.M.

Funding: This research was funded by FAPESB (Foundation for Research Support of the State of Bahia) and SENAI CIMATEC, n° TSC018/2014.

Acknowledgments: The authors are grateful to FAPESB and SENAI CIMATEC for their financial support, to the Electron Microscopy Service of the Gonçalo Moniz Research Center - FIOCRUZ (Bahia) for the Transmission Electron Microscopy analysis.

Conflicts of Interest: The authors declare no conflict of interest.

References

1. Mihindukulasuriya, S.D.F.; Lim, L.-T. Nanotechnology development in food packaging: A review. *Trends Food Sci. Technol.* **2014**, *40*, 149–167. [CrossRef]
2. Bajpai, V.K.; Kamle, M.; Shukla, S.; Mahato, D.K.; Chandra, P.; Hwang, S.K.; Kumar, P.; Huh, Y.S.; Han, Y.-K. Prospects of using nanotechnology for food preservation, safety, and security. *J. Food Drug Anal.* **2018**, *26*, 1201–1214. [CrossRef]
3. Dudefoi, W.; Villares, A.; Peyron, S.; Moreau, C.; Ropers, M.-H.; Gontard, N.; Cathala, B. Nanoscience and nanotechnologies for biobased materials, packaging and food applications: New opportunities and concerns. *Innov. Food Sci. Emerg. Technol.* **2018**, *46*, 107–121. [CrossRef]
4. Villena de Francisco, E.; García-Estepa, R.M. Nanotechnology in the agrofood industry. *J. Food Eng.* **2018**, *238*, 1–11. [CrossRef]
5. Cerqueira, M.A.; Vicente, A.A.; Pastrana, L.M. Nanotechnology in Food Packaging: Opportunities and Challenges. *Nanomater. Food Packag.* **2018**, 1–11. [CrossRef]

6. Liu, J.; Liu, S.; Zhang, X.; Kan, J.; Jin, C. Effect of gallic acid grafted chitosan film packaging on the postharvest quality of white button mushroom (Agaricus bisporus). *Postharvest Biol. Technol.* **2019**, *147*, 39–47. [CrossRef]
7. Soni, B.; Mahmoud, B.; Chang, S.; El-Giar, E.M.; Hassan, E.B. Physicochemical, antimicrobial and antioxidant properties of chitosan/TEMPO biocomposite packaging films. *Food Packag. Shelf Life* **2018**, *17*, 73–79. [CrossRef]
8. Wu, Z.; Huang, X.; Li, Y.-C.; Xiao, H.; Wang, X. Novel chitosan films with laponite immobilized Ag nanoparticles for active food packaging. *Carbohydr. Polym.* **2018**, *199*, 210–218. [CrossRef]
9. Tang, Y.; Zhang, X.; Zhao, R.; Guo, D.; Zhang, J. Preparation and properties of chitosan/guar gum/nanocrystalline cellulose nanocomposite films. *Carbohydr. Polym.* **2018**, *197*, 128–136. [CrossRef] [PubMed]
10. Dasan, Y.K.; Bhat, A.H.; Ahmad, F. Polymer blend of PLA/PHBV based bionanocomposites reinforced with nanocrystalline cellulose for potential application as packaging material. *Carbohydr. Polym.* **2017**, *157*, 1323–1332. [CrossRef]
11. Machado, B.A.S.; Nunes, I.L.; Druzian, J.I.; Pereira, F.V. Development and evaluation of the efficacy of biodegradable cassava starch films with nanocellulose as reinforcement and with erva-mate extract as an antioxidant additive. *Ciênc. Rural* **2012**, *42*, 2085–2091. [CrossRef]
12. Dilarri, G.; Rosai Mendes, C.; Otavio Martins, A. Synthesis of Chitosan biofilms crosslinked with Tripolyphosphate acting as chelating agent in the fixation of Silver nanoparticles. *Sci. Eng. J.* **2016**, *25*, 97–103.
13. Cerqueira, J.C.; Penha, S.; Oliveira, R.S.; Lefol, L.; Guarieiro, N.; Melo, S.; Viana, J.D.; Aparecida, B.; Machado, S. Production of biodegradable starch nanocomposites using cellulose nanocrystals extracted from coconut fibers. *Polímeros* **2017**, *27*, 320–329. [CrossRef]
14. Mujtaba, M.; Salaberria, A.M.; Andres, M.A.; Kaya, M.; Gunyakti, A.; Labidi, J. Utilization of flax (Linum usitatissimum) cellulose nanocrystals as reinforcing material for chitosan films. *Int. J. Biol. Macromol.* **2017**, *104*, 944–952. [CrossRef]
15. Silva, R.; Haraguchi, S.K.; Muniz, E.C.; Rubira, A.F. Applications of lignocellulosic fibers in polymer chemistry and composites. *Quim. Nova* **2009**, *32*, 661–671. [CrossRef]
16. De Lemos, A.L.; de Martins, R.M. Development and Characterization of Polymeric Composites Based on Poly (Lactic Acid) and Natural Fibers. *Polímeros Ciência e Tecnol.* **2014**, *24*, 190–197.
17. Alves, J.S.; Reis, K.C.; Menezes, E.G.T.; Pereira, F.V.; Pereira, J. Effect of cellulose nanocrystals and gelatin in corn starch plasticized films. *Carbohydr. Polym.* **2015**, *115*, 215–222. [CrossRef] [PubMed]
18. Mazzaglia, A.; Torre, L.; Puglia, D.; Luzi, F.; Del Buono, D.; Balestra, G.M.; Benincasa, P.; Fortunati, E. Revalorization of barley straw and husk as precursors for cellulose nanocrystals extraction and their effect on PVA_CH nanocomposites. *Ind. Crops Prod.* **2016**, *92*, 201–217.
19. Garcia-Garcia, D.; Lopez-Martinez, J.; Balart, R.; Strömberg, E.; Moriana, R. Reinforcing capability of cellulose nanocrystals obtained from pine cones in a biodegradable poly(3-hydroxybutyrate)/poly(ε-caprolactone) (PHB/PCL) thermoplastic blend. *Eur. Polym. J.* **2018**, *104*, 10–18. [CrossRef]
20. Fortunati, E.; Luzi, F.; Jiménez, A.; Gopakumar, D.A.; Puglia, D.; Thomas, S.; Kenny, J.M.; Chiralt, A.; Torre, L. Revalorization of sunflower stalks as novel sources of cellulose nanofibrils and nanocrystals and their effect on wheat gluten bionanocomposite properties. *Carbohydr. Polym.* **2016**, *149*, 357–368. [CrossRef]
21. Orasugh, J.T.; Saha, N.R.; Sarkar, G.; Rana, D.; Mishra, R.; Mondal, D.; Ghosh, S.K.; Chattopadhyay, D. Synthesis of methylcellulose/cellulose nano-crystals nanocomposites: Material properties and study of sustained release of ketorolac tromethamine. *Carbohydr. Polym.* **2018**, *188*, 168–180. [CrossRef]
22. Dungani, R.; Khalil, A.; Aprilia, N.A.S.; Sumardi, I.; Aditiawati, P.; Darwis, A.; Karliati, T.; Sulaeman, A.; Rosamah, E.; Riza, M. Bionanomaterial from agricultural waste and its application. In *Cellulose-Reinforced Nanofibre Composites*; Elsevier: Amsterdam, The Netherlands, 2017; pp. 45–88.
23. Islam, M.S.; Kao, N.; Bhattacharya, S.N.; Gupta, R.; Choi, H.J. Potential aspect of rice husk biomass in Australia for nanocrystalline cellulose production. *Chin. J. Chem. Eng.* **2018**, *26*, 465–476. [CrossRef]
24. Orasugh, J.T.; Saha, N.R.; Sarkar, G.; Rana, D.; Mondal, D.; Ghosh, S.K.; Chattopadhyay, D. A facile comparative approach towards utilization of waste cotton lint for the synthesis of nano-crystalline cellulose crystals along with acid recovery. *Int. J. Biol. Macromol.* **2018**, *109*, 1246–1252. [CrossRef]

25. Ghani, S.; Bakochristou, F.; ElBialy, E.M.A.A.; Gamaledin, S.M.A.; Rashwan, M.M.; Abdelhalim, A.M.; Ismail, S.M. Design challenges of agricultural greenhouses in hot and arid environments—A review. *Eng. Agric. Environ. Food* **2019**, *12*, 48–70. [CrossRef]
26. Akitt, J.W. Some observations on the greenhouse effect at the Earth's surface. *Spectrochim. Acta Part A Mol. Biomol. Spectrosc.* **2018**, *188*, 127–134. [CrossRef]
27. Tuckett, R. Greenhouse Gases. In *Reference Module in Chemistry, Molecular Sciences and Chemical Engineering*; Elsevier: Amsterdam, The Netherlands, 2018.
28. Silva, D.D.J.; D'Almeida, M.L.O. Cellulose whiskers. *O Papel.* **2009**, *70*, 34–52.
29. Mesquita, J.P. Cellulose Nanocrystals for the Preparation of Bionanocomposites with Chitosan and Nanostructured Carbonates for Technological and Environmental Applications. Ph.D. Thesis, Federal University of Minas Gerais (UFMG), Belo Horizonte, Brazil, 2012.
30. Azevedo, V.V.C.; Chaves, S.A.; Bezerra, D.C.; Fook, M.V.L.; Costa, A.C.F.M. Chitin and Chitosan: Applications as biomaterials. *Rev. Eletrôn. Mater. Process.* **2007**, *2*, 27–34.
31. Nery, T.B.R.; José, N.M. Study of Pre-treated and in natura Banana Fibers as Possible Raw Material for Reinforcement in Polymer Composites. *Rev. Virtual Quim.* **2018**, *10*, 313–322. [CrossRef]
32. Mali, S.; Grossmann, M.V.E.; Yamashita, F. Starch films: Production, properties and potential of use. *Semin. Agrar.* **2010**, *31*, 137–156. [CrossRef]
33. Fráguas, R.M.; Simão, A.A.; Faria, P.V.; Queiroz, E.D.R.; de Oliveira Junior, Ê.N.; de Abreu, C.M.P. Preparation and characterization of chitosan edible films. *Polímeros* **2015**, *25*, 48–53. [CrossRef]
34. Yassue-Cordeiro, P.H.; Zandonai, C.H.; da Silva, C.F.; Fernandes-Machado, N.R.C. Development and characterization of composite films of chitosan and zeolites with silver. *Polímeros* **2015**, *25*, 492–502. [CrossRef]
35. Rinaudo, M. Chitin and chitosan: Properties and applications. *Prog. Polym. Sci.* **2006**, *31*, 603–632. [CrossRef]
36. Ali-Komi, D.; Hamblin, M. Chitin and Chitosan: Production and Application of Versatile Biomedical Nanomaterials. *Int. J. Adv. Res.* **2016**, *4*, 411–427.
37. Grifoll-Romero, L.; Pascual, S.; Aragunde, H.; Biarnés, X.; Planas, A. Chitin deacetylases: Structures, specificities, and biotech applications. *Polymers* **2018**, *10*, 352. [CrossRef]
38. Dunlop, M.J.; Acharya, B.; Bissessur, R. Isolation of nanocrystalline cellulose from tunicates. *J. Environ. Chem. Eng.* **2018**, *6*, 4408–4412. [CrossRef]
39. Salari, M.; Sowti Khiabani, M.; Rezaei Mokarram, R.; Ghanbarzadeh, B.; Samadi Kafil, H. Development and evaluation of chitosan based active nanocomposite films containing bacterial cellulose nanocrystals and silver nanoparticles. *Food Hydrocoll.* **2018**, *84*, 414–423. [CrossRef]
40. Rodrigues, S.; Dionísio, M.; López, C.R.; Grenha, A. Biocompatibility of Chitosan Carriers with Application in Drug Delivery. *J. Funct. Biomater.* **2012**, *3*, 615–641. [CrossRef]
41. Ahsan, S.M.; Thomas, M.; Reddy, K.K.; Sooraparaju, S.G.; Asthana, A.; Bhatnagar, I. Chitosan as biomaterial in drug delivery and tissue engineering. *Int. J. Biol. Macromol.* **2017**, *110*, 97–109. [CrossRef]
42. Marín-Silva, D.A.; Rivero, S.; Pinotti, A. Chitosan-based nanocomposite matrices: Development and characterization. *Int. J. Biol. Macromol.* **2019**, *123*, 189–200. [CrossRef]
43. Celebi, H.; Kurt, A. Effects of processing on the properties of chitosan/cellulose nanocrystal films. *Carbohydr. Polym.* **2015**, *133*, 284–293. [CrossRef]
44. Corsello, F.A.; Bolla, P.A.; Anbinder, P.S.; Serradell, M.A.; Amalvy, J.I.; Peruzzo, P.J. Morphology and properties of neutralized chitosan-cellulose nanocrystals biocomposite films. *Carbohydr. Polym.* **2017**, *156*, 452–459. [CrossRef]
45. Azeredo, H.M.C.; Mattoso, L.H.C.; Avena-Bustillos, R.J.; Filho, G.C.; Munford, M.L.; Wood, D.; McHugh, T.H. Nanocellulose reinforced chitosan composite films as affected by nanofiller loading and plasticizer content. *J. Food Sci.* **2010**, *75*, 1–7. [CrossRef]
46. Pereda, M.; Dufresne, A.; Aranguren, M.I.; Marcovich, N.E. Polyelectrolyte films based on chitosan/olive oil and reinforced with cellulose nanocrystals. *Carbohydr. Polym.* **2014**, *101*, 1018–1026. [CrossRef]
47. Coupland, J.N.; Shaw, N.B.; Monahan, F.J.; Dolores O'Riordan, E.; O'Sullivan, M. Modeling the effect of glycerol on the moisture sorption behavior of whey protein edible films. *J. Food Eng.* **2000**, *43*, 25–30. [CrossRef]
48. Madaleno, E.; Rosa, D.D.S.; Zawadzki, S.F.; Pedrozo, T.H.; Ramos, L.P. Study of the Use of Plasticizer from Renewable Sources in PVC Compositions. *Polímeros* **2009**, *19*, 263–270. [CrossRef]

49. Guilbert, S.; Gontard, N.; Cuq, B. Tecnology and Application of Edible Protective Films. *Packag. Technol. Sci.* **1995**, *8*, 339–346. [CrossRef]
50. Veiga-Santos, P.; Scamparini, A.R.P.; Alves, A.J.; Cereda, M.P.; Oliveira, L.M. Mechanical properties, hydrophilicity and water activity of starch-gum films: Effect of additives and deacetylated xanthan gum. *Food Hydrocoll.* **2004**, *19*, 341–349. [CrossRef]
51. Machado, B.A.S.; Reis, J.H.D.O.; Cruz, L.S.; Leal, I.L.; Azevedo, J.B.; Barbosa, J.D.V.; Druzian, J.I. Characterization of cassava starch films plasticized with glycerol and strengthened with nanocellulose from green coconut fibers. *Afr. J. Biotechnol.* **2017**, *16*, 1567–1578.
52. Van Soest, P.J.; Robertson, J.B.; Lewis, B.A. Methods for Dietary Fiber, Neutral Detergent Fiber, and Nonstarch Polysaccharides in Relation to Animal Nutrition. *J. Dairy Sci.* **1991**, *74*, 3583–3597. [CrossRef]
53. Samir, A.S.A.; Alloin, F.; Dufresne, A. Reviews Review of Recent Research into Cellulosic Whiskers, Their Properties and Their Application in Nanocomposite Field. *Biomacromolecules* **2005**, *6*, 612–626. [CrossRef]
54. de Souza, V.C. Cellulose Nanocrystals as a Reinforcement Phase for Chitosan Films: Obtaining, Characterization and Application. Ph.D. Thesis, Federal University of Santa Catarina (UFSC), Florianópolis, Brazil, 2015.
55. Flauzino Neto, W.P.; Mariano, M.; da Silva, I.S.V.; Silvério, H.A.; Putaux, J.-L.; Otaguro, H.; Pasquini, D.; Dufresne, A. Mechanical properties of natural rubber nanocomposites reinforced with high aspect ratio cellulose nanocrystals isolated from soy hulls. *Carbohydr. Polym.* **2016**, *153*, 143–152. [CrossRef]
56. Pessanha, K.L.F.; Farias, M.G.; Carvalho, C.W.P.; Godoy, R.L.D.O. Starch Films Added of Açaí Pulp (Euterpe oleracea Martius). *Braz. Arch. Biol. Technol.* **2018**, *61*, e18170824. [CrossRef]
57. Da Silva, J.B.A.; Nascimento, T.; Costa, L.A.S.; Pereira, F.V.; Machado, B.A.; Gomes, G.V.P.; Assis, D.J.; Druzian, J.I. Effect of Source and Interaction with Nanocellulose Cassava Starch, Glycerol and the Properties of Films Bionanocomposites. *Mater. Today Proc.* **2015**, *2*, 200–207. [CrossRef]
58. Nagy, S.; Csiszár, E.; Kun, D.; Koczka, B. Cellulose nanocrystal/amino-aldehyde biocomposite films. *Carbohydr. Polym.* **2018**, *194*, 51–60. [CrossRef]
59. Pereira, F.V.; De Paula, E.L.; De Mesquita, J.P.; De Almeida Lucas, A.; Mano, V. Bionanocomposites prepared by the incorporation of cellulose nanocrystals into biodegradable polymers by means of solvent evaporation, self-assembly or electro-spinning. *Quim. Nova* **2014**, *37*, 1209–1219.
60. De Souza Lima, M.M.; Borsali, R. Rodlike cellulose microcrystals: Structure, properties, and applications. *Macromol. Rapid Commun.* **2004**, *25*, 771–787. [CrossRef]
61. Beck-Candanedo, S.; Roman, M.; Gray, D.G. Effect of Reaction Conditions on the Properties and Behavior of Wood Cellulose Nanocrystal Suspensions. *Biomacromolecules* **2005**, *6*, 1048–1054. [CrossRef]
62. Salazar, R.F.S.; Silva, G.L.P.; Silva, M.L.C.P. Study of the Composition of Corn Straw for Later Use as a Support in the Preparation of Composites. In *VI Congresso Brasileiro de Engenharia Química em Iniciação Científica* **2005**, *1*, 1–6. Available online: https://www.researchgate.net/profile/Rodrigo_Salazar/publication/235645089_Estudo_da_composicao_da_palha_de_milho_para_posterior_utilizacao_como_suporte_na_preparacao_de_compositos/links/0fcfd51242db818444000000.pdf (accessed on 10 April 2019).
63. Ziglio, B.R.; Bezerra, J.R.M.V.; Branco, I.G.; Bastos, R.; Rigo, M. Bread-making with corncob flower. *Rev. Ciências Exatas e Nat.* **2007**, *9*, 115–128.
64. Dantas, R.D.L.; Silva, G.D.S.; Rocha, A.P.T. Characterization and Technological Assessment of Stabilized Mixed Pasta. In *Encontro Nacional de Educação, Ciência e Tecnologia UEPB* **2012**, *1*, 1–10. Available online: https://editorarealize.com.br/revistas/enect/trabalhos/2d268d7f8f09e15b37f35ce1f7fc5132_586.pdf (accessed on 10 April 2019).
65. De Souza, E.E.; Vale, R.D.S.; Vieira, J.G.; Ribeiro, S.D.; Rodrigues Filho, G.; Marques, F.A.; de Assunção, R.M.N.; Meireles, C.D.S.; Barud, H.D.S.; de Souza, E.E.; et al. Preparation and Characterization of Regenerated Cellulose Membranes Using Cellulose Extracted from Agroindustrial Residues for Application in Separation Processes. *Quim. Nova* **2014**, *38*, 202–208. [CrossRef]
66. Lertwattanaruk, P.; Suntijitto, A. Properties of natural fiber cement materials containing coconut coir and oil palm fibers for residential building applications. *Constr. Build. Mater.* **2015**, *94*, 664–669. [CrossRef]
67. Merali, Z.; Collins, S.R.A.; Elliston, A.; Wilson, D.R.; Käsper, A.; Waldron, K.W. Characterization of cell wall components of wheat bran following hydrothermal pretreatment and fractionation. *Biotechnol. Biofuels* **2015**, *8*, 23. [CrossRef]

68. Mendes, D.E.C.; Antônio, F.; Adnet, D.E.O.; Christina, M.; Moreira, A.; Russi, C.; Furtado, G.; Maria, A.N.A.; Sousa, F.D.E.; De Janeiro, R. Chemical, physical, mechanical, thermal and morphological characterization of corn husk residue. *Cellul. Chem. Technol. Chem.* **2015**, *49*, 727–735.
69. Klemm, D.; Heublein, B.; Fink, H.-P.; Bohn, A. Cellulose: Fascinating Biopolymer and Sustainable Raw Material. *Angew. Chem. Int. Ed.* **2005**, *44*, 3358–3393. [CrossRef]
70. Lagerwall, J.P.F.; Schütz, C.; Salajkova, M.; Noh, J.; Hyun Park, J.; Scalia, G.; Bergström, L. Cellulose nanocrystal-based materials: From liquid crystal self-assembly and glass formation to multifunctional thin films. *NPG Asia Mater.* **2014**, *6*, e80. [CrossRef]
71. Naduparambath, S.; Jinitha, T.V.; Shaniba, V.; Sreejith, M.P.; Balan, A.K. Isolation and characterisation of cellulose nanocrystals from sago seed shells. *Carbohydr. Polym.* **2018**, *180*, 13–20. [CrossRef]
72. Martínez-Sanz, M.; Lopez-Rubio, A.; Lagaron, J.M. Optimization of the nanofabrication by acid hydrolysis of bacterial cellulose nanowhiskers. *Carbohydr. Polym.* **2011**, *85*, 228–236. [CrossRef]
73. Souza-Lima, M.M.; Borsali, R. Static and dynamic light scattering from polyelectrolyte microcrystal cellulose. *Langmuir* **2001**, 992–996.
74. Machado, B.A.S.; Reis, J.H.O.; da Silva, J.B.; Cruz, L.S.; Nunes, I.L.; Pereirae, F.V.; Druzian, J.I. Obtaining Nanocelulose from Green Coconut Fiber and Incorporation in Biodegradable Films of Starch Plasticized with Glycerol. *Quim. Nova* **2014**, *37*, 1275–1282.
75. Rosa, M.F.; Medeiros, E.S.; Malmonge, J.A.; Gregorski, K.S.; Wood, D.F.; Mattoso, L.H.C.; Glenn, G.; Orts, W.J.; Imam, S.H. Cellulose nanowhiskers from coconut husk fibers: Effect of preparation conditions on their thermal and morphological behavior. *Carbohydr. Polym.* **2010**, *81*, 83–92. [CrossRef]
76. Sarwar, M.S.; Niazi, M.B.K.; Jahan, Z.; Ahmad, T.; Hussain, A. Preparation and characterization of PVA / nanocellulose/Ag nanocomposite films for antimicrobial food packaging. *Carbohydr. Polym.* **2018**, *184*, 453–464. [CrossRef]
77. De Oliveira, T.M. Cellulose Nanocrystals: Obtaining, Characterization and Modification of Surface. Master's Thesis, Campinas State University (UNICAMP), Campinas, Brazil, 2012.
78. Silvério, H.A. Extraction and Characterization of Cellulose Nanocrystals from Corn Sabugo and Its Application as a Strengthening Agent in Polymer Nanocomposites Using Polyvinyl Alcohol as a Matrix. Master's Thesis, Federal University of Uberlândia (UFU), Uberlândia, Brazil, 2013.
79. Costa, S.S.; Silva, R.P.D.; Alves, A.R.C.; Guarieiro, L.L.N.; Machado, B.A.S. Prospective Study on the Collection and Incorporation of Cellulose Nanocrystals in Biodegradable Films. *Rev. Virtual Quim.* **2016**, *8*, 1104–1114. [CrossRef]
80. Mattos, A.L.A.; Rosa, M.D.F.; Crisóstomo, L.A.; Bezerra, F.C.; Correia, D. Benefit of the green coconut shell. In *Embrapa Agroindústria Tropical, Fortaleza*; Ceinfo EMBRAPA: Fortaleza, Brazil, 2011; p. 37.
81. Rosa, M.D.F.; Santos, F.J.D.S.; Teles, A.A.M.; de Abreu, F.A.P.; Correia, D.; de Araújo, F.B.S.; Norões, E.R.D.V. *Characterization of Green Coconut Peel Powder Used as Agricultural Substrate*; Ceinfo EMBRAPA: Fortaleza, Brazil, 2001; Volume 54, p. 6.
82. Nascimento, V.; França, C.; Hernández-Montelongo, J.; Machado, D.; Lancellotti, M.; Cotta, M.; Landers, R.; Beppu, M. Influence of pH and ionic strength on the antibacterial effect of hyaluronic acid/chitosan films assembled layer-by-layer. *Eur. Polym. J.* **2018**, *109*, 198–205. [CrossRef]
83. Sharmin, M.; Das Banya, P.; Paul, L.; Chowdhury, F.F.K.; Afrin, S.; Acharjee, M.; Rahman, T.; Noor, R. Study of microbial proliferation and the in vitro antibacterial traits of commonly available flowers in Dhaka Metropolis. *Asian Pac. J. Trop. Dis.* **2015**, *5*, 91–97. [CrossRef]
84. Heinrich, K.; Leslie, D.J.; Jonas, K. Modulation of Bacterial Proliferation as a Survival Strategy. *Adv. Appl. Microbiol.* **2015**, *92*, 127–171.
85. Mathlouthi, M. Water content, water activity, water structure and the stability of foodstuffs. *Food Control* **2001**, *12*, 409–417. [CrossRef]
86. Riggio, G.M.; Wang, Q.; Kniel, K.E.; Gibson, K.E. Microgreens—A review of food safety considerations along the farm to fork continuum. *Int. J. Food Microbiol.* **2019**, *290*, 76–85. [CrossRef]
87. Smigic, N.; Djekic, I.; Martins, M.L.; Rocha, A.; Sidiropoulou, N.; Kalogianni, E.P. The level of food safety knowledge in food establishments in three European countries. *Food Control* **2016**, *63*, 187–194. [CrossRef]
88. Alvarez-Ordóñez, A.; Broussolle, V.; Colin, P.; Nguyen-The, C.; Prieto, M. The adaptive response of bacterial food-borne pathogens in the environment, host and food: Implications for food safety. *Int. J. Food Microbiol.* **2015**, *213*, 99–109. [CrossRef]

89. Seabra, A.B.; Bernardes, J.S.; Fávaro, W.J.; Paula, A.J.; Durán, N. Cellulose nanocrystals as carriers in medicine and their toxicities: A review. *Carbohydr. Polym.* **2018**, *181*, 514–527. [CrossRef]
90. Hänninen, A.; Sarlin, E.; Lyyra, I.; Salpavaara, T.; Kellomäki, M. Nanocellulose and chitosan based films as low cost, green piezoelectric materials. *Carbohydr. Polym.* **2018**, *202*, 418–424. [CrossRef]
91. Benini, K.C.C.C.; Voorwald, H.J.C.; Cioffi, M.O.H. Mechanical properties of HIPS/sugarcane bagasse fiber composites after accelerated weathering. *Procedia Eng.* **2011**, *10*, 3246–3251. [CrossRef]
92. Ljungberg, N.; Bonini, C.; Bortolussi, F.; Boisson, C.; Heux, L.; Cavaillé, J.-Y. New Nanocomposite Materials Reinforced with Cellulose Whiskers in Atactic Polypropylene: Effect of Surface and Dispersion Characteristics. *Biomacromolecules* **2005**, *6*, 2732–2739. [CrossRef]
93. Rubentheren, V.; Ward, T.A.; Chee, C.Y.; Tang, C.K. Processing and analysis of chitosan nanocomposites reinforced with chitin whiskers and tannic acid as a crosslinker. *Carbohydr. Polym.* **2015**, *115*, 379–387. [CrossRef]

© 2019 by the authors. Licensee MDPI, Basel, Switzerland. This article is an open access article distributed under the terms and conditions of the Creative Commons Attribution (CC BY) license (http://creativecommons.org/licenses/by/4.0/).

Article

Effect of Unbleached Rice Straw Cellulose Nanofibers on the Properties of Polysulfone Membranes

Mohammad Hassan [1,2,*], Ragab E. Abou Zeid [1], Wafaa S. Abou-Elseoud [1], Enas Hassan [1], Linn Berglund [3] and Kristiina Oksman [3,4,*]

1. Cellulose and Paper Department & Centre of Excellence for Advanced Sciences, National Research Centre, 33 El-Behouth street, Dokki, Giza 12622, Egypt; r_abouzeid2002@yahoo.com (R.E.A.Z.); ws.abouelseoud@nrc.sci.eg (W.S.A.-E.); ea.elgarhy@nrc.sci.eg (E.H.)
2. Egypt Nanotechnology Centre, Cairo University, El-Shiekh Zayed, 6th October City 12588, Egypt
3. Department of Engineering Sciences and Mathematics, Luleå University of Technology, SE 97187 Luleå, Sweden; linn.berglund@ltu.se
4. Department of Mechanical and Industrial Engineering, University of Toronto, 5 Kings College Road, Toronto, ON M5S 3G8, Canada
* Correspondence: ml.hassan@nrc.sci.eg (M.H.); kristiina.oksman@ltu.se (K.O.)

Received: 5 May 2019; Accepted: 27 May 2019; Published: 29 May 2019

Abstract: In addition to their lower cost and more environmentally friendly nature, cellulose nanofibers isolated from unbleached pulps offer different surface properties and functionality than those isolated from bleached pulps. At the same time, nanofibers isolated from unbleached pulps keep interesting properties such as hydrophilicity and mechanical strength, close to those isolated from bleached pulps. In the current work, rice straw nanofibers (RSNF) isolated from unbleached neutral sulfite pulp (lignin content 14%) were used with polysulfone (PSF) polymer to make membrane via phase inversion. The effect of RSNF on microstructure, porosity, hydrophilicity, mechanical properties, water flux, and fouling of PSF membranes was studied. In addition, the prepared membranes were tested to remove lime nanoparticles, an example of medium-size nanoparticles. The results showed that using RSNF at loadings from 0.5 to 2 wt.% can significantly increase hydrophilicity, porosity, water flux, and antifouling properties of PSF. RSNF also brought about an increase in rejection of lime nanoparticles (up to 98% rejection) from their aqueous suspension, and at the same time, with increasing flux across the membranes. Tensile strength of the membranes improved by ~29% with addition of RSNF and the maximum improvement was obtained on using 0.5% of RSNF, while Young's modulus improved by ~40% at the same RSNF loading. As compared to previous published results on using cellulose nanofibers isolated from bleached pulps, the obtained results in the current work showed potential application of nanofibers isolated from unbleached pulps for improving important properties of PSF membranes, such as hydrophilicity, water flux, rejection, and antifouling properties.

Keywords: rice straw; cellulose nanofibers; unbleached pulp; polysulfone; membrane

1. Introduction

Polysulfone (PSF) is one of the attractive polymers due to its good mechanical properties, thermal stability, chemical resistance, transparency, and flexibility [1]. PSF is used in different kinds of membranes such as anion-exchange membranes [2]; hemodialyzer membranes [3]; gas separation membranes [4]; pressure-driven membranes, i.e., ultra- and nanofiltration membranes [5]; osmotically-driven membranes for desalination and ions removal [6]; proton-exchange membranes fuel cells [7]; membranes for use in artificial organs [8]; and membranes used in dehydration of solvents by pervaporation [9].

To obtain membrane with sufficient porosity, the phase inversion technique, which depends on precipitation of PSF from its solution in a nonsolvent, is usually practiced. Nevertheless, the high hydrophobic nature of PSF leads to fouling problems when used in membranes [1]. To overcome this shortcoming, PSF-based polymers have been developed such as polyether sulfone and sulfonated PSF. This of course adds to the cost of the produced membranes. Some inorganic additives have been used to improve the hydrophilicity and performance of PSF membranes such as Al_2O_3 [10], TiO_2 [11], and silica [12,13]. Another route to overcome the hydrophobicity of PSF and to produce high-flux membranes is to blend it with other relatively higher hydrophilic polymers. Compatibility between PSF and polymers used and their solubility are important factors which limit the number of candidates for that purpose. Different synthetic polymers such as polyvinylpyrrolidone [14], polyethylene glycol [15], and polyaniline-polyvinylpyrrolidone [16] have been used with PSF to improve the aforementioned shortcomings.

Cellulose, as a natural polymer, either in form of microfibers and nanocellulose, e.g., cellulose nanofibers and cellulose nanocrystals, have been studied to improve the hydrophilicity and water flux properties of PSF since they could be dispersed in the same solvents of PSF, such as dimethyl acetamide. The presence of S=O polar groups in polysulfones helps in forming hydrogen bonding with hydroxyl groups of cellulose. For example, cellulose nanocrystals have been used to improve hydrophilicity, mechanical properties, and performance of polysulfone [17–23] and polyether sulfone [24]. Regarding use of cellulose nanofibers (CNF), CNF isolated from bleached cellulose fibers have been used to improve mechanical and hydrophilicity of PSF [23], PSF/sulfonated PSF membrane [25], and polyether sulfone [26]. Also, methacryloxypropyltrimethoxy silane-modified CNF have been used to improve mechanical properties and performance of PSF membranes [27].

In addition to cellulose, the use of lignin, which produced as a byproduct from pulping processes, is another attractive route for improving hydrophilicity of polysulfones due to the presence of polar groups in the lignin structure in addition to the aromatic skeleton, which helps in achieving good compatibility between lignin and polysulfones [28,29].

Motivated by the desirable effect of both lignin and cellulose on properties of polysulfones, Ding et al. [30] studied the use of what they called "lignin/cellulose nanofibers" complex prepared by sulfuric acid hydrolysis of unbleached pulp followed by submerging the neutralized treated pulp in dimethylacetamide and high pressure homogenization, for improving hydrophilicity and mechanical properties of polyethersulfone; the ratio of lignin to cellulose nanofibers was up to 1.2 wt.%.

Furthermore, there is a recent interest in isolation of cellulose nanofibers from unbleached pulp with high lignin content using ultrafine grinding, where the nanofibers are isolated only by the action of the high shear force during grinding [31]. In addition to saving chemicals used in bleaching in case of using unbleached pulp, CNF with lignin at the surface could show good compatibility with polymers that do not have sufficient compatibility with nanofibers obtained from bleached pulp.

For the best of our knowledge, all previous work on using CNF with PSF for making membranes was mainly focused on using CNF isolated from bleached pulps. In the current work, the use of cellulose nanofibers isolated from unbleached rice straw neutral sulfite pulp (containing ~14% lignin) by ultrafine grinding for improving hydrophilicity, porosity, and ultrafiltration performance of PSF membrane was investigated. In addition, the effect of the isolated nanofibers on mechanical properties was investigated.

2. Experimental

2.1. Materials

Rice straw obtained from a local farm in Qalubiyah, Egypt was washed with water to remove the dust and allowed to air dry. Sodium sulfite and sodium carbonate used for pulping were reagent grade chemicals and used as received. Dimethylacetamide (DMAc) and polysulfone with average M_w ~35,000 and, average M_n ~16,000 were used as received from Sigmaaldrich (St. Louis, MO, USA).

Lyophilised bovine serum albumin (pH 7) was purchased from Biowest Company (Biowest, Naillé, France) and used as received.

2.2. Preparation of Rice Straw Pulp

Rice straw neutral sulfite pulp was prepared by pulping the straw using 10% sodium sulfite and 2% sodium carbonate (based on weight of rice straw) solutions at 160 °C for 2 h; the liquor ratio was 1:10. The produced pulp was thoroughly washed with water, defibrillated in a Valley beater (Valley Iron Works, Appleton, Wisconsin, USA) to a 25°SR degree of freeness, dewatered, and allowed to air dry. Chemical composition of the prepared pulp: 16.63% ash content, 14.15% Klason lignin, 3.24% acid insoluble lignin, 54.12% α-cellulose, 14.34% pentosans, and degree of polymerization: 903 [31].

2.3. Xylanases Pretreatment of Unbleached Rice Straw Pulps

Neutral sulfite unbleached pulp was pretreated with xylanases in citrate buffer (pH = 5.3) for 4 h at 50 °C as previously described [31]. The concentration of xylanase used was 0.04 g/g of pulp. Chemical composition of the pretreated pulp was 16.46% ash content, 13.18% Klason lignin, 2.31% acid insoluble lignin, 58.4% α-cellulose, 10.79% pentosans, and degree of polymerization: 1097 [31].

2.4. Isolation of Cellulose Nanofibers from Xylanase-Treated Unbleached Pulp

Isolation of cellulose nanofibers from unbleached pulp was carried out similar to the previously published protocol [31]. In brief, the unbleached pulp was first disintegrated using a shear mixer (Silverson L4RT, Silverson Machines Ltd., Chesham, UK) using pulp suspension of 2 wt.% consistency. The pulp was then fibrillated using high-shear ultrafine friction grinder, or a so-called Supermasscolloider (MKCA6-2, Masuko Sangyo, Kawaguchi, Japan).The gap between the disks was gradually adjusted to −90 µm and the pulp was run through the grinder for approximately 140 min.

2.5. Preparation of PSF/RSNF Membrane

PSF solution (18 wt.%) was prepared by dissolving in DMAc. Water in the RSNF suspension was first removed by vacuum filtration then acetone was passed once through the filtered CNF, and finally DMAc was passed twice. The DMAc-wetted RSNF were kept in a closed container in fridge at 8 °C till use. The DMAc-wetted RSNF were added to PSF solution at ratios from 0.5 to 2 wt.% of dry RSNF to PSF; the mixture was homogenized by magnetic stirring for 30 min. The viscosity of the PSF/RSNF mixture was measured using a tuning-fork vibration viscometer (Vibro Viscometer SV-10, A&D Company Limited, Tokyo, Japan). The films were prepared by phase inversion in distilled water, washed thoroughly with distilled water, and left to dry in air.

2.6. Characterization of PSF/RSNF Membrane

Tensile testing was carried out on 1-cm-wide films using a Lloyd instrument (Lloyd Instruments, West Sussex, UK) with a 100-N load cell; a cross-head speed of 2 mm/min was used and the gauge length was 20 mm. Five replicates of each sample were used and the results averaged. The water contact angle of films was measured using an Attension theta lite measuring system (Biolin Scientific AB, Gothenburg, Sweden) and calculated with the drop shape analysis OneAttension Version 2.7 (r5433), using a sessile drop technique. A 4 µL water drop was placed onto the films at four separate places for calculating the average contact angles. Microscopic features of films were investigated using a FEI Quanta 200 scanning electron microscope (FEI Company, Eindhoven, The Netherlands) at an acceleration voltage of 20 kV.

2.7. Preparation and Characterization of Lime Nanoparticles Suspension

Lime nanoparticles were prepared according to the previously published method [32]. In brief, to 100 mL containing 0.3 mol/L of calcium chloride, 0.6 mol/L of sodium hydroxide was added dropwise (≈4 mL/min) at 90°C. The precipitated calcium hydroxide nanoparticles were centrifuged and washed with previously boiled distilled water. The suspension containing purified calcium hydroxide nanoparticles was flushed with nitrogen gas and kept in closed bottle until use. Transmission electron microscopy (TEM) was carried out using high-resolution transmission electron microscopy (HR-TEM) (JEM-2100 transmission electron microscope, JEOL, Tokyo, Japan). Energy-dispersive X-ray microanalysis was carried out using JEOL JXA 8040 A electron probe microanalyzer (JEOL, Tokyo, Japan).

2.8. Evaluation of Membranes Properties

2.8.1. Porosity

The porosity (ε) of membranes was determined from water absorption of the different membranes. The membranes were submerged in distilled water for 18 h then weighed after wiping excess water. Then, the wet membranes were dried at 105 °C until no change in weight. Porosity was calculated according to the following equation [33].

$$\text{Porosity }(\varepsilon) = [(m_1 - m_2)/\varrho.A.L]*100 \tag{1}$$

where m_1 and m_2 are the weight of the wet and dry films, respectively; ϱ is the water density (g/cm^3); A is the effective area of the films (cm^2); and L is the film thickness (cm).

2.8.2. Pure Water Flux and Fouling

The water flux of the membranes was measured using a dead-end stirred cell (Sterlitech HP4750, Sterlitech, Kent, WA, USA). Prior to the measurements, discs with a diameter of ~5 cm were cut out from the membranes and soaked in water for one hour to ensure equilibration of the membrane. The conditioned membranes were placed in the dead-end cell on a stainless steel porous support disc and water was passed through the membranes at 25 °C at a differential pressure of 30 Mpa, maintained using pressure water pump. The quantity of water that passed through the membrane for a defined time interval was weighed accurately and the flux was calculated (L/h/m^2/MPa) for the active filtration area (14.6 cm^2). To avoid reduction of water flux as a result of membrane compaction, back pressure was applied every 10 min.

To test fouling of the membranes, bovine serum albumin solution (1 g/L) was passed through the membrane for 10 min under pressure of 30 MPa and flux was calculated. Then, the membranes were washed briefly with distilled water and pure water was passed through the membranes under the same pressure and protein solution was passed again; the cycle was repeated for one hour.

2.8.3. Removing Lime Nanoparticles

To test the efficiency of the prepared membranes in removing lime, suspensions containing 1 g/L was passed through the membrane under the same conditions mentioned above for testing pure water flux. The concentration of the nanoparticles in the filtrate was measured from following the turbidity using UV–Visible spectrometer (Shimadzu, Tokyo, Japan) at wavelength of 600 nm. Turbidity (T) was calculated from measuring absorbance at 500 nm: Turbidity = (A * 2.302)/L, where A is the absorbance and l is the path length (0.01 m). The rejection of the prepared membranes to remove the nanoparticles was calculated using the following formula.

$$\text{Rejection (\%)} = [(\text{Control } T_{500} - \text{sample } T_{500})/\text{Control } T_{500}] \times 100 \tag{2}$$

3. Results and Discussion

In a previous publication, it has been shown that RSNF with high lignin content (~14 wt.%) could be isolated from rice straw xylanase-treated unbleached sulfite pulp with a width of ~14 ± 7 nm, as seen in Figure 1 [31]. Water in the isolated RSNF could be easily exchanged by filtration and simple washing using different non-aqueous aprotic solvent [34].

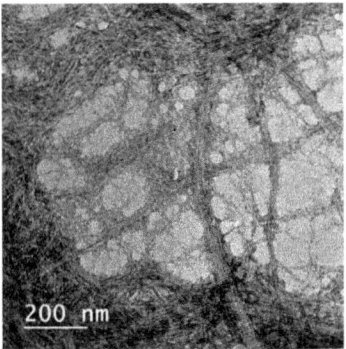

Figure 1. TEM image of cellulose nanofibers isolated from xylanase-treated rice straw unbleached neutral sulfite pulp.

3.1. Effect of RSNF on Viscosity of PSF Solution

The effect of the addition of RSNF on viscosity of the PSF solution was studied as an indication of good dispersion of RSNF in the solution; the results are presented in Table 1. As shown in the figure, the viscosity of PSF solution increased with the addition of RSNF, in spite of the small concentrations of RSNF added (2% max.). The increase in viscosity ranged from 23% to 128% at the different RSNF loadings with respect to neat PSF solution.

Table 1. Effect of rice straw nanofibers (RSNF) content on viscosity of polysulfone (PSF) solution.

Sample	Viscosity (Pa.s)
PSF	434 ± 0.71
PSF + 0.5% RSNF	533 ± 0.82
PSF + 1.0% RSNF	634 ± 1.41
PSF + 2.0% RSNF	988 ± 5.65

3.2. Effect of RSNF on Microscopic Structure of PSF Membranes

Microscopic structure of polymeric membranes is very important since it determines both mechanical and filtration properties. The effect of RSNF on the microscopic structure of PSF membranes formed by phase inversion was studied by SEM; images of both surface and cross-section are shown in Figures 2 and 3.

Regarding the cross-section, a slight change in the size of voids started to occur at RSNF loading of 1%, while at 2%, formation of larger and elongated slender-like voids was observed. On the other hand, the microscopic structure of PSF film' surfaces clearly affected by the addition of RSNF. While neat PSF film had a few and very tiny pores at the surface, adding RSNF resulted in more and wider pores, especially at RSNF loading 1–2%. At higher magnification, SEM images of the surface of membrane sample containing 2% RSNF showed that although wide pores formed at the surface, the inner pores were narrower and with mesh-like structure.

Table 2 shows the average diameter of the pores at the surface of the different PSF samples seen from the images. Increasing the loading of RSNF resulted in larger pore diameter at the surface. Similar trend was observed in membranes prepared from cellulose nanofibers isolated from bleached pulp with polysulfone/sulfonated polysulfone mixture [24]. These changes in the microscopic features due to addition of cellulose nanofibers were interpreted by the accelerated phase inversion process by the hydrophilic cellulose nanofibers, which results in formation of more pores and better pores connectivity across the membrane. In addition, due to the large aspect ratio of cellulose nanofibers, they are easily aggregated during phase inversion and resulted in formation of pore defects in the membranes [18,25]. The same findings were also reported when methacryloxypropyltrimethoxy silane-modified cellulose nanofibers were used with PSF [27].

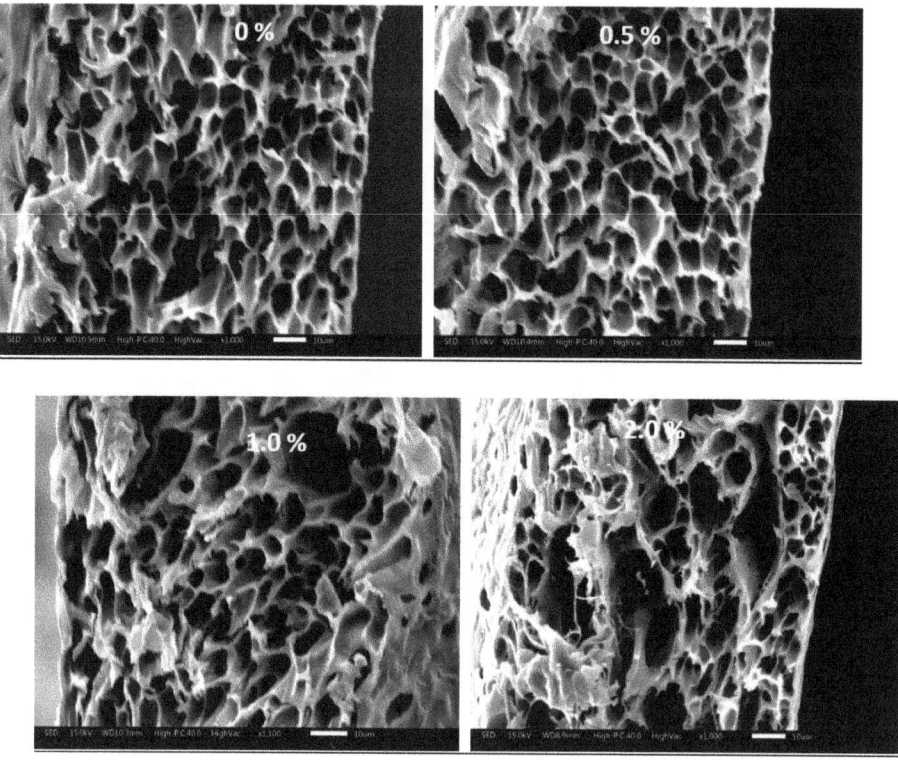

Figure 2. SEM of cross-section of polysulfone membranes with different RSNF loadings.

Table 2. Diameter of pores at the surface of PSF/RSNF films.

Sample	Diameter of Pores at Surface (µm)
PSF	2.9 ± 1.2
PSF + 0.5% RSNF	5.9 ± 3.4
PSF + 1.0% RSNF	5.8 ± 0.8
PSF + 2.0% RSNF	15.3 ± 8.6

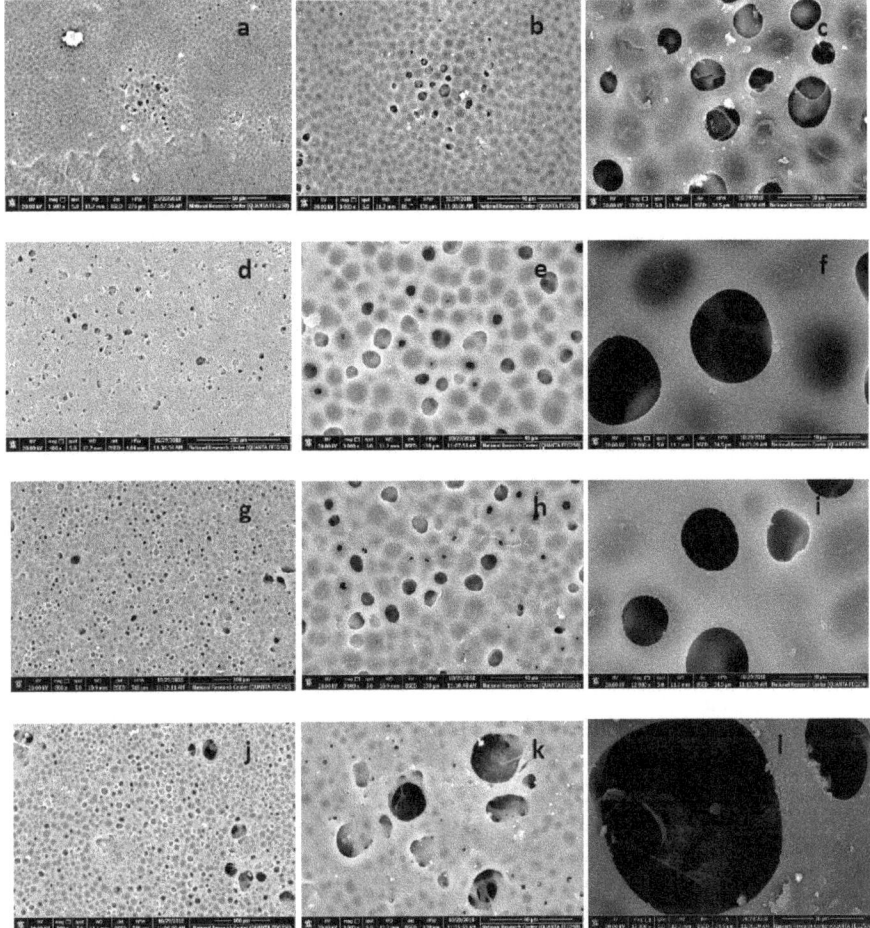

Figure 3. SEM image of surfaces of PSF (**a–c**), PSF/0.5% RSNF (**d–f**), PSF/1% RSNF (**g–i**), and PSF/2% RSNF (**j–l**) membranes at 800, 3000, and 12000x magnification, respectively (scale bar is 50, 40, and 10 μm, respectively).

3.3. Effect of RSNF on Mechanical Properties of PSF Films

The effect of RSNF on tensile strength properties of PSF was studied and results are listed in Table 3. As the results show, adding RSNF resulted in a moderate increase in maximum tensile strength (~29%) at 0.5% RSNF loading followed by a decrease at higher RSNF loading, but still as high as that of neat PSF. This could be attributed to the more pores formed at the surface and pore defects across the thickness. The high standard deviations values at 2% RSNF loading can be attributed to the high porosity of the films, and thus failure of films was non-reproducible from one sample to another. On the other hand, the Young's modulus of PSF films more obviously increased upon adding RSNF, especially at 0.5% loading of RSNF where the increase in modulus was ~40%. Increasing RSNF to higher loadings resulted in decreasing the modulus values, but they were still higher than that of the neat PSF membrane. The improvement in mechanical properties of the films indicates good compatibility between PSF matrix and the surface of RSNF which are rich in lignin with its aromatic ring structures. In addition, formation of hydrogen bonding between hydroxyl groups at the surface of the RSNF and the S=O bonds of PSF could improve mechanical properties of the films. Regarding strain at break, it generally tended to decrease with addition of RSNF; the decrease ranged from 14.5 to 20.7%.

Comparing the improvement in tensile strength properties in the current work to that found in previous work where the so-called "lignin/cellulose nanofibers" complex [30], prepared by sulfuric acid hydrolysis of unbleached pulp and high pressure homogenization in dimethylacetamide, was used with PSF, the maximum improvement in tensile strength was about 50%. Strain at break increased by ~22%, which is in contrast to the strain decrease found in the current work. The tensile modulus was not measured in that work. In another work where cellulose nanofibers from bleached pulp were used with PSF and sulfonated PSF, the maximum improvement in tensile strength and tensile modulus were approximately 20% and 38%, respectively [25]. Comparing the improvement in tensile strength and tensile modulus achieved in the current work to the aforementioned findings clearly shows the advantage of using RSNF isolated from unbleached pulp by direct grinding.

Table 3. Tensile strength properties of polysulfone (PSF)/rice straw nanofibers (RSNF) membranes.

Sample	Tensile Strength (MPa)	Tensile Modulus (MPa)	Strain at Maximum Load (%)
PSF	3.88 ± 0.40	164.8 ± 16.2	24.1 ± 2.3
PSF + 0.5% RSNF	5.00 ± 0.51	229.9 ± 27.3	20.6 ± 3.9
PSF + 1.0% RSNF	3.50 ± 0.47	186.9 ± 18.9	19.1 ± 1.7
PSF + 2.0% RSNF	3.84 ± 1.16	176.5 ± 26.6	19.3 ± 2.4

3.4. Effect of RSNF on Hydrophilicity and Porosity of PSF Membranes

One of the important reasons for using cellulose in membranes is to increase their hydrophilicity; this affects both water flux across the membrane and its resistance to fouling. The RSNF used still has strong hydrophilicity in spite of the presence of lignin. As shown in Table 4, addition of RSNF to PSF resulted in a decrease in water contact angle even at the smallest RSNF loading where water contact angle was ~90° for the neat PSF membrane and ranged from 80.1° to 82.7° for the different PSF/RSNF samples. The effect of RSNF on water absorption, and thus the calculated porosity, of PSF film was rather more pronounced at 1–2% of RSNF loading. PSF film has poor hydrophilic character and water absorption occurred is mainly due to permeation of water into the porous structure. Increasing porosity as a result of adding RSNF could be due to the higher hydrophilicity, as it is clear from the contact angle values, and wider pores formed at the surface of PSF as a result of presence of RSNF as seen from the SEM images.

Table 4. Water contact angle of the different PSF/RSNF membranes.

Sample	Contact Angle (°)	Water Absorption (%)	Porosity (%)
PSF	89.9 ± 0.4	84.2 ± 2.7	45.5 ± 2.9
PSF + 0.5% RSNF	80.1 ± 5.2	83.6 ± 2.4	54.3 ± 4.6
PSF + 1.0 % RSNF	81.5 ± 4.2	125.7 ± 5.9	69.8 ± 4.8
PSF + 2.0 % RSNF	82.7 ± 1.6	119.6 ± 6.8	66.6 ± 6.1

3.5. Effect of RSNF on Water Flux and Fouling of PSF Films

Water flux across PSF and PSF/RSNF membranes was tested and the results are presented in Figure 4. Cycles of back pressure were applied every twenty minutes to minimize the decrease of flux as a result of compactness of the film by water pressure. As it is clear in the figure, presence of RSNF with loading 1% and higher had significant effect on water flux. After one hour of the experiment, water flux values of 50 ± 5, 57 ± 5, 111 ± 18, and 131 ± 21 L/h/m^2/MPa were recorded for neat PSF, PSF/0.5%RSNF, PSF/1%RSNF, and PSF/2%RSNF, respectively. The increase in water flux could be attributed to the higher hydrophilicity, porosity, and water absorption of membrane with higher RSNF contents. The increase in water flux from cycle to another could be due to generation of more pores at the surface by the action of water pressure applied. This is could be seen from the SEM images of membranes' surface before and after the water flux test (Figure 5). As indicated by the red arrows in the image, the skin layer at the surface was ruptured by the action of water pressure applied generating more pores at the surface.

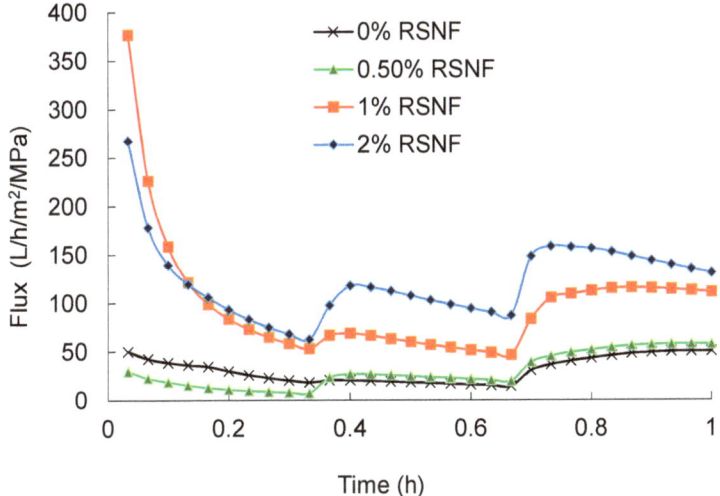

Figure 4. Water flux of PSF and PSF/RSNF membranes with different loadings of RSNF.

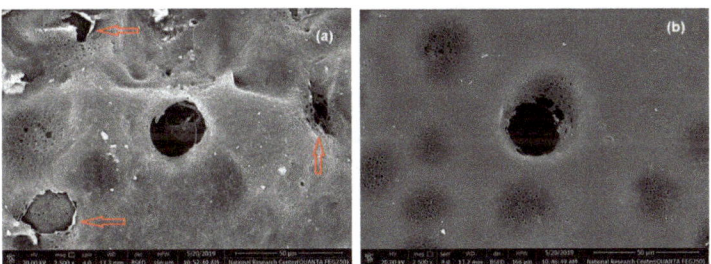

Figure 5. SEM of the PSF/2%RSNF membrane surface: (**a**) after water flux test and (**b**) before water flux test.

Regarding fouling of the membranes, as shown in Figure 6, after three cycles of passing protein solution and pure water across the membranes for one hour, water flux values of 2.7 ± 0.26, 8.9 ± 1.75, 18.6 ± 2.48, and 61.3 ± 4.4 L/h/m^2/MPa were recorded for neat PSF, PSF/0.5%RSNF, PSF/1%RSNF, and PSF/2%RSNF, respectively.

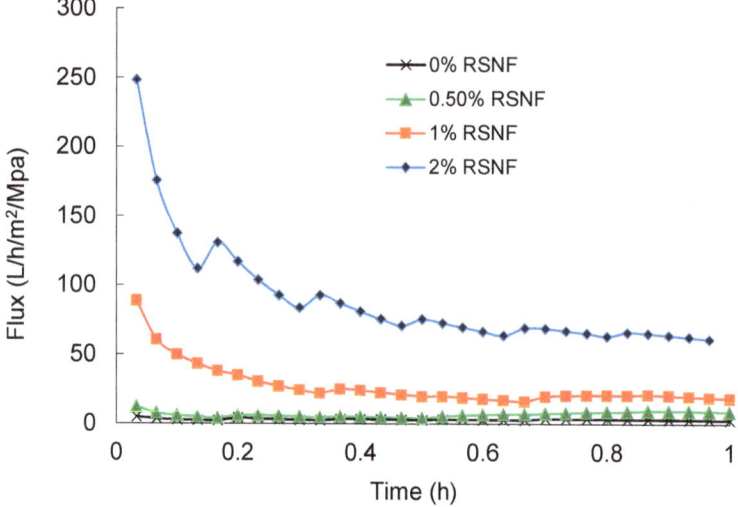

Figure 6. Fouling of PSF and PSF/RSNF membranes with different loadings of RSNF when passing 1% bovine albumin solution.

3.6. Rejection of Lime Nanoparticles

The prepared membranes were tested for removing lime nanoparticles from dilute suspension containing 2 g/L of lime nanoparticles. The XRD pattern (Figure 7) showed that structure of calcium hydroxide (lime) with hexagonal crystal structure [35]; TEM image showed nanoparticles with diagonal ranged from as low as 100 nm up to several hundred nanometers. The energy-dispersive X-ray spectroscopy (EDS) spectrum proved the purity of the prepared lime nanoparticles.

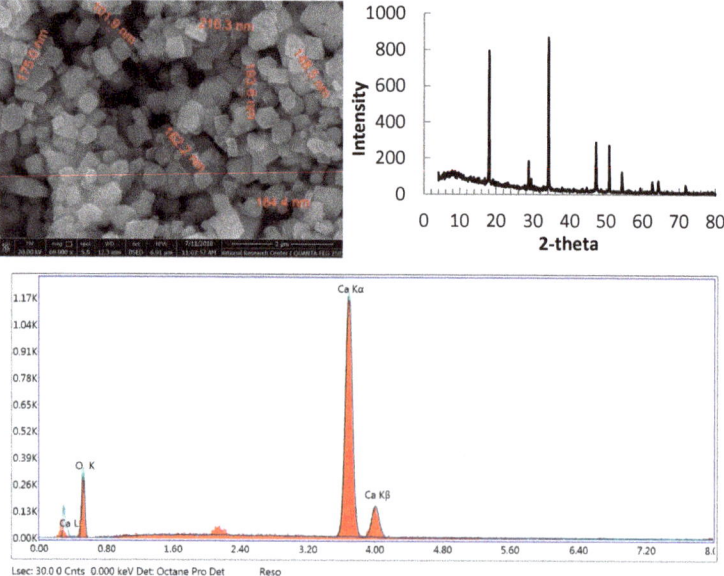

Figure 7. TEM, XRD pattern, and energy-dispersive X-ray spectroscopy (EDS) spectrum of lime nanoparticles.

The water flux of the lime nanoparticle suspension after 2 h for the different membranes is shown in Table 5. As shown in the table, adding the RSNF resulted in increasing the flux of lime nanoparticles suspension. The maximum increase of water flux was achieved on adding 1–2% of RSNF and was ~850% as compared to that of the neat PSA membrane.

Figure 8 shows visible light absorbance of the filtrate solution after passing through PSF membranes, as well as that of lime nanoparticles suspension with 1 g/L concentration. Interestingly, as shown in the figure, the rejection of lime nanoparticles increased with increasing RSNF content in spite of the increase of diameter of the pores at the surface and porosity of the membranes containing RSNF. The filtrate obtained from filtration using the neat PSF membrane showed a 60% decrease of light absorbance at 600 nm as compared to the blank lime suspension; filtrates obtained from filtration using PSF/0.5%RSNF, PSF/1%RSNF, and PSF/2% RSNF showed 76%, 97%, and 98% decrease in light absorbance, respectively. According to standard curve of lime nanoparticles with different concentrations (not shown), the concentration of lime nanoparticles in the filtrate was 17.4, 11.1, 1.3, and 0.9 mg/L when using PSF, PSF/0.5%RSNF, PSF/1%RSNF, and PSF/2% RSNF membranes, respectively. The increase in rejection of lime nanoparticles with increasing porosity and pore radius at the surface of membranes as the RSNF content increase means that the rejection of lime nanoparticles is not simply by mechanical filtration but involves interaction between the nanoparticles and the functional groups at the RSNF surfaces. RNSF with its very high surface area has plenty of hydroxyl and carboxylic functional groups at their surface which can attract the charged lime nanoparticles. Similar observation was found in previous work in case of using microcrystalline cellulose, cellulose nanocrystals or cellulose nanofibers isolated from bleached pulps when rejection of bovine serum albumin (BSA) was studied [25,27,36]. Although addition of cellulose nanocrystals or nanofibers to PSF resulted in increasing porosity and water flux, rejection of BSA (protein with charged surface) did not significantly affect by the increased porosity or water flux.

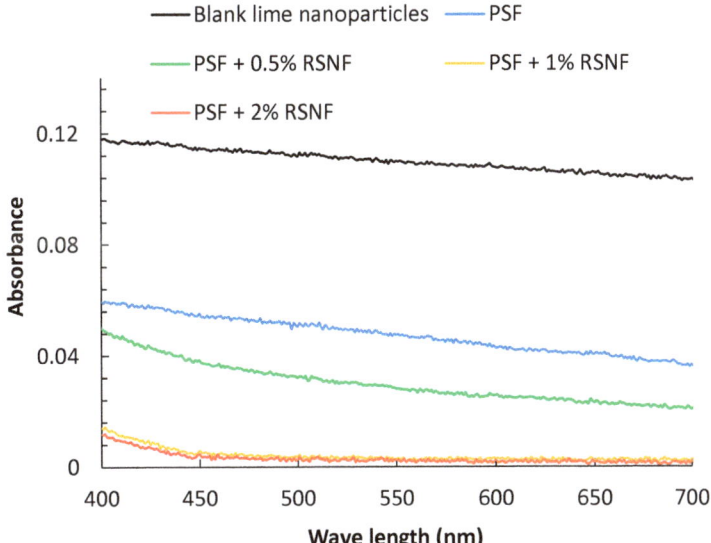

Figure 8. Absorbance spectra of lime nanoparticles suspension and filtrate produced after passing the suspension through the different PSF membranes.

Table 5. Flux of lime nanoparticles suspension through the different membranes after 2 h.

Samples	Flux rate (L/h/m^2/MPa)
Polysulfone (PSF)	1.027 ± 0.12
PSF/0.5 % RSNF	4.00 ± 0.40
PSF/1 % RSNF	10.27 ± 1.06
PSF/2 % RSNF	10.78 ± 1.67

4. Conclusions

RSNF containing lignin (isolated from unbleached pulp) could be used for improving porosity, hydrophilicity, water flux, and antifouling of PSF membranes at low loadings of RSNF (0.5–2%). Mechanical properties (tensile stress and Young' modulus) only improved at the lowest RSNF loading (0.5%); higher loadings did not enhance these properties. PSF containing 1–2% of RSNF could be successfully used for removing lime nanoparticles from aqueous suspension at a 10-fold faster rate than in case of using neat PSF, which could only partially remove the lime nanoparticles. The study showed the advantages of using cellulose nanofibers isolated with lower cost than that isolated from bleached pulps for improving properties of PSF related to their use as ultrafiltration membranes.

Author Contributions: Conceptualization, M.H. and K.O.; Funding Acquisition, M.H. and K.O.; Investigation, M.H., R.E.A.Z., W.S.A.-E., E.H., L.B., and K.O.; Methodology, M.H., R.E.A.Z., W.S.A.-E., E.H., and L.B.; Project Administration, K.O.; Supervision, M.H. and K.O.; Writing—Original Draft, M.H.; Writing—Review & Editing, M.H. and K.O.

Funding: The authors acknowledge funding of the current research by the Swedish Research Council (project no. 2015-05847), as well as the financial support received from Bio4Energy, a strategic research environment appointed by the Swedish government.

Conflicts of Interest: The authors declare no conflict of interest.

References

1. Mark, H.F. *Encyclopedia of Polymer Science and Technology*; John Wiley & Sons: Hoboken, NJ, USA, 2002. [CrossRef]
2. Cho, M.K.; Lim, A.; Lee, S.Y.; Kim, H.-J.; Yoo, S.J.; Sung, Y.-E.; Park, H.S.; Jang, J.H. A Review on Membranes and Catalysts for Anion Exchange Membrane Water Electrolysis Single Cells. *J. Electrochem. Sci. Technol.* **2017**, *8*, 183–196. [CrossRef]
3. Roy, A.; De, S. State-of-the-Art Materials and Spinning Technology for Hemodialyzer Membranes. *Sep. Purif. Rev.* **2017**, *46*, 216–240. [CrossRef]
4. Jamshidi, M.; Pirouzfar, V.; Abedini, R.; Pedram, M.Z. The influence of nanoparticles on gas transport properties of mixed matrix membranes: An experimental investigation and modeling. *Korean J. Chem. Eng.* **2017**, *34*, 829–843. [CrossRef]
5. Mehta, R.; Manna, P.; Bhattacharya, A. Sulfonated polysulfone-preparative routes and applications in membranes used for pressure driven techniques. *J. Macromol. Sci. Part A* **2016**, *53*, 644–650. [CrossRef]
6. Alsvik, I.L.; Hagg, M.-B. Pressure Retarded Osmosis and Forward Osmosis Membranes: Materials and Methods. *Polymers* **2013**, *5*, 303–327. [CrossRef]
7. Iojoiu, C.; Sanchez, J.-Y. Polysulfone-based Ionomers for Fuel Cell Applications. *High Perform. Polym.* **2009**, *21*, 673–692. [CrossRef]
8. Kawakami, H. Polymeric membrane materials for artificial organs. *J. Artif. Organs* **2008**, *11*, 177–181. [CrossRef]
9. Chapman, P.D.; Oliveira, T.; Livingston, A.G.; Li, K. Membranes for the dehydration of solvents by pervaporation. *J. Membr. Sci.* **2008**, *318*, 5–37. [CrossRef]
10. Liu, F.; Abed, M.M.; Li, K. Preparation and characterization of poly(vinylidene fluoride) (PVDF) based ultrafiltration membranes using nano γ-Al2O3. *J. Membr. Sci.* **2011**, *366*, 97–103. [CrossRef]
11. Yuliwati, E.; Ismail, A. Effect of additives concentration on the surface properties and performance of PVDF ultrafiltration membranes for refinery produced wastewater treatment. *Desalination* **2011**, *273*, 226–234. [CrossRef]

12. Merkel, T.C.; Merkel, T.C.; Freeman, B.D.; Spontak, R.J.; He, Z.; Pinnau, I. Ultrapermeable, Reverse-Selective Nanocomposite Membranes. *Science* **2002**, *296*, 519–522. [CrossRef] [PubMed]
13. Huang, Y.; Jin, H.; Yu, P.; Luo, Y. Polyamide thin-film composite membrane based on nano-silica modified polysulfone microporous support layer for forward osmosis. *Desalin. Water Treat.* **2016**, *57*, 20177–20187. [CrossRef]
14. Ahmad, A.L.; Sarif, M.; Ismail, S. Development of an integrally skinned ultrafiltration membrane for wastewater treatment: effect of different formulations of PSf/NMP/PVP on flux and rejection. *Desalination* **2005**, *179*, 257–263. [CrossRef]
15. Yunos, M.Z.; Harun, Z.; Basri, H.; Ismail, A.F. Studies on fouling by natural organic matter (NOM) on polysulfone membranes: Effect of polyethylene glycol (PEG). *Desalination* **2014**, *333*, 36–44. [CrossRef]
16. Zhao, B.; Yang, S.; Zhao, S.; Wang, Z.; Wei, X.; Wang, J.; Wang, S. Performance Improvement of Polysulfone Ultrafiltration Membrane Using Well-Dispersed Polyaniline–Poly(vinylpyrrolidone) Nanocomposite as the Additive. *Ind. Eng. Chem.* **2012**, *51*, 4661–4672. [CrossRef]
17. Noorani, S.; Simonsen, J.; Atre, S. Nano-enabled microtechnology: polysulfone nanocomposites incorporating cellulose nanocrystals. *Cellulose* **2007**, *14*, 577–584. [CrossRef]
18. Bai, H.; Wang, X.; Sun, H.; Zhang, L. Permeability and morphology study of polysulfone composite membrane blended with nanocrystalline cellulose. *Desalin. Water Treat.* **2015**, *53*, 2882–2896. [CrossRef]
19. Daraei, P.; Ghaemi, N.; Ghari, H.S.; Norouzi, M. Mitigation of fouling of polyethersulfone membranes using an aqueous suspension of cellulose nanocrystals as a nonsolvent. *Cellulose* **2016**, *23*, 2025–2037. [CrossRef]
20. Yang, X.; Liu, H.; Zhao, Y.; Liu, L. Preparation and characterization of polysulfone membrane incorporating cellulose nanocrystals extracted from corn husks. *Fibers Polym.* **2016**, *17*, 1820–1828. [CrossRef]
21. Zhou, Y.; Zhao, H.; Bai, H.; Zhang, L.; Tang, H. Papermaking Effluent Treatment: A New Cellulose Nanocrystalline/Polysulfone Composite Membrane. *Procedia Environ. Sci.* **2012**, *16*, 145–151. [CrossRef]
22. Li, S.; Gao, Y.; Bai, H.; Zhang, L.; Qu, P.; Bai, L. Preparation and characteristics of polysulfone dialysis composite membranes modified with nanocrystalline cellulose. *BioResources* **2011**, *6*, 1670–1680.
23. Daria, M.; Fashandi, H.; Zarrebini, M.; Mohamadi, Z. Contribution of polysulfone membrane preparation parameters on performance of cellulose nanomaterials. *Mater. Res. Express* **2019**, *6*, 015306. [CrossRef]
24. Daraei, P.; Ghaemi, N.; Sadeghi Ghari, H. An ultra-antifouling polyethersulfone membrane embedded with cellulose nanocrystals for improved dye and salt removal from water. *Cellulose* **2017**, *24*, 915–929. [CrossRef]
25. Zhong, L.; Ding, Z.; Li, B.; Zhang, L. Preparation and Characterization of Polysulfone/Sulfonated Polysulfone/Cellulose Nanofibers Ternary Blend Membranes. *BioResources* **2015**, *10*, 2936–2948. [CrossRef]
26. Qu, P.; Tang, H.; Gao, Y.; Zhang, L.P.; Wang, S. Polyethersulfone composite membrane blended With cellulose fibrils. *BioResources* **2010**, *5*, 2323–2336.
27. Zhang, W.; Zhong, L.; Wang, T.; Jiang, Z.; Gao, X.; Zhang, L. Surface modification of cellulose nanofibers and their effects on the morphology and properties of polysulfone membranes. *IOP Conf. Ser. Mater. Sci. Eng.* **2018**, *397*, 012016. [CrossRef]
28. Vilakati, G.D.; Hoek, E.M.; Mamba, B.B. Probing the mechanical and thermal properties of polysulfone membranes modified with synthetic and natural polymer additives. *Polym. Test.* **2014**, *34*, 202–210. [CrossRef]
29. Zhang, X.; Benavente, J.; Garcia-Valls, R. Lignin-based membranes for electrolyte transference. *J. Power Sources* **2005**, *145*, 292–297. [CrossRef]
30. Ding, Z.; Liu, X.; Liu, Y.; Zhang, L.; Li, K. Enhancing the Compatibility, Hydrophilicity and Mechanical Properties of Polysulfone Ultrafiltration Membranes with Lignocellulose Nanofibrils. *Polymers* **2016**, *8*, 349. [CrossRef]
31. Hassan, M.; Berglund, L.; Hassan, E.; Abou-Zeid, R.; Oksman, K. Effect of xylanase pretreatment of rice straw unbleached soda and neutral sulfite pulps on isolation of nanofibers and their properties. *Cellulose* **2018**, *25*, 2939–2953. [CrossRef]
32. Taglieri, G.; Mondelli, C.; Daniele, V.; Pusceddu, E.; Trapananti, A. Synthesis and X-Ray Diffraction Analyses of Calcium Hydroxide Nanoparticles in Aqueous Suspension. *Adv. Mater. Phys. Chem.* **2013**, *3*, 108–112. [CrossRef]
33. Yu, H.; Zhang, Y.; Sun, X.; Liu, J.; Zhang, H. Improving the antifouling property of polyethersulfone ultrafiltration membrane by incorporation of dextran grafted halloysite nanotubes. *Chem. Eng. J.* **2014**, *237*, 322–328. [CrossRef]

34. Hassan, M.; Berglund, L.; Abou-Zeid, R.; Hassan, E.; Abou-Elseoud, W.; Oksman, K. Nanocomposite Film Based on Cellulose Acetate and Lignin-Rich Rice Straw Nanofibers. *Materials* **2019**, *12*, 595. [CrossRef] [PubMed]
35. Desgranges, L.; Grebille, D.; Calvarin, G.; Chevrier, G.; Floquet, N.; Niepce, J.-C. Hydrogen thermal motion in calcium hydroxide: Ca(OH)$_2$. *Acta Crystallogr. Sect. B Struct. Sci.* **1993**, *49*, 812–817. [CrossRef]
36. Zhang, L.; Chen, G.; Tang, H.; Cheng, Q.; Wang, S. Preparation and characterization of composite membranes of polysulfone and microcrystalline cellulose. *J. Appl. Sci.* **2009**, *112*, 550–556. [CrossRef]

© 2019 by the authors. Licensee MDPI, Basel, Switzerland. This article is an open access article distributed under the terms and conditions of the Creative Commons Attribution (CC BY) license (http://creativecommons.org/licenses/by/4.0/).

Article

Poly(3-hydroxybutyrate) Modified by Plasma and TEMPO-Oxidized Celluloses

Denis Mihaela Panaitescu [1,*], Sorin Vizireanu [2,*], Sergiu Alexandru Stoian [1,3], Cristian-Andi Nicolae [1], Augusta Raluca Gabor [1], Celina Maria Damian [3], Roxana Trusca [4], Lavinia Gabriela Carpen [2,5] and Gheorghe Dinescu [2,5]

[1] Polymer Department, National Institute for Research and Development in Chemistry and Petrochemistry, 202 Spl. Independentei, 060021 Bucharest, Romania; stoian.sergiu@gmail.com (S.A.S.); ca_nicolae@yahoo.com (C.-A.N.); ralucagabor@yahoo.com (A.R.G.)
[2] National Institute for Laser, Plasma and Radiation Physics, Atomistilor 409, Magurele-Bucharest, 077125 Ilfov, Romania; lavinia.carpen@infim.ro (L.G.C.); dinescug@infim.ro (G.D.)
[3] Advanced Polymers Materials Group, University Politehnica of Bucharest, 1-7 Polizu Street, 011061 Bucharest, Romania; c.damian@tsocm.pub.ro
[4] Science and Engineering of Oxide Materials and Nanomaterials, University Politehnica of Bucharest, 1-7 Gh. Polizu Street, 011061 Bucharest, Romania; truscaroxana@yahoo.com
[5] Faculty of Physics, Bucharest University, 405 Atomistilor Street, Magurele-Bucharest, 077125 Ilfov, Romania
* Correspondence: panaitescu@icechim.ro (D.M.P.); s_vizi@infim.ro (S.V.)

Received: 22 June 2020; Accepted: 6 July 2020; Published: 7 July 2020

Abstract: Microcrystalline cellulose (MCC) was surface modified by two approaches, namely a plasma treatment in liquid using a Y-shaped tube for oxygen flow (MCC-P) and a TEMPO mediated oxidation (MCC-T). Both treatments led to the surface functionalization of cellulose as illustrated by FTIR and XPS results. However, TEMPO oxidation had a much stronger oxidizing effect, leading to a decrease of the thermal stability of MCC by 80 °C. Plasma and TEMPO modified celluloses were incorporated in a poly(3-hydroxybutyrate) (PHB) matrix and they influenced the morphology, thermal, and mechanical properties of the composites (PHB-MCC-P and PHB-MCC-T) differently. However, both treatments were efficient in improving the fiber–polymer interface and the mechanical properties, with an increase of the storage modulus of composites by 184% for PHB-MCC-P and 167% for PHB-MCC-T at room temperature. The highest increase of the mechanical properties was observed in the composite containing plasma modified cellulose although TEMPO oxidation induced a much stronger surface modification of cellulose. This was due to the adverse effect of more advanced degradation in this last case. The results showed that Y-shaped plasma jet oxidation of cellulose water suspensions is a simple and cheap treatment and a promising method of cellulose functionalization for PHB and other biopolymer reinforcements.

Keywords: cellulose; plasma in liquids; polyhydroxyalkanoate; polymer composites; thermal properties; DMA

1. Introduction

Biopolymers are seen as a viable alternative to synthetic polymers in many applications and they are unparalleled in the medical field [1–3]. Poly(3-hydroxybutyrate) (PHB) is an aliphatic polyester obtained by microbial synthesis and, for the moment, is the most affordable of polyhydroxyalkanoates (PHAs) family. The known shortcomings of PHB are the high cost and brittleness along with the low thermal stability during melt processing [4,5]. Adding plasticizers, elastomers, and organic or inorganic fillers are the most used strategies to improve PHB properties [6–8]. Among the fillers, micro- and nano-celluloses were most intensely studied to modify PHB and other aliphatic polyesters [9–12].

Microfibrillated cellulose is obtained from cellulose, a sustainable and low-cost naturally occurring biopolymer. This is the most abundant organic polymer on Earth and comprises 33% of all the plants on the planet [13,14]. Microfibrillated cellulose is biodegradable and biocompatible and has reduced carbon dioxide emissions in the environment. It shows an adaptable surface chemistry and a unique spectrum of properties: high water-uptake capability, low density, high crystallinity and Young's modulus, and good thermal stability [12–14]. The research, production, and application of microfibrillated cellulose have experienced a real boost in the last twenty years. It can be extracted from a multitude of sources, such as wood, plants, algae, by-products, and waste of the agro-food industry by mechanical disintegration [13–15]. Refining, homogenization, high power ultrasonication, or microfluidization, with or without additional enzymatic or chemical pretreatments are generally the most used methods to obtain long flexible microfibrils [12–15].

TEMPO-mediated oxidation of cellulose consists of surface oxidation of cellulose in aqueous suspension using sodium hypochlorite as oxidant and 2,2,6,6-tetramethylpiperidine-1-oxyl radical (TEMPO) and sodium bromide as catalysts at an alkaline pH [16–18]. This pre-treatment proved to be a successful method for ensuring the selective conversion of C6 hydroxyl group of cellulose to carboxylate group. Previous studies have shown the great advantage of this pre-treatment which reduces the energy consumption and the number of cycles in the subsequent stages of mechanical defibrillation [18,19].

Recently, plasma treatment was found to be efficient for the defibrillation and functionalization of cellulose [20,21]. Plasma treatment was mainly studied to improve the surface of cellulose textiles or membranes but, in certain conditions, plasma treatment may ensure the defibrillation and chemical functionalization of cellulose [21]. In particular, a plasma torch immersed in the water suspension of cellulose in the presence of different reactive gases induced physical and chemical changes of cellulose [21] and is proposed as an alternative to the mechano-chemical methods usually employed for the modification of cellulose and fabrication of nanocellulose. Moreover, plasma treated cellulose showed a good reinforcing effect in PHB/cellulose composites, increasing both the Young's modulus and tensile strength [21].

Plasma treatment of nanocellulose water suspensions using a filamentary plasma jet based on dielectric barrier discharge (DBD) proved to be very advantageous in the surface functionalization of nanocellulose [20]. Only 0.2 wt % of filamentary jet plasma functionalized nanocellulose increased the tensile strength and Young's modulus of PHB nanocomposite by 10–15% compared to the nanocomposite with the same amount of untreated nanocellulose. The surface functionalization of cellulose in water suspensions using DBD plasma is an environmentally friendly, simple, cheap, and low-energy consuming treatment that overcomes the disadvantages of complex, time-consuming, and energy-intensive chemical-mechanical treatments. Moreover, toxic products that must be deactivated are generally released from chemical treatments. However, DBD filamentary plasma sources cannot work with higher fluxes of reactive gases. Our preliminary attempts have shown that a flux of reactive gas (O_2 or NH_3) greater than 20 sccm (3.33×10^{-7} standard cubic meter per second) led to the extinction of the discharge submersed in the water suspensions of cellulose. For this reason, in this work, the DBD plasma source was modified for the addition of the reactive gas laterally in the discharge. This new method makes it possible to work with much higher fluxes of oxygen or other reactive gases, even greater than 500 sccm.

In this study, a microcrystalline cellulose (MCC) was treated with a filamentary Y-shaped DBD plasma jet generated in argon-oxygen gas mixture and immersed in the liquid suspension of cellulose. For a comparison with chemical treatments, MCC was chemically modified using TEMPO mediated oxidation, an eco-friendlier chemical method compared to acid hydrolysis. The changes induced by the two treatments (plasma and TEMPO) were highlighted by Fourier transform infrared (FTIR) spectroscopy, X-ray photoelectron spectroscopy (XPS), and thermogravimetric analysis (TGA). Plasma and TEMPO modified celluloses (MCC-P and MCC-T) were used to reinforce a PHB matrix and the composites were characterized by thermal analyses, dynamic mechanical analysis, and scanning electron microscopy (SEM).

2. Materials and Methods

2.1. Materials

Microcrystalline cellulose (MCC) was purchased from Sigma-Aldrich (Saint Louis, MO, USA). MCC is rod-shaped and has a mean diameter of 20 µm, a bulk density of 0.5 g/cm^3 and a small aspect ratio. PHB type P304 from Biomer (Schwalbach am Taunus, Germany) was used to prepare the composites. A tensile strength of 28 MPa, determined according to ISO 527:2012 standard with 50 mm/min [22], was provided for PHB type P304 by the producer. Acetonitrile 99% (ACN) purchased from Fluka Chemie AG (Buchs, Switzerland) was used as received. Further, 2,2,6,6-tetramethylpiperidine-1-oxyl radical, 98% (TEMPO) was purchased from Sigma-Aldrich (Saint Louis, MO, USA), sodium hypochlorite 10% from Oltchim (Rm-Valcea, Romania), sodium bromide, hydrochloric acid, and sodium hydroxide, all of analytical grade, were obtained from Merck KGaA (Darmstadt, Germany).

2.2. Plasma Treatment of Cellulose Suspensions

The cellulose suspension (5 wt %) was prepared by dispersing 5 g MCC in 95 mL of distilled water by ultrasonication using a Elmasonic S 15 H bath (Elma, Singen, Germany) for 1 h. The treatment of MCC suspensions was carried out using a DBD plasma source with floating electrode, coupled with a radiofrequency power supply [23,24]. The discharge tube was modified in a Y-shape geometry (Figure 1). The Y-shaped configuration allows for the introduction of large amounts of reactive gases without extinguishing the discharge.

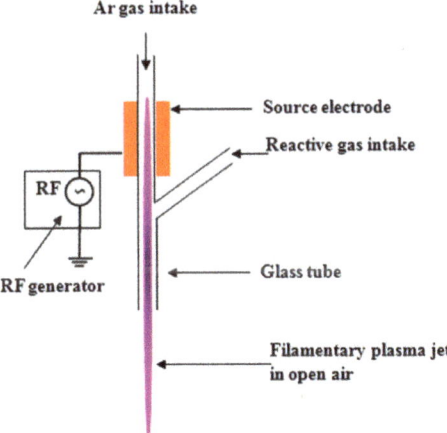

Figure 1. Configuration of the DBD Y-shaped source.

The plasma jet was initiated in argon (RF power 100 W, flow rate 3000 sccm), in open ambient conditions, and then immersed in the water suspension of MCC. Sccm, standard cubic centimeters per minute, is a measure for the flow rate of particles, 1 sccm denotes that in one minute in the chamber entered a number of particles corresponding to those existing in a volume of 1 cubic centimeter at standard temperature (273 K) and pressure (1 atm). An oxygen flux of 500 sccm was introduced through the lateral branch of the Y-tube and the discharge was maintained for 1 h of treatment. Then, ACN (30%) was mixed in the water suspension of MCC for ensuring better conditions for surface functionalization of cellulose and the suspension was plasma treated with Ar for 30 min using the same source. Plasma treated MCC suspension was ultrasonicated for 30 min and then freeze dried for 72 h using a FreeZone 2.5 L Benchtop Freeze Dry System (Labconco, Kansas City, MO, USA).

2.3. TEMPO Mediated Oxidation of MCC

MCC was oxidized using 2,2,6,6-tetramethylpiperidine-1-oxyl radical as a catalyst and sodium hypochlorite as an oxidant. Five grams of MCC were dispersed in 195 mL of distilled water by magnetic stirring at room temperature for 24 h. TEMPO (0.08 g corresponding to 0.1 mmol/g cellulose) was dispersed in 100 mL of distilled water until complete dissolution using an Elmasonic S40 ultrasonic bath (Elma, Germany). Meanwhile, 0.52 g NaBr (1.0 mmol/g cellulose) were added over the TEMPO solution and dispersed using the same bath. The TEMPO/NaBr solution was added to the cellulose suspension in small portions under magnetic stirring. Then, 10% NaClO solution (29.80 g corresponding to 8.0 mmol/g cellulose) was added dropwise for half an hour and stirred for 3h from the moment of the first drop of NaClO. The pH of the solution was kept constant (pH 10–11) during the reaction by the addition of 1M solutions of HCl and NaOH. After the completion of the reaction, the TEMPO treated cellulose suspension was dialyzed using Spectra/Por molecular porous membrane (MWCO 3.5 kDa, SpectrumLabs (Rancho Dominguez, CA USA) for removing the unreacted products. After the treatments, the MCC suspension was ultrasonicated for 30 min and then freeze dried for 72 h using a FreeZone 2.5 L Benchtop Freeze Dry System (Labconco, Kansas City, MO, USA). Plasma treated and TEMPO oxidized MCC were denoted as MCC-P and MCC-T.

2.4. Preparation of PHB Composites with Untreated, Plasma and TEMPO Oxidized MCC

PHB pellets, MCC-P and MCC-T were dried in vacuum ovens at 60 °C for 4 h. Different treated MCC (2 wt %) was melt blended with PHB using a Brabender LabStation (Duisburg, Germany) with a mixing chamber of 30 cm^3, at 165 °C for 8 min at a rotor speed of 50 min^{-1}. For morphological and mechanical characterization, composites sheets of 0.5 mm in thickness were obtained by compression molding using an electrically heated press (P200E, Dr. Collin, Maitenbeth, Germany). The composites were pressed at 175 °C for 120 s preheating (without pressure), 75 s under pressure (100 bar), and cooling for 1 min in a cooling cassette. The sheets were kept at room temperature for at least 2 weeks before characterization. PHB composites with untreated MCC, MCC-T, and MCC-P were denoted as PHB-MCC, PHB-MCC-T, and PHB-MCC-P. The schematic representation of operations for obtaining PHB composites with different treated MCC is shown in Figure 2.

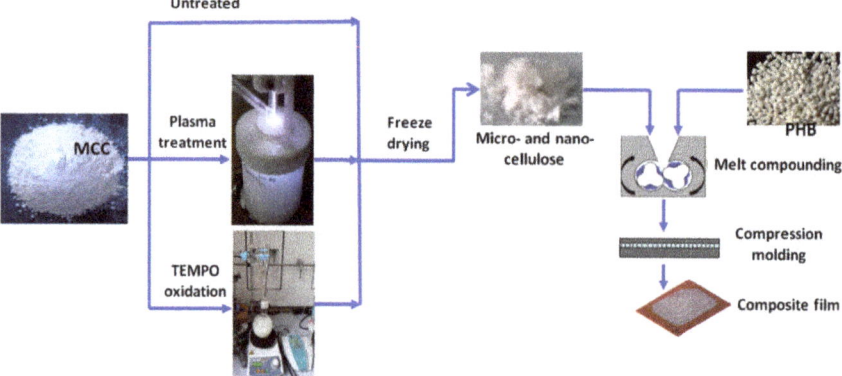

Figure 2. The sequence of main operations for the fabrication of PHB composites with untreated, plasma treated and TEMPO oxidized MCC.

2.5. Characterization

2.5.1. SEM Investigation

The compression molded sheets of PHB–MCC composites were fractured in liquid nitrogen and then sputter-coated with a thin layer of gold. The surface of fractured composites was analyzed by SEM using a Quanta Inspect F scanning electron microscope (FEI-Philips, Hillsboro, OR, USA) equipped with a field emission gun. The microscope was operated in high vacuum at 30 kV with a resolution of 1.2 nm.

2.5.2. Chemical Characterization by FTIR and XPS

The FTIR spectra of untreated, plasma, and TEMPO treated cellulose were recorded on a Tensor 37 spectrometer from Bruker Optics (Ettlingen, Germany) with an attenuated total reflectance (ATR) accessory. The spectra were collected in duplicate, at room temperature, from 400 to 4000 cm^{-1}, with 16 scans at a resolution of 4 cm^{-1}. Plasma treated and TEMPO oxidized cellulose suspensions were dropwise deposited on silicon wafers and dried several days at room temperature to obtain films. The surface of cellulose films was characterized by X-ray photoelectron spectroscopy (XPS) using ESCALAB™ XI+ spectrometer (Thermo Scientific, Waltham, MA, USA) with a monochromatic Al Kα source at 1486.6 eV. Both survey spectra (step of 1 eV) at a pass energy of 100 eV and high-resolution spectra in the C 1s and O 1s regions (step of 0.1 eV) at pass energy of 20 eV were recorded.

2.5.3. Thermal Characterization

Thermogravimetric analysis (TGA) was used to characterize the thermal stability of MCC samples before and after the treatments and the thermal behavior of PHB composites with different treated celluloses. TGA was carried out on duplicate cellulose samples sealed in aluminium pans using a SDT Q600 V20.9 (TA Instruments Inc., New Castle, DE, USA) with helium as the purge gas (100 mL/min). TGA measurements of PHB composites were carried out on a TGA Q500 V20.10 (TA Instruments Inc., New Castle, DE, USA) in platinum pans, using nitrogen as purge gas at a flow rate of 90 mL/min. All the samples, celluloses or composites, were heated from 25 to 700 °C at a heating rate of 10 °C/min.

2.5.4. Differential Scanning Calorimetry (DSC)

DSC measurements were carried out on composites using a DSC Q2000 V24.9 from TA Instruments (New Castle, DE, USA) under helium flow (30 mL/min). Samples of 6–7 mg were heated from 30 to 200 °C and equilibrated for 3 min for erasing the thermal history (first heating cycle), cooled down to −50 °C, isothermal for 2 min (cooling cycle) and reheated to 200 °C (second heating cycle). A heating/cooling rate of 10 °C/min was used in all cycles. The degree of crystallinity (X_c) was determined from the second heating cycle by dividing the total melting enthalpy ΔH_m (of the peak and shoulder) by the amount of PHB in the composites (W_{PHB}) and ΔH_0, the melting enthalpy of 100% crystalline PHB (146 J/g [25]):

$$X_c(\%) = \frac{\Delta H_m}{\Delta H_0} \cdot \frac{100}{W_{PHB}} \tag{1}$$

2.5.5. DMA Characterization

The dynamic mechanical properties of composites were determined using a DMA Q800 (TA Instruments, New Castle, DE, USA) in multi-frequency-strain mode. Bar specimens of 12 × 6 × 0.5 mm^3 were cut from the composite sheets and cooled to −35 °C, kept isothermally at this temperature for 5 min, and heated to 165 °C with a heating rate of 3 °C/min. Storage modulus, showing the recoverable stored energy and tan δ or damping factor, which is the ratio of the loss modulus to the storage modulus, were plotted against temperature.

3. Results and Discussion

3.1. Characterization of MCC, MCC-P and MCC-T

The changes induced in cellulose by the two treatments (plasma and TEMPO) were highlighted by TGA, FTIR, and XPS.

3.1.1. Effect of the Treatments on the Thermal Stability of MCC

TGA and derivative thermogravimetric (DTG) curves of MCC, MCC-T, and MCC-P are shown in Figure 3. Pure MCC showed one major degradation step with the onset degradation temperature (T_{on}) at 309.3 °C and the temperature of the maximum degradation rate (T_d) at 331.5 °C, similar to other observations [21,26]. The highest loss of weight was noticed between 280 and 360 °C, when the reactions of decomposition, depolymerization, and decarboxylation overlap [26].

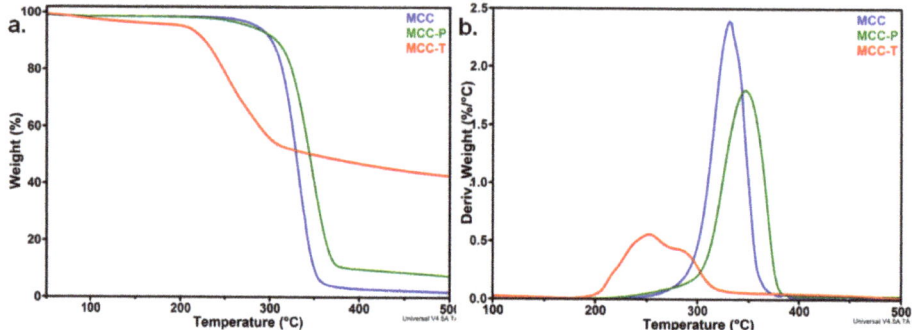

Figure 3. TGA (**a**) and DTG curves (**b**) for microcrystalline cellulose (MCC), TEMPO oxidized cellulose (MCC-T) and plasma treated cellulose (MCC-P).

The thermal stability of MCC was significantly influenced by TEMPO-mediated oxidation and only small changes were observed after the plasma treatment (Figure 3). The degradation started at a much lower temperature of 222 °C in the case of MCC-T instead of 309 °C for pristine MCC. Similarly, the T_d decreased by 80 °C, from 331.5 to 251.9 °C. These changes are due to the oxidation reactions induced by the TEMPO treatment [27,28]. A similar decrease of the onset degradation temperature, with about 100 °C, was reported by Fukuzumi et al. in the case of TEMPO oxidized bleached kraft pulp fibers and explained by the formation of sodium carboxylate groups at the C6 primary hydroxyls on the surface of cellulose crystals and in the disordered regions [27]. Saito and Isogai reported an important decrease of the degree of polymerization (DP) in the case of TEMPO-modified cellulose cotton linter [16]. The decrease of DP, related to a lower average molecular weight and a lower thermal stability, was explained by the oxidation of all hydroxyl groups of cellulose and the formation of C6 carboxylate or aldehyde groups and C2 or C3 ketones [29]. The high broadness of the DTG peak in the case of MCC-T shows the overlap of at least two degradation processes. The first peak, at about 222 °C, was determined by the degradation of more unstable anhydroglucuronate units resulting from TEMPO-oxidation, and the second at a temperature of 285 °C, lower than the T_d of unmodified cellulose, may be ascribed to the decomposition of unmodified or slightly oxidized cellulose influenced by the adjacent sodium carboxylate groups [28].

A completely different behavior was observed after the plasma treatment of MCC (Figure 3): T_{on} increased by 9 °C and T_d by 15 °C compared to untreated MCC. A similar increase of the maximum degradation temperature was reported for nanocellulose treated with a DBD plasma source in Ar or Ar/N_2 flow [20]. The better thermal stability of MCC-P compared to untreated MCC resulted from the removal of low molecular weight fractions or impurities from the MCC suspended in water under the influence of submerged plasma discharge [20]. The different charred residue values at 500 °C (R_{500}), 7.6% for MCC-P, 42.7% for MCC-T, and only 1.9% for MCC, also suggest different degradation processes after the treatments. Cellulose decomposed almost completely in inert atmosphere and the residue is very low for untreated MCC [29]. The large charred residue value of MCC-T, 22 times higher than that of MCC, may be related to the decomposition products of sodium anhydroglucuronate units. Furthermore, carboxylate, aldehyde, or ketone groups of MCC-T may be involved in a great number of condensation, grafting, or crosslinking reactions, resulting in an important amount of highly condensed polycyclic aromatic structures [21,30]. Moreover, the decreased DP supposed in the case of MCC-T may lead to an increased number of chain ends, which decompose at a lower temperature giving an increased amount of char [31]. Therefore, the smaller R_{500} value of MCC-P compared to MCC-T, only four times higher than that of MCC, may be related to a much lower amount of oxidized cellulose after the plasma treatment. FTIR analysis may give more information on the chemical modifications induced by the treatments on the surface of cellulose.

3.1.2. Surface Chemical Properties of Modified Celluloses

The FTIR spectra of different treated MCC are shown in Figure 4a. FTIR spectrum of MCC shows the characteristic vibrations of cellulose: (i) 3000–3600 cm^{-1} due to the O–H stretching vibrations of hydrogen bonded hydroxyl groups, (ii) 2800–3000 cm^{-1} due to asymmetric and symmetric stretching vibrations of C–H, (iii) at 1645 cm^{-1} due to the O–H bending vibration in bound water, (iv) multiple bands between 1275 and 1435 cm^{-1} due to the bending and deformation vibrations of C–H and CH$_2$, (v) at 1162 and 897 cm^{-1} due to the C–O–C stretching vibrations at the β-glucosidic linkage, and (vi) at 1054 and 1030 cm^{-1} due to the C–O stretching vibrations [26,32,33].

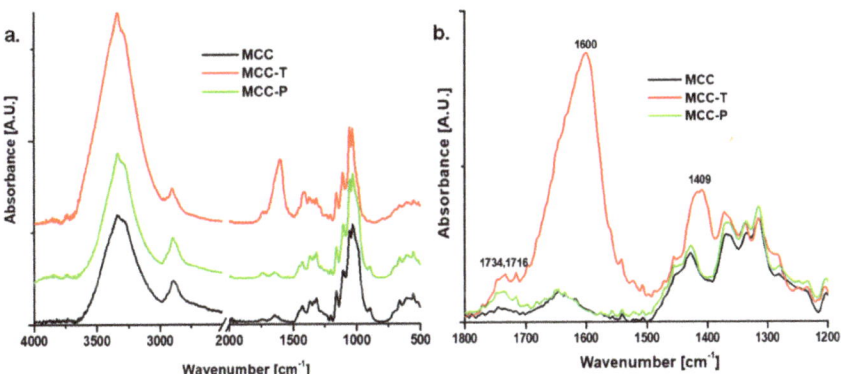

Figure 4. FTIR spectra of MCC, MCC-T and MCC-P (**a**); FTIR spectra in the range from 1800 to 1200 cm^{-1} (**b**).

Major changes were noticed in the FTIR spectrum of MCC-T (Figure 4). A new and intense peak appeared at 1600 cm^{-1} which is characteristic of asymmetric C=O stretching vibrations in COO$^-$ of sodium carboxylate [34,35]. A new and broad band appeared at 1409 cm^{-1} due to the symmetric stretching vibrations of COO^{-1} [34]. In addition, two new bands were noticed in MCC-T at 1716 and 1734 cm^{-1} (Figure 4b), which may be associated to the C=O stretching vibrations in carboxylic acid [27,36,37].

Small changes were observed between the FTIR spectra of MCC-P and untreated MCC (Figure 4a,b). A broad band at 1737 cm^{-1} and a second band at 1716 cm^{-1} were noticed in MCC-P similar to MCC-T, and they are assigned to the C=O stretching vibrations in carboxylic acid. The appearance of the new bands around 1734/1737 cm^{-1} indicates the oxidation of MCC by both TEMPO and plasma treatments [36]. The degree of oxidation (DO) may be estimated from the FTIR spectra by comparing the intensity of the band near 1735 cm^{-1} with that of the band at 1050 cm^{-1} which comes from the cellulose backbone [36]. Different DO values were obtained for MCC-T and MCC-P, 0.06% and 0.12%. They correlate with the different oxidation intensity induced by the treatments. Similar DO values were reported for TEMPO-mediated oxidation of cellulose whiskers and cellulose from pulp residue in comparable reaction conditions [34,36].

For a better understanding of the chemical changes on the surface of MCC after the treatments, untreated and treated cellulose was also characterized by XPS. Survey spectra are shown in Figure 5. XPS survey scans indicate the presence of carbon at 286.3 eV and oxygen at 533.3 eV in all samples and, in addition, chlorine (199.3 eV) and sodium (1072.3 eV) [38] in MCC-T. The O KLL Auger peak was also noticed in all survey spectra at 978 eV. The O/C ratio differed slightly before and after treatments (Table 1) and it was close to the theoretical ratio for pure cellulose (0.83). It can be supposed that NaCl is present as an impurity from the TEMPO treatment, however the exceeding Na could come from sodium carboxylate, also highlighted by FTIR.

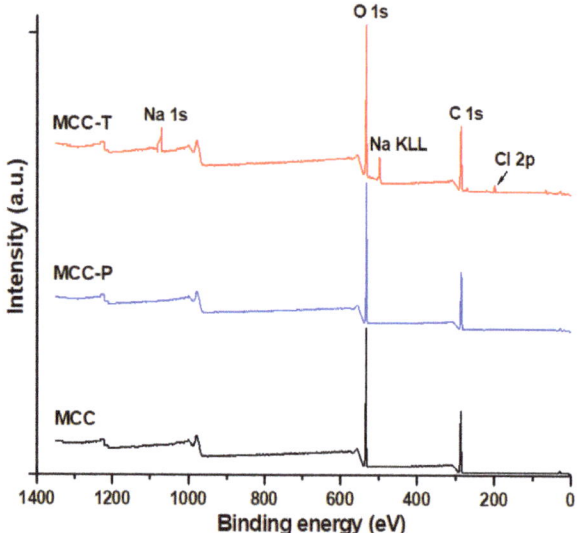

Figure 5. XPS survey spectra of different treated MCC.

Table 1. The elemental surface composition, in atomic %, from XPS survey spectra before (MCC) and after the treatments (MCC-T and MCC-P).

Samples	C1s (%)	O1s (%)	Na1s (%)	Cl 2p (%)	O/C
MCC	57.3	42.7	-	-	0.75
MCC-T	52.8	39.4	5.3	2.5	0.75
MCC-P	56.9	43.1	-	-	0.76

High resolution XPS spectra for C1s and O1s regions are shown in Figure 6. The C1s peak was fitted with 3 components in the case of MCC and 4 components for treated cellulose (Table 2): C_1 corresponding to C–C or C–H at a binding energy of 284.6 eV, C_2 assigned to a carbon bonded to a single oxygen in alcohol or ether groups (C–O) at a binding energy of 286.3 eV, C_3 ascribed to O–C–O or C=O at 288.0 eV, and C_4 in O–C=O at 289.7 eV [39,40]. The C3/C2 ratio, 0.29, 0.41 and 0.31 for MCC, MCC-T and MCC-P, is higher than the theoretical value (0.2), showing the presence of carbonyl groups and a higher degree of oxidation, especially for MCC-T. In contrast to MCC, both treated celluloses contain C_4 species (i.e., carboxyl or carboxylate groups) but MCC-T in a higher proportion. In general, the TEMPO treatment of cellulose led to a much higher degree of oxidation than the plasma one, which is obvious from the higher increase of C_3 and C_4 proportion at the expense of C_2 in this case. Similar trends were reported by Coseri et al. for differently oxidized microcrystalline cellulose [38].

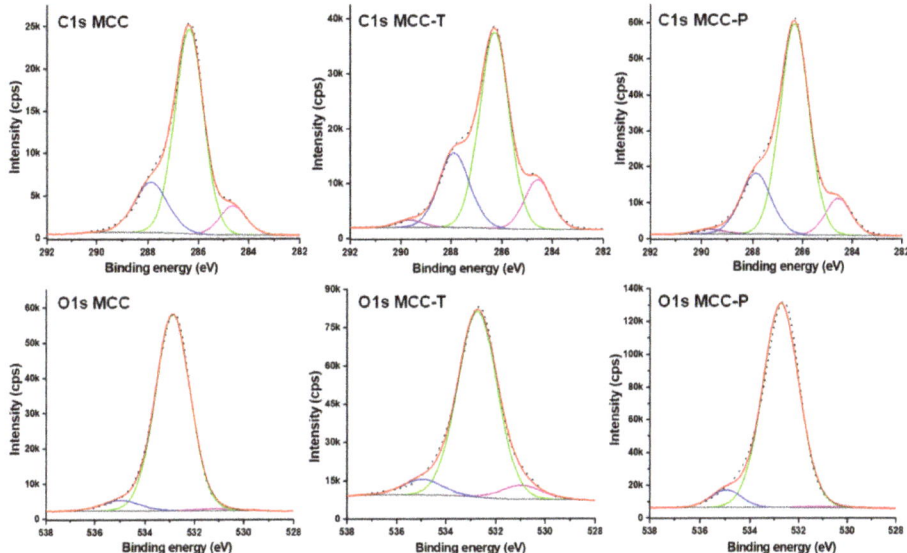

Figure 6. High resolution XPS spectra and deconvolution in the C1s and O1s regions.

Table 2. The relative atomic concentrations of differently bound carbon and oxygen atoms as determined from high resolution XPS spectra.

Samples	C1s				O1s		
	C_1 (%)	C_2 (%)	C_3 (%)	C_4 (%)	O_1 (%)	O_2 (%)	O_3 (%)
MCC	9.4	70.3	20.3	-	1.1	93.4	5.5
MCC-T	14.0	59.6	24.3	2.1	6.7	85.7	7.6
MCC-P	10.5	67.4	20.6	1.4	0.6	92.7	6.8

Some more information can be obtained from the deconvolution of O1s in high resolution XPS spectra (Figure 6). To O1s region contributed: O_1, double bonded O (C=O) in sodium carboxylate and other oxidized groups (aldehydes, ketones) at 531.0 eV, O_2, single bonded O from C-OH and C–O–C of pyranose ring at 532.9 eV and O_3, which can be assigned to carboxylic acid functional groups and water, at 535.0 eV [40–42]. The relative atomic concentrations of these components (Table 2) undoubtedly show that the plasma treatment induced a small increase of carboxylic groups on cellulose, ensuring a slight oxidation while the TEMPO treatment determined a significant oxidation of cellulose, especially as carboxylate and carboxylic groups. The cellulose surface changes observed by XPS are in good agreement and strengthen the results obtained by FTIR.

3.2. Thermal Properties of PHB Composites with Different Treated MCC

Untreated and surface treated celluloses were used as reinforcement in PHB. Figure 7 shows the TGA and derivative (DTG) curves of PHB composites containing 2 wt % MCC, MCC-T, and MCC-P. Although the effect of different treated MCC on the thermal stability of PHB is similar to the effect of the treatments on cellulose (Figure 3), meaning an increased stability for MCC-P and a decreased one for MCC-T, the difference between the characteristic temperatures of the composites is much attenuated (Table 3).

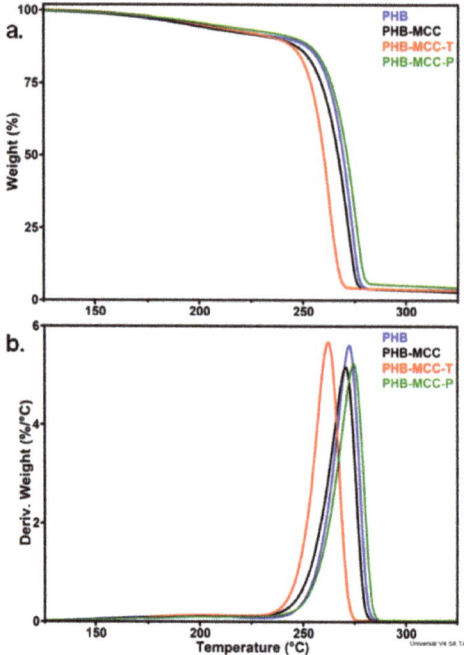

Figure 7. TGA (a) and DTG (b) curves of PHB composites with untreated and treated MCC.

Table 3. TGA results for PHB composites with different treated MCC.

Characteristic Temperatures	PHB	PHB-MCC	PHB-MCC-T	PHB-MCC-P
$T_{5\%}$, °C	195.6	195.7	200.9	206.8
T_{on}, °C	262.6	259.4	253.7	263.6
T_d, °C	272.3	270.6	262.1	274.3

The addition of MCC in PHB decreases the T_{on} and T_d, with only 2–3 °C, which is a common effect of cellulose fillers in PHB [7]. One cause of the increased degradation was supposed to be the higher thermal conductivity of cellulose compared to PHB [12]. Impurities, such as increased melt viscosity after the incorporation of cellulose in the polymer matrix or the presence of bond water in cellulose, were also considered as sources of an enhanced thermal degradation of PHB after the addition of cellulose [7].

The addition of MCC-T decreased the T_{on} and T_d of PHB composites with 9–10 °C. The presence of oxidized groups on the surface of MCC-T may enhance not only the thermal degradation of cellulose but also the degradation of PHB because they may catalyze the depolymerization reaction of PHB by interacting with the carboxyl group and favoring the chain scission reaction [43]. Considering that the melt processing took place at about 170 °C, the effect of oxidized compounds might be much enhanced at this temperature. Moreover, XPS survey spectra (Figure 5) and FTIR showed the presence of Na as sodium carboxylate on the surface of MCC-T which may catalyze the depolymerization reaction of PHB. Indeed, previous study has shown that many metal compounds but especially Na and Mg accelerate the random chain scission and the degradation of PHB [44].

In contrast to MCC and MCC-T, the addition of MCC-P slightly increased the T_{on} and T_d of PHB composites with 1–2 °C. Plasma treated MCC led to a better thermal behavior of PHB-MCC-P composite because of the mild oxidation of cellulose following the plasma treatment in contrast to the harsh TEMPO oxidation, also illustrated by the FTIR results. Moreover, the cleaning of cellulose during the submerged plasma treatment removes metals or other impurities which may catalyze the decomposition of PHB at high temperature [44]. It is remarkable that the thermal stability of composites at processing temperature, illustrated by the temperature at 5% weight loss ($T_{5\%}$), was much improved in PHB composites containing treated MCC (Table 3). An increase of $T_{5\%}$ with about 11 °C in the case of PHB-MCC-P and 5 °C in the case of PHB-MCC-T was noticed.

The DSC thermograms during the cooling and second heating cycles are shown in Figure 8 and the corresponding thermal parameters are listed in Table 4. Important differences between the crystallization behaviors of composites with different treated MCC were noticed in the cooling cycle (Figure 8a).

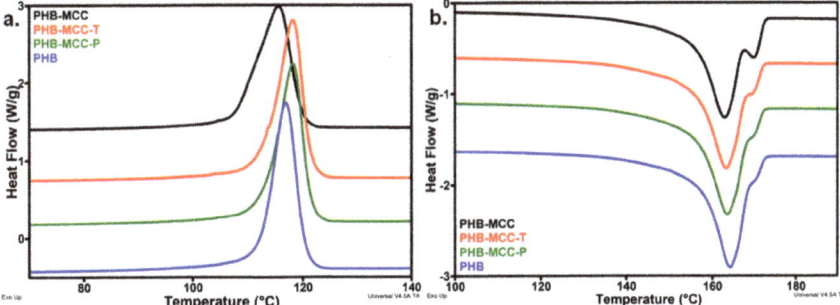

Figure 8. DSC thermograms of PHB and composites for the first (**a**) and second heating (**b**).

Table 4. DSC data collected from the heating-cooling-heating DSC curves of composites.

Composites	PHB	PHB-MCC	PHB-MCC-T	PHB-MCC-P
T_c, °C	116.9	115.5	118.2	118.2
ΔH_c	76.6	74.0	77.1	76.9
T_m, °C	164.2	163.0	163.3	163.6
ΔH_m	81.1	78.4	81.5	81.5
X_c, %	55.5	54.8	57.0	57.0

ΔH_c, T_c—crystallization enthalpy and crystallization temperature. ΔH_m, T_m—melting enthalpy and melting temperature.

As observed in Figure 8a, The T_c of PHB slightly increased in the composites with treated MCC and decreased in the composite with untreated cellulose. Although the difference between the composites with untreated and treated cellulose is at most 3 °C, this reflects the influence of MCC treatment on the crystallization of PHB. Moreover, the width at half-height of crystallization peak was almost double in the case of PHB-MCC compared to PHB and the composites with treated MCC. The broader crystallization peak and the lower T_c show that MCC hindered the crystallization of PHB and slowed down the crystallization rate. Previous study has shown that nanocellulose is a good nucleating agent in PHB, increasing the crystallization rate [7]. Therefore, the effect of MCC on the crystallization of PHB may be due to the poor interface with the polymer and the micrometric size of cellulose fibers. On the contrary, the good influence of treated fibers suggests a better interface between PHB and MCC-T or MCC-P. In addition, both plasma and TEMPO treatments contributed to the defibrillation of cellulose, reducing the size of the fibers, as pointed out by previous studies [16,21].

Similar behavior was observed in the second heating cycle, with a main melting peak and a shoulder at a higher temperature (Figure 8b). The melting temperature remained unchanged after the addition of cellulose, regardless the treatment. However, the crystallinity followed the same trend as observed in the cooling cycle, a slight decrease in the composite with untreated MCC and a slight increase in that containing treated celluloses. This may be an effect of a better dispersion of treated

MCC in PHB and better interactions between PHB and plasma or TEMPO treated cellulose. SEM analysis may enlighten this aspect.

3.3. SEM Images of PHB Composites with Treated MCC

The SEM images in Figure 9 show the surface morphology of the fractured PHB composites with different treated cellulose.

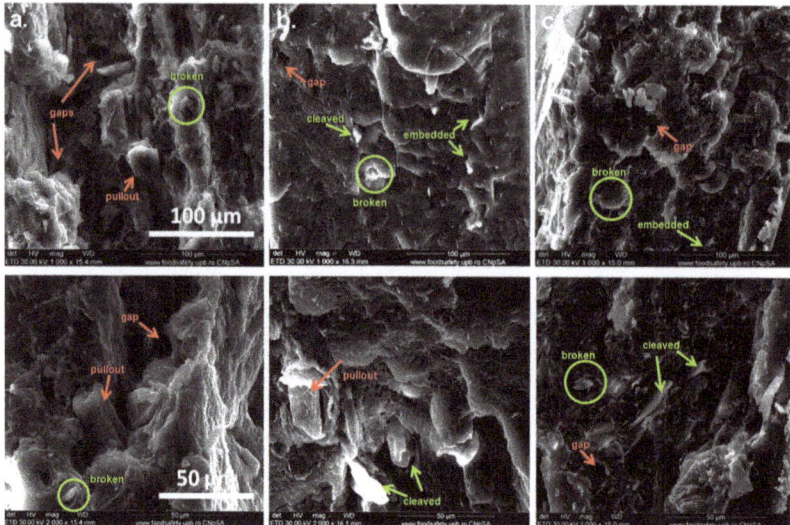

Figure 9. SEM images of PHB composites with different treated MCC: untreated MCC ((**a**); ×1000, ×2000); MCC-T ((**b**); ×1000, ×2000); MCC-P ((**c**); ×1000, ×2000).

Both well embedded broken fibers and pullout fibers and gaps were observed in all the samples but the frequency of debonded fibers and gaps was higher in PHB-MCC than in the composites with treated fibers. Notably, the presence of fibers broken on the transverse section indicates a good PHB–cellulose interface, and such fibers were frequently observed in the composites with TEMPO and plasma modified fibers. This shows good bonding between the polymer and cellulose fibers, which may be benefic to their mechanical properties. Similarly, thinner submicron fibers were mostly observed in the composites with MCC-T and MCC-P, because both TEMPO oxidation and plasma treatment decrease the size of original cellulose fibers and caused defibrillation [13,21]. A better dispersion of fibers was noticed in the case of PHB-MCC-P.

3.4. DMA Analysis of PHB Composites with Treated MCC

The stiffness of PHB composites as function of temperature was investigated by DMA (Figure 10). The addition of untreated and treated MCC in PHB led to an increase of the storage modulus (E') (Figure 10a). However, the increase was more accentuate (2–3 times) for the composites with treated MCC, especially with plasma treated fibers. An increase of the E' by 184% for PHB-MCC-P and 167% for PHB-MCC-T was noticed at room temperature (Table 5).

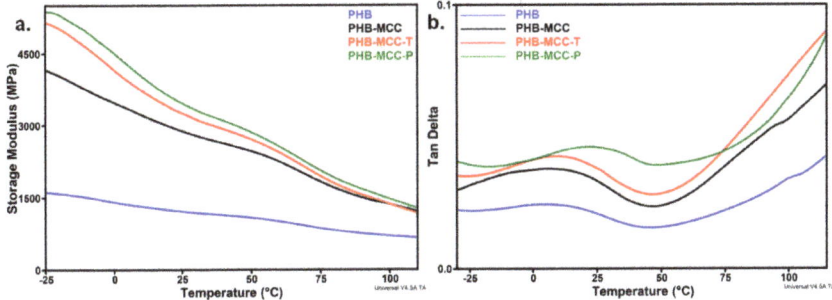

Figure 10. Storage modulus (**a**) and tan δ (**b**) vs temperature curves for PHB and composites with different treated MCC.

Table 5. DMA data for PHB and composites: PHB-MCC, PHB-MCC-T and PHB-MCC-P.

Composites	PHB	PHB-MCC	PHB-MCC-T	PHB-MCC-P
T_g, °C	3.8	5.8	9.2	21.8
E' (−25 °C), MPa	1617	4164	5160	5370
E' (25 °C), MPa	1221	2885	3264	3470
E' (75 °C), MPa	871	1830	1952	2068

The more accentuated decrease of the E', observed between about −10 °C and 25–30 °C, was determined by the transition of PHB from glassy to rubbery state and the activation of molecular segmental motions at the glass transition temperature [44]. Variation of loss factor (tan δ) with temperature (Figure 10b) emphasizes the relative contributions of the elastic and viscous components when temperature increases. A shift of the glass transition temperature (T_g) to a higher temperature was observed for all the composites compared to PHB, however the highest shift, of 18 °C, was observed for PHB-MCC-P. This shows a good interface and a uniform dispersion of MCC-P in the polymer matrix. Plasma treatment induced a mild oxidation on the surface of cellulose, as indicated by both FTIR and XPS results, and the formation of carboxylate groups may increase the compatibility between poly(hydroxyl butyric acid) matrix and COOH bearing cellulose. The influence of MCC-T was not so important, probably because of the opposite effects of surface functionalization and degradation. TEMPO oxidation induces the surface functionalization of cellulose, improving the interface and mechanical properties, but the presence of metal compounds and other impurities from the chemical treatment enhances the decomposition of PHB, thus decreasing the molecular weight and the mechanical properties [7,45]. This is supported by the lower thermal stability of this composite (Figure 7). As a result, the increase of the mechanical properties in PHB-MCC-T was not as high as in the case of PHB-MCC-P, although TEMPO oxidation induced a much strong surface modification of cellulose.

4. Conclusions

Microcrystalline cellulose was surface modified by a plasma treatment in liquid using a Y-shaped tube and TEMPO mediated oxidation. The plasma treatment of water suspensions of cellulose ensured an increased flow of the oxidative gas, being a simple and highly reproducible method for cellulose modification. FTIR and XPS results showed that both treatments led to the surface functionalization of cellulose. However, TEMPO oxidation had a much stronger effect, leading to a decrease of thermal stability of MCC by 80 °C. Plasma and TEMPO modified celluloses were incorporated in a PHB matrix and they influenced the morphology and thermal and mechanical properties of composites. Both treatments improved the fiber-polymer interface and the mechanical properties of PHB composites. An increase of the storage modulus by184% in PHB-MCC-P and 167% in PHB-MCC-T was noticed at room temperature. The increase of the mechanical properties in the composite with TEMPO oxidized cellulose was not as high as in the case of that containing plasma treated cellulose, although the chemical treatment induced a much stronger surface modification of cellulose. The main cause of this behavior is, probably, the significant degradation of MCC following TEMPO oxidation. The PHB composites containing plasma modified cellulose are promising materials for biomedical applications. In addition, the filamentary Y-shaped DBD plasma jet immersed in the liquid suspension of cellulose is an environmentally friendly, simple, and cheap treatment, which is proposed as an alternative to the mechano-chemical methods usually employed for the modification of cellulose.

Author Contributions: Conceptualization, D.M.P. and G.D.; methodology, D.M.P. (cellulose treatments and composites), S.V. (plasma experiments), and C.-A.N. (thermal experiments); formal analysis and investigation, S.V. and L.G.C. (plasma), S.A.S. (TEMPO, composites), C.-A.N. (TGA, DSC composites), A.R.G. (DMA), C.M.D. (TGA celluloses), and R.T. (SEM); writing—original draft preparation, D.M.P. and S.V.; writing—review & editing, D.M.P.; validation, G.D.; supervision, D.M.P. All authors have read and agreed to the published version of the manuscript.

Funding: This work was funded by two grants of the Romanian Ministry of Research and Innovation, CCCDI–UEFISCDI, project number PN-III-P1-1.2-PCCDI-2017-0428 Napoli19 (PCCDI 40/2018) and project number PN-III-P1-1.2-PCCDI-2017-0637 MultiMonD2 (PCCDI 33/2018) within PNCDI III.

Acknowledgments: The authors gratefully acknowledge Veronica Satulu for XPS measurements.

Conflicts of Interest: The authors declare no conflict of interest.

References

1. Gandini, A.; Belgacem, M.N. The state of the art of polymers from renewable resources. In *Handbook of Biopolymers and Biodegradable Plastics: Properties, Processing and Applications*, 1st ed.; Ebnesajjad, S., Ed.; Elsevier Inc.: Kidlington, Oxford, UK, 2013; Chapter 4; pp. 71–85.
2. Kargarzadeh, H.; Huang, J.; Lin, N.; Ahmad, I.; Marino, M.; Dufresne, A.; Thomas, S.; Gałęski, A. Recent developments in nanocellulose-based biodegradable polymers, thermoplastic polymers, and porous nanocomposites. *Prog. Polym. Sci.* **2018**, *87*, 197–227. [CrossRef]
3. Wróblewska-Krepsztul, J.; Rydzkowski, T.; Michalska-Pożoga, I.; Thakur, V.K. Biopolymers for Biomedical and Pharmaceutical Applications: Recent Advances and Overview of Alginate Electrospinning. *Nanomaterials (Basel)* **2019**, *9*, 404.
4. Albuquerque, P.B.S.; Malafaia, C.B. Perspectives on the production, structural characteristics and potential applications of bioplastics derived from polyhydroxyalkanoates. *Int. J. Biol. Macromol.* **2018**, *107*, 615–625. [CrossRef]
5. Panaitescu, D.M.; Nicolae, C.A.; Gabor, A.R.; Trusca, R. Thermal and mechanical properties of poly(3-hydroxybutyrate) reinforced with cellulose fibers from wood waste. *Ind. Crops Prod.* **2020**, *145*, 112071. [CrossRef]
6. Wang, L.; Zhu, W.; Wang, X.; Chen, X.; Chen, G.-Q.; Xu, K. Processability modifications of poly(3-hydroxybutyrate) by plasticizing, blending, and stabilizing. *J. Appl. Polym. Sci.* **2008**, *107*, 166–173. [CrossRef]
7. Srithep, Y.; Ellingham, T.; Peng, J.; Sabo, R.; Clemons, C.; Turng, L.-S.; Pilla, S. Melt compounding of poly (3-hydroxybutyrate-co-3-hydroxyvalerate)/nanofibrillated cellulose nanocomposites. *Polym. Degrad. Stab.* **2013**, *98*, 1439–1449. [CrossRef]
8. Panaitescu, D.M.; Nicolae, C.A.; Frone, A.N.; Chiulan, I.; Stanescu, P.O.; Draghici, C.; Iorga, M.; Mihailescu, M. Plasticized poly(3-hydroxybutyrate) with improved melt processing and balanced properties. *J. Appl. Polym. Sci.* **2017**, *134*, 44810. [CrossRef]
9. Frone, A.N.; Panaitescu, D.M.; Chiulan, I.; Nicolae, C.A.; Vuluga, Z.; Vitelaru, C.; Damian, M.C. The effect of cellulose nanofibers on the crystallinity and nanostructure of poly (lactic acid) composites. *J. Mater. Sci.* **2016**, *51*, 9771–9791. [CrossRef]
10. Garcia-Garcia, D.; Lopez-Martinez, J.; Balart, R.; Strömberg, E.; Moriana, R. Reinforcing capability of cellulose nanocrystals obtained from pine cones in a biodegradable poly(3-hydroxybutyrate)/poly(ε-caprolactone) (PHB/PCL) thermoplastic blend. *Eur. Polym. J.* **2018**, *104*, 10–18. [CrossRef]

11. Panaitescu, D.M.; Frone, A.N.; Chiulan, I.; Gabor, R.A.; Spataru, I.C.; Căşărică, A. Biocomposites from polylactic acid and bacterial cellulose nanofibers obtained by mechanical treatment. *BioResources* **2017**, *12*, 662–672. [CrossRef]
12. Ambrosio-Martin, J.; Fabra, M.J.; Lopez-Rubio, A.; Gorrasi, G.; Sorrentino, A.; Lagaron, J.M. Assessment of ball milling as a compounding technique to develop nanocomposites of poly(3-hydroxybutyrate-co-3-hydroxyvalerate) and bacterial cellulose nanowhiskers. *J. Polym. Environ.* **2016**, *24*, 241–254. [CrossRef]
13. Nechyporchuk, O.; Belgacem, M.N.; Bras, J. Production of cellulose nanofibrils: A review of recent advances. *Ind. Crops Prod.* **2016**, *93*, 2–25. [CrossRef]
14. Agarwal, C.; Csoka, L. *Surface-Modified Cellulose in Biomedical Engineering In Materials for Biomedical Engineering: Bioactive Materials, Properties, and Applications*; Grumezescu, V., Grumezescu, A.M., Eds.; Elsevier Radarweg: Amsterdam, The Netherlands, 2019; Chapter 6; pp. 251–261.
15. Frone, A.N.; Chiulan, I.; Panaitescu, D.M.; Cristian, A.N.; Ghiurea, M.; Galan, A.-M. Isolation of cellulose nanocrystals from plum seed shells, structural and morphological characterization. *Mater. Lett.* **2017**, *194*, 160–163. [CrossRef]
16. Saito, T.; Isogai, A. TEMPO-mediated oxidation of native cellulose. The effect of oxidation conditions on chemical and crystal structures of the water-insoluble fractions. *Biomacromolecules* **2004**, *5*, 1983–1989. [CrossRef]
17. Besbes, I.; Alila, S.; Boufi, S. Nanofibrillated cellulose from TEMPO-oxidized eucalyptus fibres: Effect of the carboxyl content. *Carbohydr. Polym.* **2011**, *84*, 975–983. [CrossRef]
18. Lavoine, N.; Desloges, I.; Dufresne, A.; Bras, J. Microfibrillated cellulose—Its barrier properties and applications in cellulosic materials: A review. *Carbohydr. Polym.* **2012**, *90*, 735–764. [CrossRef]
19. Missoum, K.; Belgacem, M.N.; Bras, J. Nanofibrillated Cellulose Surface Modification: A Review. *Materials (Basel)* **2013**, *6*, 1745–1766. [CrossRef]
20. Panaitescu, D.M.; Vizireanu, S.; Nicolae, C.A.; Frone, A.N.; Casarica, A.; Carpen, L.G.; Dinescu, G. Treatment of nanocellulose by submerged liquid plasma for surface functionalization. *Nanomaterials (Basel)* **2018**, *8*, 467. [CrossRef]
21. Vizireanu, S.; Panaitescu, D.M.; Nicolae, C.A.; Frone, A.N.; Chiulan, I.; Ionita, M.D.; Satulu, V.; Carpen, L.G.; Petrescu, S.; Birjega, R.; et al. Cellulose defibrillation and functionalization by plasma in liquid treatment. *Sci. Rep.* **2018**, *8*, 15473. [CrossRef]
22. International Organisation for Standardization (ISO). *ISO 527-2:2012: Plastics—Determination of Tensile Properties—Part. 2: Test. Conditions for Moulding and Extrusion Plastics*; International Organization for Standardization: Geneva, Switzerland, 2012.
23. Teodorescu, M.; Bazavan, M.; Ionita, E.R.; Dinescu, G. Characteristics of a long and stable filamentary argon plasma jet generated in ambient atmosphere. *Plasma Sources Sci. Technol.* **2015**, *24*, 25033. [CrossRef]
24. Carpen, L.G.; Chireceanu, C.; Teodorescu, M.; Chiriloaie, A.; Teodoru, A.; Dinescu, G. The effect of argon/oxygen and argon/nitrogen atmospheric plasma jet on stored products pests. *Rom. J. Phys.* **2019**, *64*, 503.
25. Barham, P.J.; Keller, A.; Otun, E.L.; Holmer, P.A. Crystallization and morphology of a bacterial thermoplastic: Poly-3-hydroxybutyrate. *J. Mater. Sci.* **1984**, *19*, 2781–2794. [CrossRef]
26. Baruah, J.; Deka, R.C.; Kalita, E. Greener production of microcrystalline cellulose (MCC) from Saccharum spontaneum (Kans grass): Statistical optimization. *Int. J. Biol. Macromol.* **2020**, *154*, 672–682. [CrossRef] [PubMed]
27. Fukuzumi, H.; Saito, T.; Iwata, T.; Kumamoto, Y.; Isogai, A. Transparent and high gas barrier films of cellulose nanofibers prepared by TEMPO-mediated oxidation. *Biomacromolecules* **2009**, *10*, 162–165. [CrossRef] [PubMed]
28. Fukuzumi, H.; Saito, T.; Okita, Y.; Isogai, A. Thermal stabilization of TEMPO-oxidized cellulose. *Polym. Degrad. Stab.* **2010**, *95*, 1502–1508. [CrossRef]
29. Loof, D.; Hiller, M.; Oschkinat, H.; Koschek, K. Quantitative and qualitative analysis of surface modified cellulose utilizing TGA-MS. *Materials (Basel)* **2016**, *9*, 415. [CrossRef]
30. Peng, Y.; Gardner, D.J.; Han, Y.; Kiziltas, A.; Cai, Z.; Tshabalala, M.A. Influence of drying method on the material properties of nanocellulose I: Thermostability and crystallinity. *Cellulose* **2013**, *20*, 2379–2392. [CrossRef]

31. Piskorz, J.; Radlein, D.S.A.G.; Donald, S.S.; Czernik, S. Pretreatment of wood and cellulose for production of sugars by fast pyrolysis. *J. Anal. Appl. Pyrolysis* **1989**, *16*, 127–142. [CrossRef]
32. Oh, S.Y.; Yoo, D.I.; Shin, Y.; Kim, H.C.; Kim, H.Y.; Chung, Y.S.; Park, W.H.; Youk, J.H. Crystalline structure analysis of cellulose treated with sodium hydroxide and carbon dioxide by means of X-ray diffraction and FTIR spectroscopy. *Carbohydr. Res.* **2005**, *340*, 2376–2391. [CrossRef]
33. Gwon, J.G.; Lee, S.Y.; Doh, G.H.; Kim, J.H. Characterization of chemically modified wood fibers using ftir spectroscopy for biocomposites. *J. Appl. Polym. Sci.* **2010**, *116*, 3212–3219. [CrossRef]
34. Sehaqui, H.; de Larraya, U.P.; Liu, P.; Pfenninger, N.; Mathew, A.P.; Zimmermann, T.; Tingaut, P. Enhancing adsorption of heavy metal ions onto biobased nanofibers from waste pulp residues for application in wastewater treatment. *Cellulose* **2014**, *21*, 2831–2844. [CrossRef]
35. Zhu, C.; Soldatov, A.; Mathew, A.P. Advanced microscopy and spectroscopy reveal the adsorption and clustering of Cu(II) onto TEMPO oxidized cellulose nanofibers. *Nanoscale* **2017**, *9*, 7419–7428. [CrossRef]
36. Habibi, Y.; Chanzy, H.; Vignon, M.R. TEMPO-mediated surface oxidation of cellulose whiskers. *Cellulose* **2006**, *13*, 679–687. [CrossRef]
37. Socrates, G. *Infrared and Raman Characteristic Group Frequencies*, 3rd ed.; John Wiley & Sons Ltd.: Chichester, UK, 2001; pp. 50–67.
38. Coseri, S.; Biliuta, G.; Zemljič, L.F.; Srndovic, J.S.; Larsson, P.T.; Strnad, S.; Kreže, T.; Naderi, A.; Lindstrom, T. One-shot carboxylation of microcrystalline cellulose in the presence of nitroxyl radicals and sodium periodate. *RSC Adv.* **2015**, *5*, 85889–85897. [CrossRef]
39. Clark, D.T.; Cromarty, B.J.; Dilks, A. A theoretical investigation of molecular core binding and relaxation energies in a series of oxygen-containing organic molecules of interest in the study of surface oxidation of polymers. *J. Polym. Sci. Polym. Chem. Ed.* **1978**, *16*, 3173–3184. [CrossRef]
40. Zhou, J.-H.; Sui, Z.-J.; Zhu, J.; Li, P.; Chen, D.; Dai, Y.-C.; Yuan, W.-K. Characterization of surface oxygen complexes on carbon nanofibers by TPD, XPS and FT-IR. *Carbon* **2007**, *45*, 785–796. [CrossRef]
41. Li, S.-S.; Wang, X.-L.; An, Q.-D.; Xiao, Z.-Y.; Zhai, S.-R.; Cui, L.; Li, Z.-C. Upon designing carboxyl methylcellulose and chitosan-derived nanostructured sorbents for efficient removal of Cd(II) and Cr(VI) from water. *Int. J. Biol. Macromol.* **2020**, *143*, 640–650. [CrossRef] [PubMed]
42. Smith, M.; Scudiero, L.; Espinal, J.; McEwen, J.-S.; Garcia-Perez, M. Improving the deconvolution and interpretation of XPS spectra from chars by ab initio calculations. *Carbon* **2016**, *110*, 155–171. [CrossRef]
43. Ariffin, H.; Nishida, H.; Shirai, Y.; Hassan, M.A. Highly selective transformation of poly[(R)-3-hydroxybutyric acid] into *trans*-crotonic acid by catalytic thermal degradation. *Polym. Degrad. Stab.* **2010**, *95*, 1375–1381. [CrossRef]
44. Kaiser, M.R.; Anuar, H.; Razak, S.B.A. Ductile–brittle transition temperature of polylactic acid-based biocomposite. *J. Thermoplast. Compos. Mater.* **2011**, *26*, 216–226. [CrossRef]
45. Kim, K.J.; Doi, Y.; Abe, H. Effect of metal compounds on thermal degradation behavior of aliphatic poly(hydroxyalkanoic acid)s. *Polym. Degrad. Stab.* **2008**, *93*, 776–785. [CrossRef]

© 2020 by the authors. Licensee MDPI, Basel, Switzerland. This article is an open access article distributed under the terms and conditions of the Creative Commons Attribution (CC BY) license (http://creativecommons.org/licenses/by/4.0/).

Article

Microfibrillated Cellulose Grafted with Metacrylic Acid as a Modifier in Poly(3-hydroxybutyrate)

Marius Stelian Popa [1,2], Adriana Nicoleta Frone [1], Ionut Cristian Radu [2], Paul Octavian Stanescu [2], Roxana Trușcă [3], Valentin Rădițoiu [1], Cristian Andi Nicolae [1], Augusta Raluca Gabor [1] and Denis Mihaela Panaitescu [1,*]

[1] Institute for Research & Development in Chemistry and Petrochemistry—ICECHIM, 202 SplaiulIndependentei, 060021 Bucharest, Romania; popamarius7777@gmail.com (M.S.P.); adriana.frone@icechim.ro (A.N.F.); vraditoiu@icechim.ro (V.R.); cristian.nicolae@icechim.ro (C.A.N.); raluca.gabor@icechim.ro (A.R.G)

[2] Faculty of Applied Chemistry and Materials Science, University Politehnica of Bucharest, 1–7 Gh. Polizu Street, 011061 Bucharest, Romania; ionut_cristian.radu@upb.ro (I.C.R.); paul.stanescu@upb.ro (P.O.S.)

[3] National Research Centre for Micro and Nanomaterials, University Politehnica of Bucharest, 313 Spl. Indendentei, 060042 Bucharest, Romania; truscaroxana@yahoo.com

* Correspondence: panaitescu@icechim.ro

Citation: Popa, M.S.; Frone, A.N.; Radu, I.C.; Stanescu, P.O.; Trușcă, R.; Rădițoiu, V.; Nicolae, C.A.; Gabor, A.R.; Panaitescu, D.M. Microfibrillated Cellulose Grafted with Metacrylic Acid as a Modifier in Poly(3-hydroxybutyrate). *Polymers* **2021**, *13*, 3970. https://doi.org/10.3390/polym13223970

Academic Editor: Wei Zhang

Received: 6 November 2021
Accepted: 15 November 2021
Published: 17 November 2021

Publisher's Note: MDPI stays neutral with regard to jurisdictional claims in published maps and institutional affiliations.

Copyright: © 2021 by the authors. Licensee MDPI, Basel, Switzerland. This article is an open access article distributed under the terms and conditions of the Creative Commons Attribution (CC BY) license (https://creativecommons.org/licenses/by/4.0/).

Abstract: This work proposes a new method for obtaining poly(3-hydroxybutyrate) (PHB)/microfibrillated cellulose (MC) composites with more balanced properties intended for the substitution of petroleum-based polymers in packaging and engineering applications. To achieve this, the MC surface was adjusted by a new chemical route to enhance its compatibility with the PHB matrix: (i) creating active sites on the surface of MC with γ-methacryloxypropyltrimethoxysilane (SIMA) or vinyltriethoxysilane (SIV), followed by (ii) the graft polymerization of methacrylic acid (MA). The high efficiency of the SIMA-MA treatment and the lower efficiency in the case of SIV-MA were proven by the changes observed in the Fourier transform infrared FTIR spectra of celluloses. All modified celluloses and the PHB composites containing them showed good thermal stability close to the processing temperature of PHB. SIMA-modified celluloses acted as nucleating agents in PHB, increasing its crystallinity and favoring the formation of smaller spherulites. A uniform dispersion of SIMA-modified celluloses in PHB as a result of the good compatibility between the two phases was observed by scanning electron microscopy and many agglomerations of fibers in the composite with unmodified MC. The dual role of SIMA-MA treatment, as both compatibilizer and plasticizer, was pointed out by mechanical and rheological measurements. This new method to modify MC and obtain PHB/MC composites with more balanced stiffness–toughness properties could be a solution to the high brittleness and poor processability of PHB-based materials.

Keywords: microfibrillated cellulose; polymethacrylic acid; grafting; poly(3-hydroxybutyrate); biocomposites; compatibility

1. Introduction

The poor disposal of waste combined with the wasteful mentality of humankind has led to well-known environmental problems. Currently, petroleum is used in everything that requires electricity [1], and the dependence on petroleum-based plastics increases the demand for oil, which drags along an increased need for energy, and so on. The focus of research in recent years has been to reduce this demand by exploring more natural-based materials with the aim to replace classic, fossil-based plastics [2–4]. Promising results have been obtained with poly-(3-hydroxybutyrate) (PHB), an aliphatic microbial polyester that shows similar mechanical and thermal properties to polypropylene with the added bonuses of better barrier properties and biodegradability [5]. It is biosynthesized by certain bacteria as a means of stocking energy in nitrogen- and phosphorus-deficient conditions [6]. However, in its pure form, PHB is highly brittle at room temperature and has a low processing window, which limits its possibilities of application. Thus, different approaches such as plasticization, melt blending with other polymers, copolymerization, and preparation of nanocomposites have been studied to overcome these drawbacks [7]. The introduction of organic or inorganic fillers has proven to be an attractive and versatile technique to enhance the performances of PHB. Several reinforcing agents have been incorporated into PHB for improving its properties, such as (nano)clays [8], carbon nanotubes [9], graphene [10], and cellulosic materials [11].

Cellulose is one of the most abundant and studied materials in the world and can be obtained in various sizes and shapes [12]. It can be extracted from a variety of sources, such as wood and plants, and can be biosynthesized by some bacterial strands as extracellular material [13]. Its versatility regarding geometry, from (nano)fibers to (nano)particles, and its sources allows cellulose to be used in countless applications. Cellulose micro- and nanofillers have been used to improve the thermal and mechanical properties of PHB [11,14,15]. A big issue in the case of PHB–cellulose composites is the poor compatibility between the matrix and the filler due to the hydrophobic nature of PHB and the strong hydrophilicity of cellulosic fillers [16]. The methods attempted so far to improve this compatibility, such as the treatment of cellulose fillers by acetylation, silylation, and TEMPO-mediated or plasma oxidation, have led to some improvement in PHB properties [17–19]. Reactive extrusion has also been used to improve the adhesion at the PHB–cellulose interface, leading to a moderate enhancement of mechanical properties [20].

A different route for improving the compatibility between PHB and microfibrillated cellulose was attempted in this paper. The hypothesis underpinning this work is that cellulose modification with polymethacrylic acid units improves its compatibility with PHB. Indeed, only a limited number of polymers are miscible with PHB, such as polyvinyl acetate or polymethyl acrylate [21,22]. An et al. have shown that polymethyl acrylate is fully miscible with PHB over the entire composition range [22]. This may be presumed from the similarity of the repeating units, both vinyl acetate and methyl acrylate units being isomers of 3-hydroxybutyrate (HB). Another isomer of the HB unit is methacrylic acid, which is used in this work to modify microfibrillated cellulose. Thus, polymethacrylic acid (PMA) grafts were grown on the surface of cellulose to increase the compatibility between PHB and the cellulosic filler. Although polymethacrylic acid was previously grafted on a few cellulosic substrates [23,24], especially for medical application, no attempt to modify cellulose with PMA for increased compatibility with PHB has been performed so far. In addition, this is the first attempt to pre-activate the cellulose surface with various silanes for better grafting of PMA and obtaining improved properties. In this work, Fourier transform infrared spectroscopy (FTIR) and thermogravimetric analysis (TGA) were used to highlight the grafting of silanes and the presence of polymethacrylic acid on the surface of cellulose. The influence of modified celluloses on the morphology, thermal, and mechanical properties of PHB composites was also studied.

2. Experimental Section

2.1. Materials

Microcrystalline cellulose (MCC) with an average diameter of 20 µm and purchased from Sigma Aldrich (Saint Louis, MO, USA) was used to obtain the microfibrillated cellulose (MC). Methacrylic acid (MA, purity 99%), vinyltriethoxysilane (SIV) (purity 97%), and 2,2′-azobis(2-methylpropionitrile) (AIBN) (purity > 98%) were purchased from Sigma Aldrich (Saint Louis, MO, USA). Moreover, γ-methacryloxypropyltrimethoxysilane (SIMA) (purity 98%, trade name—Xiameter OFS-6030 silane) was obtained from Dow Corning Co. (Midland, MI, USA) and acetone (purity > 99.92%) was obtained from Chimreactiv (Bucharest, Romania). All chemicals were used without further purification. PHB pellets (P304) from Biomer (Schwalbach am Taunus, Germany) with a density of 1.24 g/cm^3 were used as the polymer matrix in the composites.

2.2. Preparation of Microfibrillated Cellulose

MC was originally obtained from a 2 wt% MCC suspension in distilled water, which was maintained at room temperature for 48 h to enable MCC soaking. Then, the MCC suspension was processed by a high-pressure mechanical treatment for 12 cycles at 200 MPa using a microfluidizer LM20 (Microfluidics, Westwood, MA, US). A cellulose gel resulted from the treatment. This was first frozen at −20 °C for 48 h and then freeze-dried (FreeZone 2.5 L, Labconco, Kansas-City, MO, USA) at −85 °C and 0.006 mbar for 48 h. The dried cellulose was further milled using an ultra-centrifugal mill ZM 200 (Retsch GmbH & Co., Düsseldorf, Germany) at a speed of 6000 min^{-1}, resulting in microfibrillated cellulose (MC) powder.

2.3. Silanization Reaction

MC powder was soaked in water (10 wt%) under magnetic stirring for 8 h at room temperature. A silane solution (5 wt%) was prepared by dissolving either SIV or SIMA in water at room temperature under strong stirring for 30 min. The silane solution was acidified with glacial acetic acid until the pH value of 3.5 and allowed to hydrolyze for 30 min. Then, the hydrolyzed silane solution was poured over the MC suspension and allowed to react at 50 °C under strong magnetic stirring for another 4 h. Following this, the modified celluloses were washed three times with a plentiful amount of distilled water. The suspensions were then freeze-dried and the dried celluloses were milled using the same equipment and conditions as previously mentioned, resulting in MC-SIV or MC-SIMA. The reaction schemes for the chemical modification of cellulose by silanization with SIMA and SIV are shown in Figure 1.

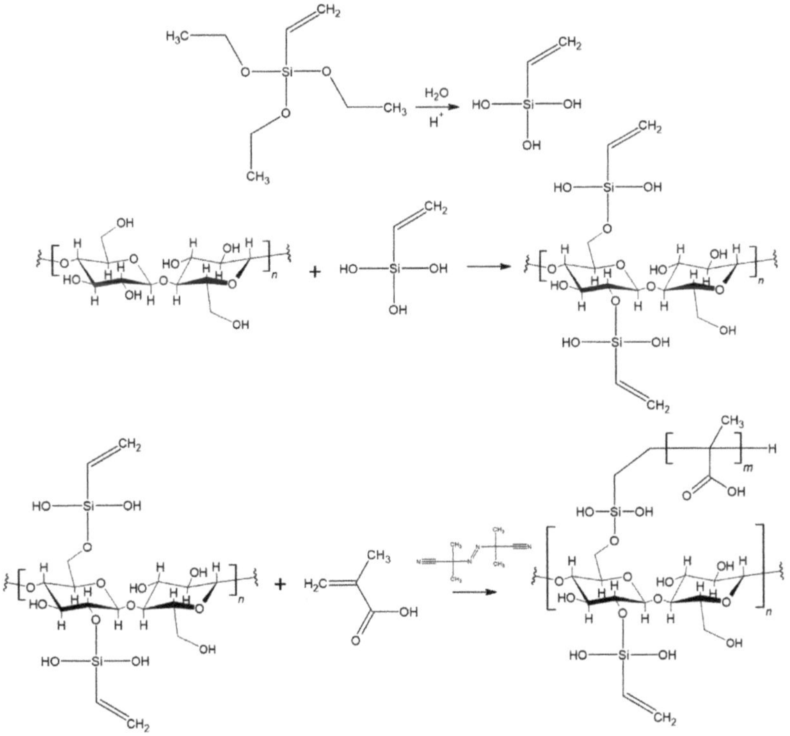

(a)

Figure 1. *Cont.*

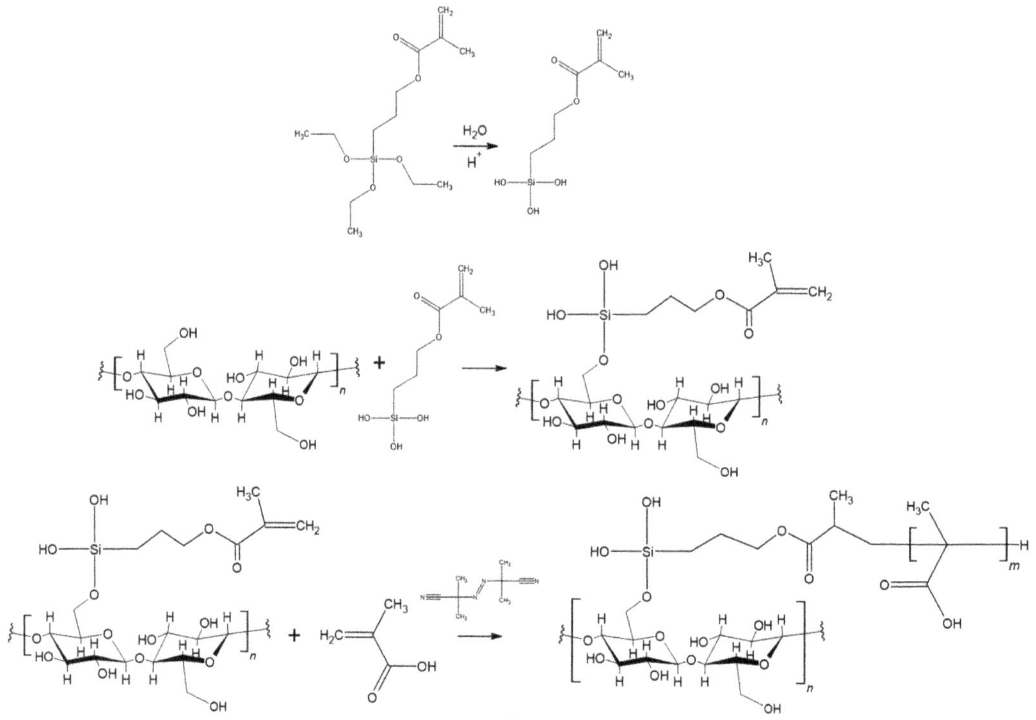

Figure 1. Reaction schemes for the chemical modification of microfibrillated cellulose for obtaining: (**a**) MC-SIV and MC-SIV-MA, (**b**) MC-SIMA and MC-SIMA-MA.

2.4. Polymerization Reaction

MC-SIV and MC-SIMA powders were soaked in acetone, an aprotic solvent that was chosen as reaction medium. In the meantime, methacrylic acid (MA) was dissolved in acetone at room temperature left under magnetic stirring for 8 h. Trace amounts of acetone were used to dissolve the initiator, AIBN. Overall, the methacrylic acid/cellulose ratio was $1/2 \, v/w$, and the AIBN concentration was 10^{-2} mol/L regarding MA volume. Next, all components were mixed together and allowed to react under reflux for 4 h. The modified celluloses were washed with distilled water, freeze-dried, and milled, similar to the previous step. Polymethacrylic acid-modified celluloses, MC-SIV-MA and MC-SIMA-MA, were thus obtained. Figure 1 presents the reaction schemes for the MA graft polymerization on the surface of silanized celluloses.

2.5. Preparation of the PHB/Modified Cellulose Composite Films

The composite films were prepared by mixing PHB and modified celluloses (2 wt%) in a 30 cm³ Brabender mixing chamber(Brabender GmbH & Co. KG, Duisburg, Germany) at 165 °C for 7 min, followed by molding on a two-rollmill to obtain sheets. These were further compression-molded in a P200E press (Dr. Collin, Ebersberg, Germany) at 175 °C with 120 s of preheating at 0.5 MPa and 60 s of compression at 10 MPa. A cooling cassette accessory was used for the rapid cooling of the films.

2.6. Characterization

2.6.1. Fourier Transform Infrared Spectroscopy

The FTIR attenuated total reflectance (ATR) analysis was carried out on a JASCO 6300 spectrophotometer (JASCO International Co., Ltd., Tokyo, Japan) equipped with a Specac ATR Golden Gate unit (Specac, Inc., Orpington, UK) with KRS5 lens. FTIR spectra were scanned from 4000 to 400 cm^{-1}, with 32 scans per spectrum at a resolution of 4 cm^{-1}.

2.6.2. Thermogravimetric Analysis

Original and differently modified MCs and the composites containing these fibers were characterized by TGA using a TA-Q5000 (TA Instruments, New Castle, DE, USA). Measurements were carried out on samples of 8–10 mg from room temperature to 700 °C with 10 °C/min in nitrogen atmosphere purged with 40 mL/min.

2.6.3. Differential Scanning Calorimetry (DSC)

Calorimetric measurements were carried out on a DSC Q2000 from TA Instruments (New Castle, DE, USA) under a helium flow of 25 mL/min using 10–13 mg from each sample and a heating/cooling rate of 10 °C/min during cycles. The applied program involved: a rapid cooling from the ambient temperature to −60 °C, equilibration for 2 min at this temperature, heating to 200 °C (first heating cycle), equilibration for 2 min, cooling to −60 °C (cooling cycle), equilibration for 2 min and heating again to 200 °C (second heating cycle). The crystallinity (X_c) was calculated from the second heating cycle with:

$$X_c\ (\%) = \frac{\Delta H}{\Delta H_0 \times w_{PHB}} \times 100 \qquad (1)$$

The formula includes the melting enthalpy (ΔH) and the weight fraction (w_{PHB}) of PHB in the composite along with the melting enthalpy of 100% crystalline PHB (ΔH_0 = 146 J/g [25]).

2.6.4. Dynamic Mechanical Analysis (DMA)

Composite specimens with the length × width × thickness of 12.7 mm × 6.8 mm × 0.8 mm were cut from the compression-molded films and analyzed with a DMA Q800 (TA Instruments, New Castle, DE, USA) in multifrequency-strain mode using a tension clamp. The specimens were cooled from the ambient temperature to −50 °C with 10 °C/min, equilibrated for 2 min at this temperature, and heated to 125 °C with a heating rate of 3 °C/min.

2.6.5. Tensile Properties

The tensile properties of PHB compression-molded films were determined according to ISO 527 using an Instron 3382 universal testing machine (Instron, Norwood, MA, USA) with a 10 kN load cell. Five specimens were tested for each sample at room temperature with a crosshead speed of 10 mm/min. The average mechanical properties, Young's modulus, and tensile strength and elongation at break were calculated as mean and standard deviation using the Bluehill 2 software.

2.6.6. Scanning Electron Microscopy (SEM)

Morphological aspects of composites in the fracture were investigated with a Quanta Inspect F Scanning Electron Microscope (Philips/FEI, Eindhoven, The Netherlands), with a resolution of 1.2 nm at an accelerating voltage of 30 kV. Prior to the measurements, the samples were frozen in liquid nitrogen and fractured, then the fractures were sputter-coated with gold for a better contrast.

The MC powder obtained after freeze-drying and milling was analyzed with the same scanning electron microscope. Before the measurements, the powder was spread on an adhesive tape and sputter-coated with gold.

2.6.7. Polarized Light Optical Microscopy

Polarized light optical microscopy (POM) analysis was carried out using an Olympus BX53F Microscope equipped with a DP23/DP28 Digital Camera (Olympus, Tokyo, Japan). Composite films of 20–30 µm in thickness were used for the measurements. The films were obtained by compression molding at 175°C (preheating for 150 s and compression for 65 s at 10 MPa). Sections of the films were melted between glass slides in an oven at 220 °C for 5 min, kept at room temperature for about 15 min, and then analyzed by POM.

3. Results and Discussion

3.1. Morphological Investigation of MC Powder

Figure 2 shows the SEM image of dried microfibrillated cellulose powder. One may observe a network of micro- and nanofibers. Most of the nanofibers had a width of less than 100 nm and a length of a couple of microns. This is better seen in the detail in the upper-left corner of Figure 2. Microfibers with a width of less than 1 µm may also be observed. It turns out that 12 cycles of high-pressure homogenization were sufficient to defibrillate microcrystalline cellulose with an initial width of about 20 µm in submicron-sized fibers.

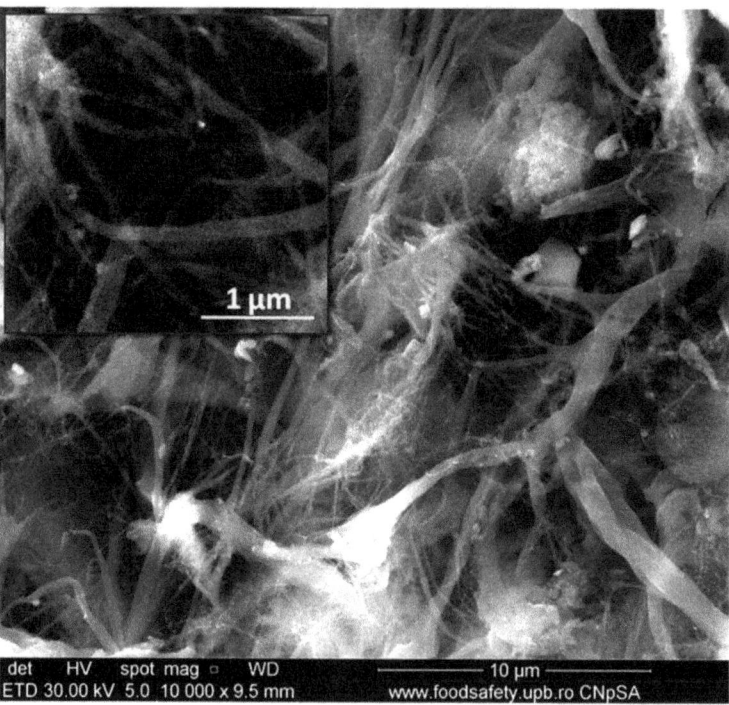

Figure 2. SEM image of microfibrillated cellulose after freeze-drying and milling.

Many entangled and agglomerated cellulose nanofibers were noticed in Figure 2. This is an effect of the removal of water during the freeze-drying process that favors the self-aggregation of individual nanofibers. However, the aggregates disentangled during the melt mixing with the polymer. A network of cellulose fibers with similar size was obtained, starting with delignified Fique tow using a TEMPO oxidation process coupled with ultrasonication [26].

3.2. Characterization of Modified Celluloses
3.2.1. FTIR Analysis

Figure 3 presents the FTIR spectra of the MC and modified celluloses. The unmodified MC spectrogram shows the expected broad peak centered on 3333 cm^{-1}, characteristic of the stretching vibration of hydrogen bonded –OH groups of the cellulose, which does not significantly change after chemical reactions.

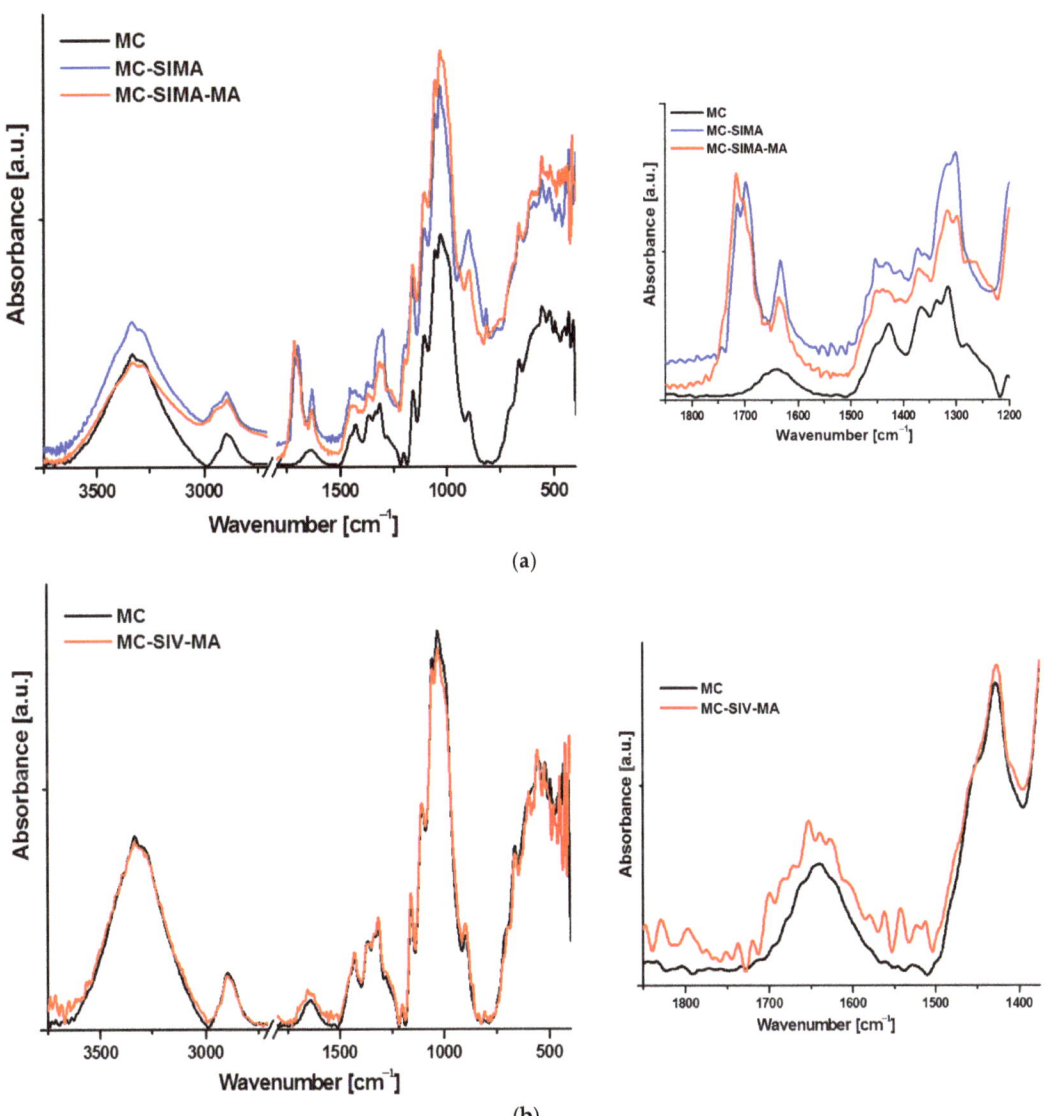

Figure 3. FTIR spectra of MC, MC-SIMA, and MC-SIMA-MA (**a**) and MC-SIV-MA (**b**).

In unmodified cellulose, the peak from 2895 cm^{-1}, which is attributed to C–H stretching vibrations, suffers changes after chemical modifications; thus, a shoulder appears in both SIMA-modified celluloses at about 2950 cm^{-1} (Figure 3a) due to the stretching of CH$_2$ groups in the oxypropyl chain of SIMA [27]. Further, the broad small peak at 1640 cm^{-1}, attributed to the bending vibration of the –OH groups in water, was overlapped by a new peak located at 1633 cm^{-1} in MC-SIMA and a lower intensity broad peak at 1637/1630 cm^{-1} in MC-SIMA-MA, which may be assigned to the stretching vibrations of unreacted C=C groups of the acrylic moiety [27,28]. New peaks with different shapes and intensities appeared between 1750 and 1650 cm^{-1} in MC-SIMA and MC-SIMA-MA (Figure 3a—detail). Two peaks were observed at 1714 cm^{-1} and 1698 cm^{-1} in MC-SIMA spectrum, deriving from the stretching vibrations of the C=O groups in the methacryloyl chains of the grafted silane [28]. A strong peak at 1716 cm^{-1} and three shoulders at 1701, 1740, and 1690 cm^{-1} were observed in the case of MC-SIMA-MA (Figure 3a—detail). As specified above, the peaks at 1716 and 1701 cm^{-1} may be ascribed to the carbonyl groups of silane, however, the shoulder at 1690 cm^{-1} can be attributed to the stretching vibration of the C=O groups in the poly(methacrylic acid) [29]. The shoulder from 1740 cm^{-1} belonged to the ester carbonyl signal and shows the formation of the cellulose ester in the case of MC-SIMA-MA [30]. Therefore, the acylation of cellulose was proved by the appearance of the new signal at 1740 cm^{-1} along with the decreased intensity of the band that is characteristic of the hydrogen bonded –OH in MC-SIMA-MA from 3333 cm^{-1}. However, the weak signal of the ester C=O group shows that this reaction was only a secondary one.

In the case of MC modified by vinyl silane and methacrylic acid, the changes observed in the FTIR spectra (Figure 3b) were minor. The small changes noticed in the band from 1600 to 1700 cm^{-1} (Figure 3b—detail) may have derived from the stretching vibration of the C=O groups in the poly(methacrylic acid) (1696 cm^{-1}) and the C=C stretching in the terminal vinyl groups of methacrylic acid oligomers, or in the unreacted vinyl silane (1637 cm^{-1}). However, the signals were so weak that it can therefore be presumed that the chemical modification of cellulose was less efficient in this case.

3.2.2. Thermogravimetric Analysis of Modified Celluloses

Figure 4 shows the thermograms of the MC and the modified celluloses. After the loss of water at up to 100 °C, MC decomposed between 270 and 370 °C when it lost 80% of its weight. A significantly increased thermal stability was observed after the treatment of microfibrillated cellulose with SIMA. Thus, the temperature at 5% weight loss (T$_{5\%}$) increased by about 34 °C after SIMA treatment. Further grafting with MA led to an increase in T$_{5\%}$ by 20 °C for MC-SIMA-MA and by 14 °C for MC-SIV-MA (Table 1). Several differences in the degradation process were noticed depending on the treatment. Thus, the temperature at the maximum decomposition rate (T$_{max}$) increased in the case of MC-SIMA by 8 °C and for MC-SIV-MA by about 5 °C, and decreased for MC-SIMA-MA by 13 °C (Table 1). The increased thermal stability after the treatment of MC with silanes was an effect of the oligomers formed by the condensation of silanes at increased temperature that act as a protective barrier and delay the degradation of cellulose, similar to other observations [31,32]. Indeed, a new peak was observed after that characteristic to the decomposition of cellulose in the derivative curves of MC-SIMA (427 °C) and MC-SIMA-MA (415 °C), which probably derived from the thermal decomposition of siloxanes [33]. The lower T$_{max}$ of MC-SIMA-MA may have been caused by the breaking of the esteric bonds in grafted SIMA or of the cellulose–silane bonds in the presence of the initiator (Figure 1b), leading to the removal of the silane and the slightly lower thermal stability of cellulose.

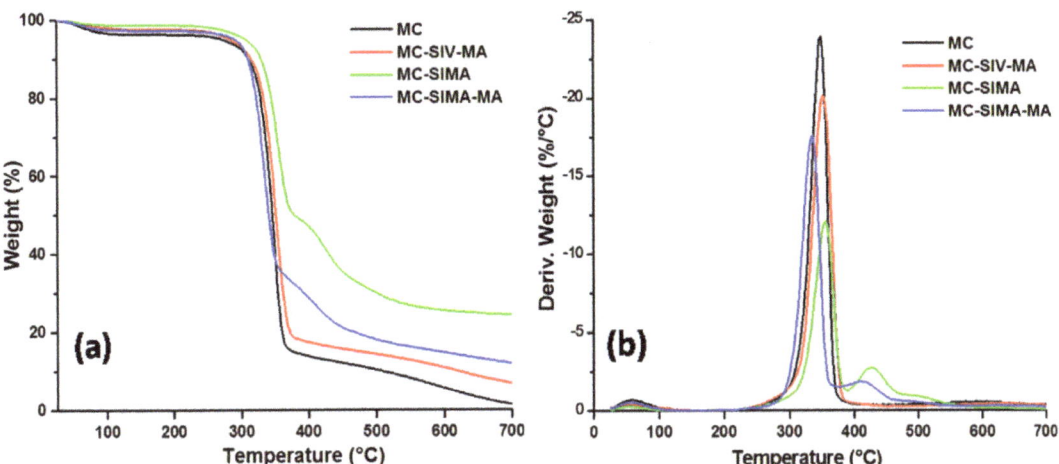

Figure 4. Thermogravimetric (**a**) and derivative (**b**) curves of MC and modified celluloses.

Table 1. Thermogravimetric data for the modified celluloses.

Sample	MC	MC-SIMA	MC-SIMA-MA	MC-SIV-MA
$T_{5\%}$, °C	270.7	305.1	290.6	284.5
T_{max}, °C	348.7	356.5	335.5	353.2
$WL_{200°C}$, %	3.8	1.4	2.7	2.5
Residue at 700 °C, %	1.4	24.4	12.0	6.8

The residue at 700 °C was very high in MC-SIMA (24%) due to the presence of the crosslinked polysiloxanes and was proof of the successful silanization of cellulose. The halved residue obtained in MC-SIMA-MA supports the hypothesis of cellulose–silane bonds breaking and the removal of the silane, also highlighted by the lower T_{max} of MC-SIMA-MA. The lower weight loss at 200 °C ($WL_{200°C}$) of modified celluloses shows their good thermal stability close to the processing temperature of PHB composites.

3.3. Characterization of Composites

3.3.1. TGA Analysis

The TGA and DTG curves of PHB composites with 2 wt% modified celluloses are shown in Figure 5. The incorporation of MC and modified celluloses had a small influence on the thermal stability of PHB. The main decomposition step occurred between 250 and 300 °C in neat PHB and in all the composites. The $T_{5\%}$ and T_{max} of PHB slightly decreased by less than 5 °C after the addition of celluloses. A similar decrease in the thermal stability of PHB was reported after the addition of 2 wt% nanofibrillated bacterial cellulose [15], and a stronger decrease was noticed in the case of poly(3-hydroxybutyrate-co-3-hydroxyvalerate) composites with 2.5–10 wt% nanofibrillated cellulose [34]. The greatest influence on the thermal stability of PHB was noticed in the case of MC-SIMA, the cellulose which had undergone the strongest modification, as demonstrated by FTIR and TGA (Figures 3 and 4). Indeed, the esteric bond of SIMA grafted on MC was labile enough to undergo breaking at a high processing temperature (165 °C), releasing free methacryloyl radicals, which can promote the thermal degradation of PHB [35]. However, these processes have a low intensity and their contribution to the PHB degradation is minor, as proven by the small changes in the $T_{5\%}$ and T_{max} (Table 2). Similarly, the residue at 700 °C of PHB shows little variation after the addition of modified celluloses (Table 2).

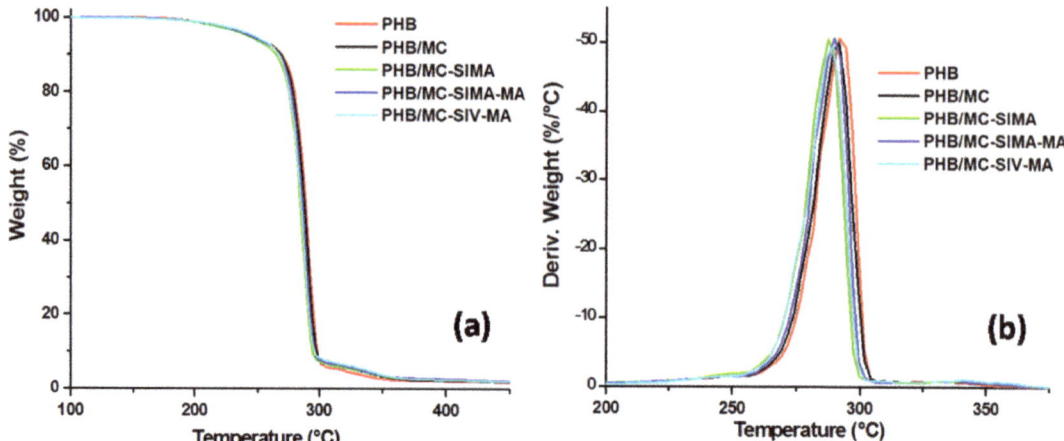

Figure 5. TGA (a) and DTG (b) curves of composites with differently modified celluloses.

Table 2. TGA data for the PHB composites with differently modified celluloses.

Composites	PHB	PHB/MC	PHB/MC-SIMA	PHB/MC-SIMA-MA	PHB/MC-SIV-MA
$T_{5\%}$, °C	246.1	245.5	242.4	244.5	246.8
T_{max}, °C	292.4	290.9	287.9	290	288.6
Residue at 700 °C, %	1.3	1.3	1.7	2.0	1.3

3.3.2. Differential Scanning Calorimetry

Figure 6 presents the behavior of the composites upon heating and cooling, while the crystallinity degree (X_c), melting (T_{m1}, T_{m2}), and crystallization (T_c) temperatures along with corresponding enthalpies (ΔH_{m1}, ΔH_{m2}, ΔH_c) are listed in Table 3. Double endothermic melting peaks were observed in neat PHB and composites during the first heating cycle (Figure 6a). The phenomenon is generally ascribed to the melt–recrystallization mechanism [11,36]: the first peak arises from the melting of PHB fraction that was formerly crystallized during the compression molding of the films, while the second peak from 168 °C can be related to the melting of the recrystallized PHB fraction during heating. One can observe that the addition of modified celluloses only influenced the peak from the lower temperature (Figure 6a). Thus, a slight shift of this peak to a higher temperature along with an increase in intensity was observed in all the composites. This behavior can be associated with the interactions between the cellulose fibers and PHB, which restrict the flexibility of the polymer chain and increase the melting temperature. The higher intensity of the lower temperature peak in composites shows a higher proportion of smaller and less perfect crystallites, possibly an effect of the nucleating effect of the cellulose fibers. No obvious effect of modified celluloses on the crystallization temperature of the composites was noticed, regardless of the treatment, meaning that all the composites had a crystallization rate similar to that of neat PHB (Figure 6b, Table 3).

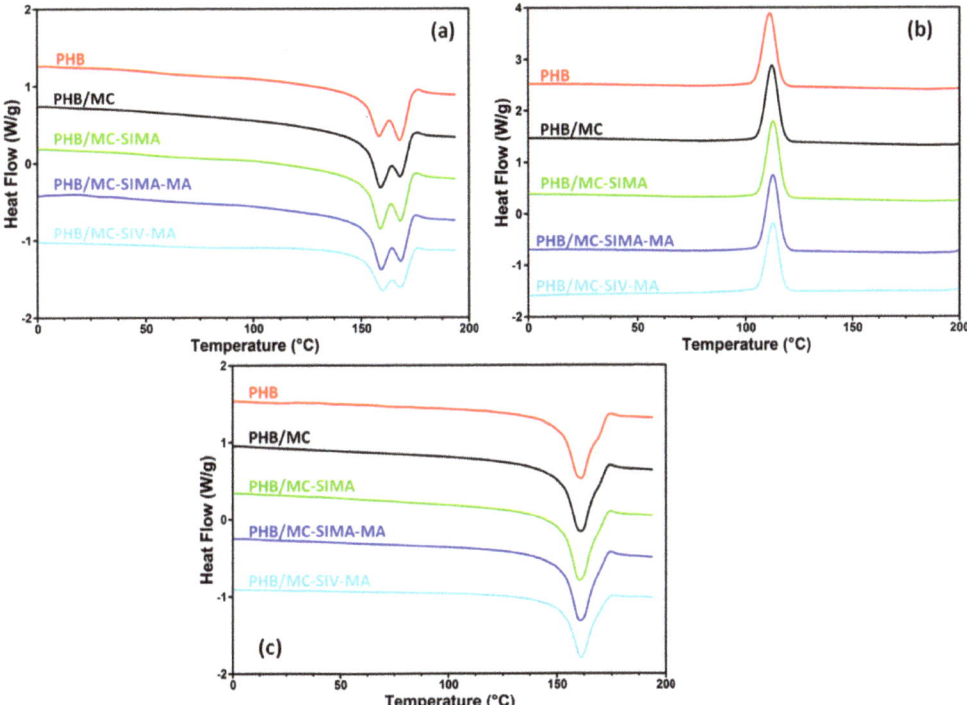

Figure 6. DSC first heating (**a**) and cooling (**b**), and second heating (**c**) of the composites.

Table 3. DSC data for the composites.

Composites	First Heating		Cooling		Second Heating		X_c (%)
	T_{m1} (°C)	ΔH_{m1} (J/g)	T_c (°C)	ΔH_c (J/g)	T_{m2} (°C)	ΔH_{m2} (J/g)	
PHB	158.4/168.0	67.3	111.3	62.5	160.9	69.6	47.7
PHB/MC	159.0/168.2	67.3	112.3	64.2	160.9	72.3	50.3
PHB/MC-SIMA	158.9/168.2	69.1	112.8	64.9	160.2	73.6	51.4
PHB/MC-SIMA-MA	159.4/168.5	67.8	112.8	64.2	160.7	72.5	50.7
PHB/MC-SIV-MA	160.1/168.2	56.2	113.0	57.2	161.0	61.5	42.9

A single melting (T_{m2}) event was observed in the second heating cycle for all the composites and the small influence of cellulose on it, regardless of the treatment (Figure 6c). However, the degree of crystallinity was higher in composites than in PHB, except for PHB/MC-SIV-MA (Table 3), showing the nucleating effect of MC and modified celluloses, in agreement with previous results [11,34]. The increase in X_c in PHB composites may contribute to an improvement in their mechanical properties relative to the reference matrix.

The different influence of SIMA- and SIV-modified celluloses upon PHB crystallinity and its thermal behavior can be explained by the different chemical compositions of the two modifiers; methacrylic and polymethacrylic pendant units have better interaction with the PHB matrix as opposed to the less reactive vinyl silane group. In addition, the action of unreacted SIV as a crosslinker in PHB cannot be excluded. This is in agreement with the slightly higher melting temperature and lower crystallinity of the PHB/MC-SIV-MA composite. The crystallization behavior of PHB composites was also investigated by polarized light microscopy (POM), as discussed below.

3.3.3. Polarized Optical Microscopy

The formation of spherulites in PHB and in composites and their shape and size can be observed from the images obtained with the polarized optical microscope presented in Figure 7.

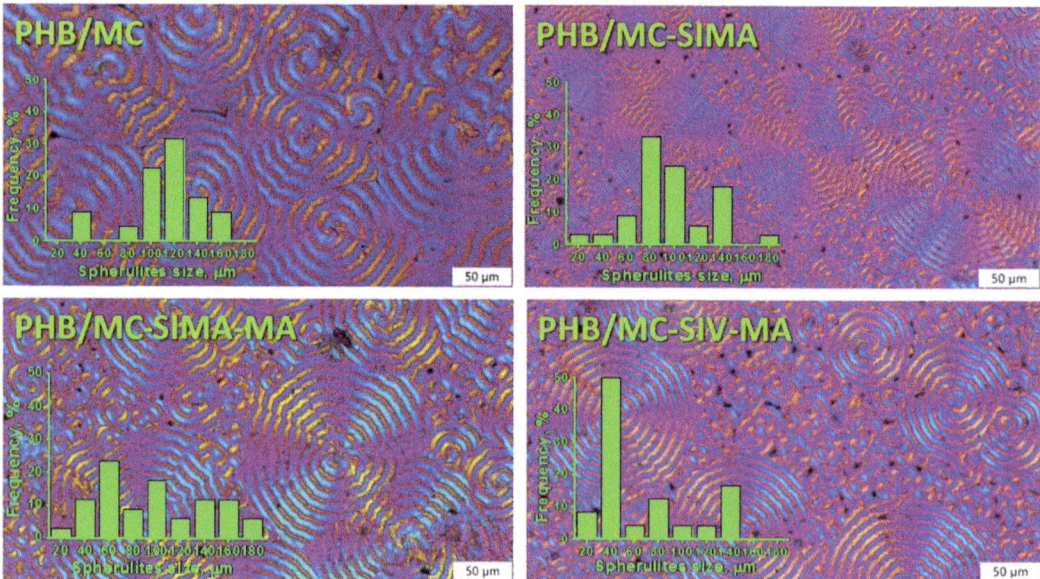

Figure 7. Polarized optical micrographs of PHB and composites with differently modified celluloses (×40); scale bar 50 µm.

POM images show ring-banded spherulites with the characteristic Maltese cross in both PHB and its composite. In PHB, the spherulites' size varied between 60 and 140 µm and in composites between 20 and 180 µm, with a greater proportion of smaller spherulites in the composites with treated celluloses, as observed in the histograms attached to the representative POM images (Figure 7). This supports the nucleating effect of celluloses, and of the modified celluloses especially, which is in agreement with the DSC results.

3.3.4. Dynamical Mechanical Analysis

The mechanical behavior of the composite films was investigated by DMA and the storage modulus (E') and tan δ variation with temperature are presented in Figure 8. The glass transition temperature of PHB (T_g), determined from the tan δ vs. temperature curve, was not changed by the addition of untreated MC but increased by 3–5 °C in the composites with modified celluloses (Table 4). The shift of the T_g to a higher temperature was due to the restriction of polymer chain movements. This may be caused by the interactions between

PHB and the modified celluloses [19,37], which are also indicated by the lower intensity of the tan δ peak in composites compared to PHB, except for PHB/MC-SIV-MA (Figure 8). The increased damping in the composite with MC-SIV-MA was a result of the higher content of amorphous phase in PHB/MC-SIV-MA, with 42.9% crystallinity instead of 50–51% for the other composites (Table 3). In addition, the breadth of the tan δ peak was larger for PHB/MC-SIMA-MA because of the difference in PHB chains' mobility; the movements of the PHB chains close to the modified cellulose fibers being much more restrained than those of the farthest ones [14]. Indeed, the methacrylate groups and polymethacrylate grafts on modified celluloses show good interactions with the PHB matrix thanks to their compatibility [22].

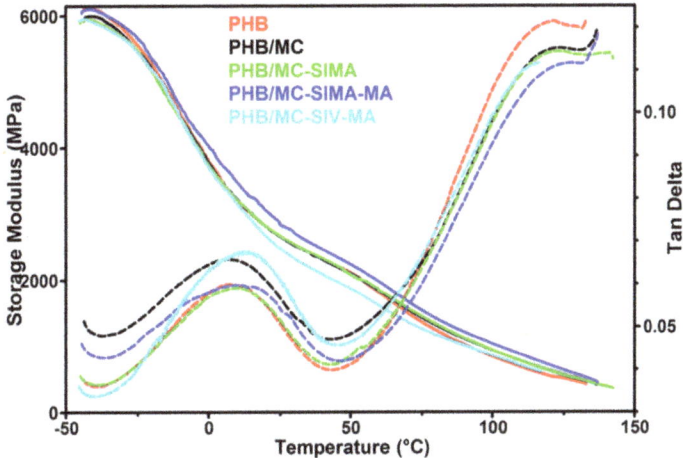

Figure 8. Storage modulus and tan δ of the composites vs. temperature.

Table 4. Storage modulus (E') of the composites at different temperatures, glass transition temperature (T_g) determined from tan δ vs. temperature curve and tan δ value at T_g.

Composites	PHB	PHB/MC	PHB/MC-SIMA	PHB/MC-SIMA-MA	PHB/MC-SIV-MA
$E'_{-25°C}$, MPa	5569	5484	5488	5674	5460
$E'_{0°C}$, MPa	3817	3783	3741	4041	3722
$E'_{25°C}$, MPa	2660	2659	2671	2844	2486
$E'_{50°C}$, MPa	2116	2099	2087	2242	1880
$E'_{100°C}$, MPa	844	938	943	1040	863
T_g, °C	6.8	6.9	9.3	10.9	12.2
tan δ	0.060	0.066	0.059	0.059	0.067
C	-	0.886	0.882	0.827	0.959

Microfibrillated celluloses had a small influence on the storage modulus of PHB, except for MC-SIMA-MA, which led to higher E' values on the whole tested temperature range (Table 4). Thus, an increase in the storage modulus by up to 23% was noticed in PHB/MC-SIMA-MA compared to the PHB reference. The reinforcing efficiency of MC and modified celluloses in PHB composites was assessed by the effectiveness coefficient (C), which is the ratio of the storage modulus values in the glassy and rubbery regions for the composite reported inthe similar ratio for the matrix [11]. The values of E' at −25 °C and 100 °C were used for the storage modulus in the glassy and rubbery regions. The lowest C values, corresponding to the highest reinforcing effectiveness of cellulose fibers, were obtained for the PHB/MC-SIMA-MA composites (Table 4). Therefore, the treatment of MC with SIMA and MA improved the compatibility of cellulose fibers with the PHB matrix and increased the mechanical properties. In contrast, the treatment of MC with SIV and MA led to an opposite effect. Indeed, the lowest E' values of almost the entire temperature range were obtained for PHB/MC-SIV-MA. The poor polymerization of methacrylic acid on the SIV-modified MC, as demonstrated by FTIR, may explain this behavior.

3.3.5. Tensile Properties of the Composites

The mechanical properties of PHB and composites, elongation at break, tensile strength at break (σ), and Young's modulus (M) are presented in Table 5, and the representative stress–strain curves in Figure 9. Without any surface treatment, MC had a poor effect on the mechanical properties of PHB; σ increased by 6%, which is in the limit of the experimental error, and M by 10%. A higher increase in the tensile strength was noticed in the composites with modified celluloses, in PHB/MC-SIMA by 13% and in PHB/MC-SIMA-MA by 18%. In the second composite, the Young's modulus increased by almost 30%. The increase in the mechanical properties was higher than that reported for a PHBV/2.5% nanofibrillated cellulose composite [34]. Thus, the reinforcing effect observed in the composite containing MC-SIMA-MA proved the effectiveness of this surface treatment of cellulose fibers, which increased the interfacial bonding between PHB and cellulose. Indeed, the polymerization reaction of methacrylic acid on the SIMA-modified cellulose resulted in a compatibilization with the PHB matrix. A drastic decrease in all mechanical properties was observed in PHB/MC-SIV-MA. The opposite effect for MC-SIV-MA may be due to the ineffective treatment of cellulose when vinyl silane groups were involved, as also demonstrated by FTIR. The degree of crystallinity also has a strong influence on the mechanical properties. The increase in crystallinity, determined by the nucleating effect of cellulose fibers, was similar in all composites with the exception of PHB/MC-SIV-MA (Table 3). In this case, X_c decreased by more than 10%, and this can be considered as an important cause of the drop in the mechanical properties and of inefficient treatment.

Table 5. Tensile properties data of composites.

Composites	PHB	PHB/MC	PHB/MC-SIMA	PHB/MC-SIMA-MA	PHB/MC-SIV-MA
Elongation at break, %	5.3 ± 0.6	4.9 ± 0.4	4.5 ± 0.6	5.0 ± 0.2	3.0 ± 0.5
Tensile strength at break, MPa	18.7 ± 1.9	19.8 ± 1.6	21.1 ± 0.7	22.0 ± 0.3	17.4 ± 2.0
Young's modulus, MPa	868 ± 58	954 ± 42	946 ± 61	1116 ± 12	966 ± 17

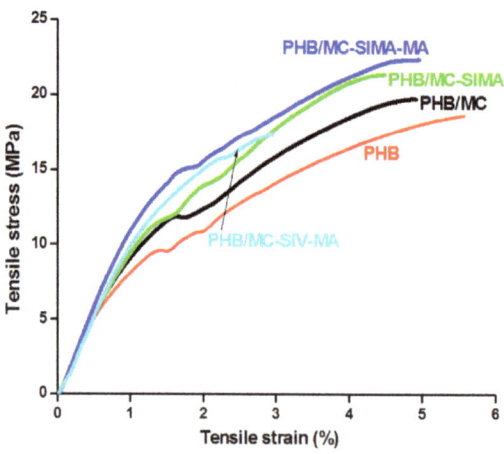

Figure 9. Representative stress–strain curves of PHB and composites.

Looking at the stress–strain curves of composites (Figure 9), one may observe that the reinforcing effect of MC-SIMA-MA and MC-SIMA was not followed by a strong decrease in elongation at break, as is the case for MC-SIV-MA and in the literature [34]. This behavior may result from a plasticizing effect of SIMA and polymethacrylic acid grafts. For verifying this hypothesis, the variation in time of the torque during the melt processing of the samples was analyzed (Figure 10).

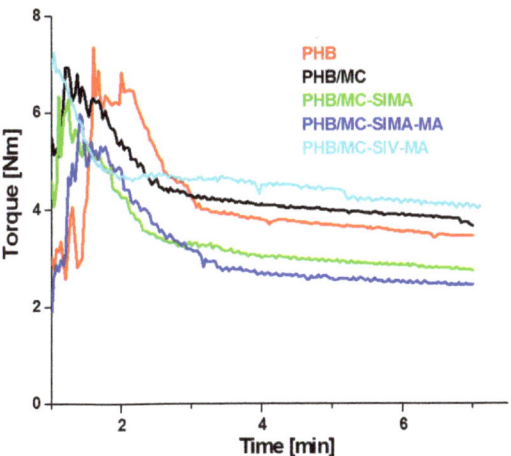

Figure 10. Torque vs. time diagrams recorded during the melt processing of PHB and composites.

The two composites, PHB/MC-SIMA and PHB/MC-SIMA-MA, showed lower viscosity than PHB and PHB/MC. Therefore, these treatments for the surface modification of MC not only have a compatibilizing effect in PHB, but also a plasticizing one. This is an important finding because the addition of fillers in PHB generally increases its brittleness, which is already large and deteriorates its processability. Therefore, the double role of SIMA-MA treatment, as both compatibilizer and plasticizer, could better solve the issues related to PHB application. Moreover, the overlap of the plasticizing effect of modified celluloses with the reinforcing effect can, to a certain extent, diminish the increase in tensile strength and modulus, leading to much more balanced stiffness–toughness properties.

3.3.6. Morphological Investigation of Composites

The SEM images in the fracture of PHB and composite plates are shown in Figure 11. In fracture, PHB shows pores of different sizes and several impurities. These are probably additives with different purposes than commercial PHB. The PHB composite with untreated MC shows many agglomerations of fibers, encircled with a red line in Figure 11 (PHB/MC).

In the SEM images of the composites with modified celluloses, the dispersion of cellulose fibers was much better, especially in PHB/MC-SIMA-MA. Many individual fibers, marked with green arrows, can be seen in these images. Their thickness was less than 100 nm. It should be mentioned that in the composites with modified celluloses the pores appeared only scarcely. The morphological features of the composites with modified celluloses support their improved mechanical behavior.

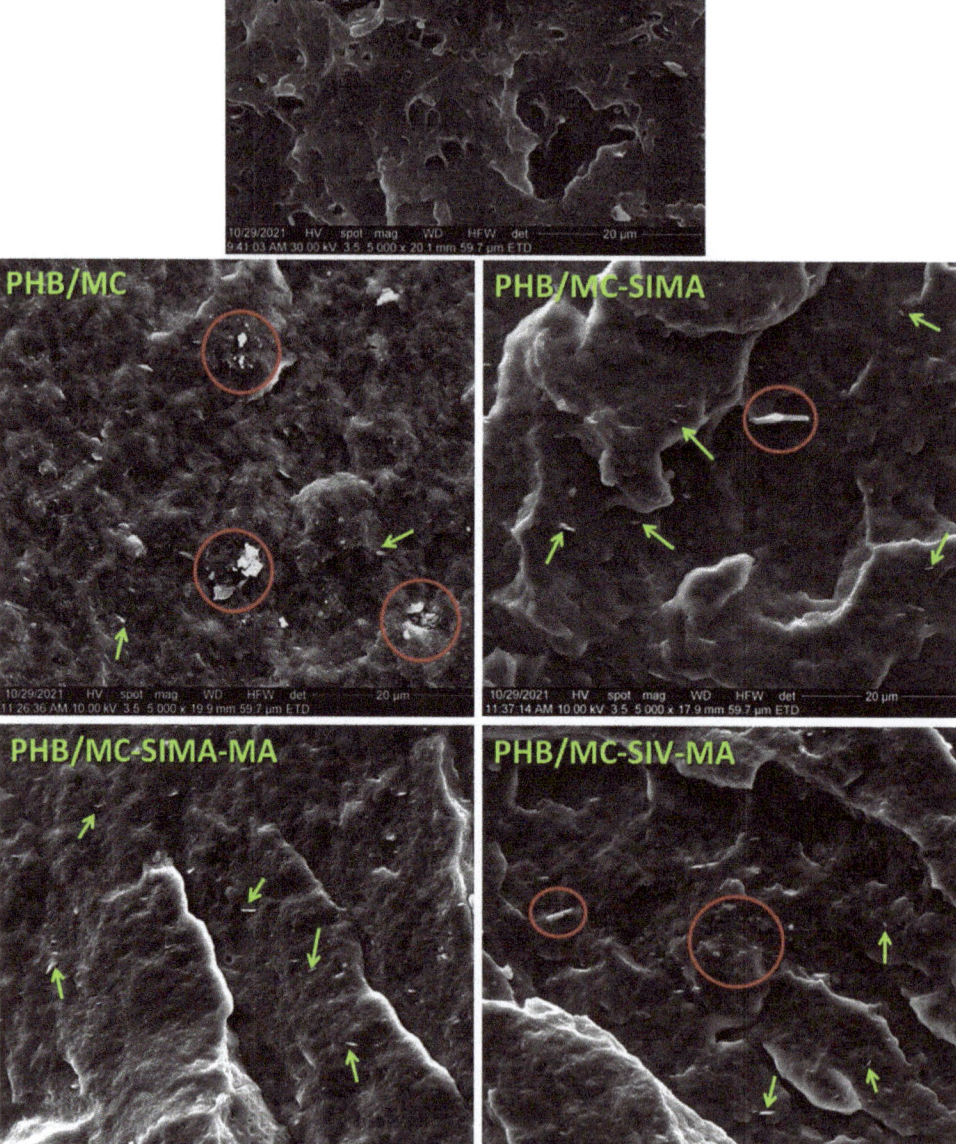

Figure 11. SEM images of PHB and composites frozen in liquid nitrogen and fractured.

4. Conclusions

This study was the first attempt to use a different strategy for the chemical modification of the surface of cellulose by firstly using silanes to create active sites, followed by the graft polymerization of methacrylic acid. FTIR spectroscopy confirmed the high efficiency of the SIMA and SIMA-MA treatments and the low efficiency of the SIV-MA treatment. The good adhesion between SIMA-modified celluloses and PHB and their good dispersion, highlighted by SEM, led to a significant improvement in the tensile strength and modulus in these composites; an opposite effect was noticed for PHB/MC-SIV-MA. Furthermore, the effectiveness coefficient, calculated from the storage modulus data, highlighted the reinforcing effect of MC-SIMA-MA and the unfavorable effect of MC-SIV-MA. The beneficial effect of SIMA-modified celluloses in PHB was also supported by the increased crystallinity and the smaller spherulites formed in this composite, as observed from the differential scanning calorimetry and polarized optical microscopy analyses. In addition, SIMA and SIMA-MA treatments had a dual role in achieving both compatibilization and plasticization. Therefore, this new method to modify cellulose fibers proved to be facile and effective when considering the improvements in the thermal and overall mechanical properties of the PHB matrix.

Author Contributions: Writing—original draft preparation, formal analysis, M.S.P.; funding acquisition, project administration, investigation, writing—review and editing, A.N.F.; formal analysis, I.C.R.; formal analysis, resources, P.O.S.; formal analysis, R.T.; formal analysis, V.R.; formal analysis, methodology, C.A.N.; formal analysis, A.R.G.; conceptualization, data curation, supervision, writing—review and editing, D.M.P. All authors have read and agreed to the published version of the manuscript.

Funding: This research and APC were funded by a grant from the Ministry of Research, Innovation and Digitization, CNCS/CCCDI—UEFISCDI, project number PN-III-P2-2.1-PED-2019-5002, no. 530/PED—EPOCEL, within PNCDI III.

Institutional Review Board Statement: Not applicable.

Informed Consent Statement: Not applicable.

Data Availability Statement: The data presented in this study are available on request from the corresponding author.

Conflicts of Interest: The authors declare no conflict of interest.

References

1. Friedlingstein, P.; Grasso, M.; Staiano, G.; Martorelli, M. Global carbon budget. *Earth Syst. Sci. Data* **2019**, *11*, 1783–1838. [CrossRef]
2. Payne, J.; Mc Keown, P.; Jones, M.D. A circular economy approach to plastic waste. *Polym. Degrad. Stab.* **2019**, *165*, 170–181. [CrossRef]
3. Khalid, M.Y.; Al Rashid, A.; Arif, Z.U.; Ahmed, W.; Arshad, H. Recent advances in nanocellulose-based different biomaterials: Types, properties, and emerging applications. *J. Mater. Res. Technol.* **2021**, *14*, 2601–2623. [CrossRef]
4. Panaitescu, D.M.; Frone, A.N.; Ghiurea, M.; Chiulan, I. Influence of storage conditions on starch/PVA films containing cellulose nanofibers. *Ind. Crops Prod.* **2015**, *70*, 170–177. [CrossRef]
5. Raza, Z.A.; Abid, S.; Banat, I.M. Polyhydroxyalkanoates: Characteristics, production, recent developments and applications. *Int. Biodeter. Biodegr.* **2018**, *126*, 45–56. [CrossRef]
6. Kamravamanesh, D.; Pflügl, S.; Nischkauer, W.; Limbeck, A.; Lackner, M.; Herwig, C. Photosynthetic poly-β-hydroxybutyrate accumulation in unicellular cyanobacterium Synechocystis sp. PCC 6714. *AMB Express* **2017**, *7*, 143. [CrossRef]
7. Yeo, J.C.C.; Muiruri, J.K.; Thitsartarn, W.; Li, Z.; He, C. Recent advances in the development of biodegradable PHB-based toughening materials: Approaches, advantages and applications. *Mat. Sci. Eng. C* **2018**, *92*, 1092–1116. [CrossRef]
8. El-Hadi, A. Investigation of the effect of nanoclay type on the non-isothermal crystallization kinetics and morphology of poly(3(R)-hydroxybutyrate) PHB/clay nanocomposites. *Polym. Bull.* **2014**, *71*, 1449–1470. [CrossRef]
9. Cai, Z.; Xiong, P.; He, S.; Zhu, C. Improved piezoelectric performances of highly orientated poly(β-hydroxybutyrate) electrospun nanofiber membrane scaffold blended with multiwalled carbon nanotubes. *Mater. Lett.* **2019**, *240*, 213–216. [CrossRef]

10. Vadahanambi, S.; Lee, I.; Chun, H.; Park, H. Graphene reinforced biodegradable poly (3-hydroxybutyrate-co-4-hydroxybutyrate) nano-composites. *Express Polym. Lett.* **2013**, *7*, 320–328.
11. Panaitescu, D.M.; Nicolae, C.A.; Gabor, A.R.; Trusca, R. Thermal and mechanical properties of poly(3-hydroxybutyrate) reinforced with cellulose fibers from wood waste. *Ind. Crops Prod.* **2020**, *145*, 112071. [CrossRef]
12. Omran, A.A.B.; Mohammed, A.A.B.A.; Sapuan, S.M.; Ilyas, R.A.; Asyraf, M.R.M.; RahimianKoloor, S.S.; Petru, M. Micro- and nanocellulose in polymer composite materials: A review. *Polymers* **2021**, *13*, 231. [CrossRef] [PubMed]
13. Moon, R.J.; Martini, A.; Nairn, J.; Simonsen, J.; Youngblood, J. Cellulose nanomaterials review: Structure, properties and nanocomposites. *Chem. Soc. Rev.* **2011**, *40*, 3941–3994. [CrossRef]
14. Ten, E.; Turtle, J.; Bahr, D.; Jiang, L.; Wolcott, M. Thermal and mechanical properties of poly(3-hydroxybutyrate-co-3-hydroxyvalerate)/cellulose nanowhiskers composites. *Polymer* **2010**, *51*, 2652–2660. [CrossRef]
15. Panaitescu, D.M.; Frone, A.N.; Chiulan, I.; Nicolae, C.A.; Trusca, R.; Ghiurea, M.; Gabor, A.R.; Mihailescu, M.; Casarica, A.; Lupescu, I. Role of bacterial cellulose and poly (3-hydroxyhexanoate-co-3-hydroxyoctanoate) in poly (3-hydroxybutyrate) blends and composites. *Cellulose* **2018**, *25*, 5569–5591. [CrossRef]
16. Nagarajan, V.; Misra, M.; Mohanty, A.K. New engineered biocomposites from poly(3-hydroxybutyrate-co-3-hydroxyvalerate) (PHBV)/poly(butylene adipate-co-terephthalate) (PBAT) blends and switch grass: Fabrication and performance evaluation. *Ind. Crops Prod.* **2013**, *42*, 461–468. [CrossRef]
17. Ghasemlou, M.; Daver, F.; Ivanova, E.P.; Habibi, Y.; Adhikari, B. Surface modifications of nanocellulose: From synthesis to high-performance nanocomposites. *Prog. Polym. Sci.* **2021**, *119*, 101418. [CrossRef]
18. Oprea, M.; Panaitescu, D.M.; Nicolae, C.A.; Gabor, A.R.; Frone, A.N.; Raditoiu, V.; Trusca, R.; Casarica, A. Nanocomposites from functionalized bacterial cellulose and poly(3-hydroxybutyrate-co-3-hydroxyvalerate). *Polym. Degrad. Stab.* **2020**, *179*, 109203. [CrossRef]
19. Panaitescu, D.M.; Vizireanu, S.; Stoian, S.A.; Nicolae, C.-A.; Gabor, A.R.; Damian, C.M.; Trusca, R.; Carpen, L.G.; Dinescu, G. Poly(3-hydroxybutyrate) modified by plasma and TEMPO-oxidized celluloses. *Polymers* **2020**, *12*, 1510. [CrossRef]
20. Sánchez-Safont, E.L.; Aldureid, A.; Lagarón, J.M.; Cabedo, L.; Gámez-Pérez, J. Study of the compatibilization effect of different reactive agents in PHB/natural fiber-based composites. *Polymers* **2020**, *12*, 1967. [CrossRef]
21. Chiu, H.J.; Chen, H.L.; Lin, T.L.; Lin, J.S. Phase structure of poly(3-hydroxy butyrate)/poly(vinyl acetate) blends probed by small-angle X-ray scattering. *Macromolecules* **1999**, *32*, 4969–4974. [CrossRef]
22. An, Y.; Dong, L.; Li, G.; Mo, Z.; Feng, Z. Miscibility, crystallizationkinetics, and morphology of poly(β-hydroxybutyrate) andpoly(methylacrylate) blends. *J. Polym. Sci. B Polym. Phys.* **2000**, *38*, 1860–1867. [CrossRef]
23. Vitta, S.B.; Stahel, E.P.; Stannett, V.T. The preparation and properties of acrylic and methacrylic acid grafted cellulose prepared by ceric ion initiation. Part I. Preparation of the grafted cellulose. *J. Macromol. Sci. A* **1985**, *22*, 579–590. [CrossRef]
24. Eldin, M.S.M.; Rahman, S.A.; Fawal, G.F.E. Preparationandcharacterization of grafted cellophane membranes for affinity separation of His-tagChitinase. *Adv. Polym. Technol.* **2011**, *30*, 191–202. [CrossRef]
25. Barham, P.J.; Keller, A.; Otun, E.L.; Holmer, P.A. Crystallization and morphology of a bacterial thermoplastic: Poly-3-hydroxybutyrate. *J. Mater. Sci.* **1984**, *19*, 2781. [CrossRef]
26. Ovalle-Serrano, S.A.; Gómez, F.N.; Blanco-Tirado, C.; Combariza, M.Y. Isolation and characterization of cellulose nanofibrils from Colombian Fique decortication by-products. *Carbohydr. Polym.* **2018**, *189*, 169–177. [CrossRef]
27. Pantoja, M.; Díaz-Benito, B.; Velasco, F.; Abenojar, J.; del Real, J.C. Analysis of hydrolysis process of γ-methacryloxypropyltrimethoxysilane and its influence on the formation of silane coatings on 6063 aluminum alloy. *Appl. Surf. Sci.* **2009**, *255*, 6386–6390. [CrossRef]
28. Abdelmouleh, M.; Boufi, S.; Belgacem, M.N.; Dufresne, A. Short natural-fibre reinforced polyethylene and natural rubber composites: Effect of silane coupling agents and fibres loading. *Compos. Sci. Technol.* **2007**, *67*, 1627–1639. [CrossRef]
29. Lumbreras-Aguayo, A.; Meléndez-Ortiz, H.I.; Puente-Urbina, B.; Alvarado-Canché, C.; Ledezma, A.; Romero-García, J.; Betancourt-Galindo, R. Poly(methacrylic acid)-modified medical cotton gauzes with antimicrobial and drug delivery properties for their use as wound dressings. *Carbohydr. Polym.* **2019**, *205*, 203–210. [CrossRef]
30. Kakko, T.; King, A.W.T.; Kilpeläinen, I. Homogenous esterification of cellulose pulp in DBNH.OAc. *Cellulose* **2017**, *24*, 5341–5354. [CrossRef]
31. Frone, A.N.; Panaitescu, D.M.; Chiulan, I.; Nicolae, C.A.; Casarica, A.; Gabor, A.R.; Trusca, R.; Damian, C.M.; Purcar, V.; Alexandrescu, E.; et al. Surface Treatment of Bacterial Cellulose in Mild, Eco-Friendly Conditions. *Coatings* **2018**, *8*, 221. [CrossRef]
32. Castro Cabrera, I.; Berlioz, S.; Fahs, A.; Louarn, G.; Carriere, P. Chemical functionalization of nano fibrillated cellulose by glycidyl silane coupling agents: A grafted silane network characterization study. *Int. J. Biol. Macromol.* **2020**, *165*, 1773–1782. [CrossRef]
33. Sarmento, V.H.V.; Schiavetto, M.G.; Hammer, P.; Benedetti, A.V.; Fugivara, C.S.; Suegama, P.H.; Pulcinelli, S.H.; Santilli, C.V. Corrosion protection of stainless steel by polysiloxane hybrid coatings prepared using the sol-gel process. *Surf. Coat. Technol.* **2010**, *204*, 2689–2701. [CrossRef]
34. Srithep, Y.; Ellingham, T.; Peng, J.; Sabo, R.; Clemons, C.; Turng, L.S.; Pilla, S. Melt compounding of poly (3-hydroxybutyrate-co-3-hydroxyvalerate)/nanofibrillated cellulose nanocomposites. *Polym. Degrad. Stab.* **2013**, *98*, 1439–1449. [CrossRef]
35. Saïdi, L.; Vilela, C.; Oliveira, H.; Silvestre, A.J.D.; Freire, C.S.R. Poly(N-methacryloyl glycine)/nanocellulose composites as pH-sensitive systems for controlled release of diclofenac. *Carbohydr. Polym.* **2017**, *169*, 357–365. [CrossRef] [PubMed]

36. Seoane, I.T.; Cerrutti, P.; Vazquez, A.; Cyras, V.P.; Manfredi, L.B. Ternary nanocomposites based on plasticized poly(3-hydroxybutyrate) and nanocellulose. *Polym. Bull.* **2019**, *76*, 967–988. [CrossRef]
37. Iggui, K.; Le Moigne, N.; Kaci, M.; Cambe, S.; Degorce-Dumas, J.R.; Bergeret, A. A biodegradation study of poly(3-hydroxybutyrate-co-3-hydroxyvalerate)/organoclay nanocomposites in various environmental conditions. *Polym. Degrad. Stab.* **2015**, *119*, 77–86. [CrossRef]

Communication

Biosynthesis of Polyhydroxybutyrate with Cellulose Nanocrystals Using *Cupriavidus necator*

Giyoung Shin [1], Da-Woon Jeong [1], Hyeri Kim [1], Seul-A Park [1], Semin Kim [1], Ju Young Lee [1], Sung Yeon Hwang [1,2,*], Jeyoung Park [1,2,*] and Dongyeop X. Oh [1,2,*]

[1] Research Center for Bio-Based Chemistry, Korea Research Institute of Chemical Technology (KRICT), Ulsan 44429, Korea; sky77@krict.re.kr (G.S.); dawoon@krict.re.kr (D.-W.J.); hr0962@krict.re.kr (H.K.); seula@krict.re.kr (S.-A.P.); seminkim@krict.re.kr (S.K.); juylee@krict.re.kr (J.Y.L.)

[2] Advanced Materials and Chemical Engineering, University of Science and Technology (UST), Daejeon 34113, Korea

* Correspondence: crew75@krict.re.kr (S.Y.H.); jypark@krict.re.kr (J.P.); dongyeop@krict.re.kr (D.X.O.)

Citation: Shin, G.; Jeong, D.-W.; Kim, H.; Park, S.-A.; Kim, S.; Lee, J.Y.; Hwang, S.Y.; Park, J.; Oh, D.X. Biosynthesis of Polyhydroxybutyrate with Cellulose Nanocrystals Using *Cupriavidus necator*. *Polymers* **2021**, *13*, 2604. https://doi.org/10.3390/polym13162604

Academic Editors: Denis Mihaela Panaitescu and Adriana Nicoleta Frone

Received: 22 June 2021
Accepted: 3 August 2021
Published: 5 August 2021

Publisher's Note: MDPI stays neutral with regard to jurisdictional claims in published maps and institutional affiliations.

Copyright: © 2021 by the authors. Licensee MDPI, Basel, Switzerland. This article is an open access article distributed under the terms and conditions of the Creative Commons Attribution (CC BY) license (https://creativecommons.org/licenses/by/4.0/).

Abstract: Polyhydroxybutyrate (PHB) is a natural polyester synthesized by several microorganisms. Moreover, it has excellent biodegradability and is an eco-friendly material because it converts water and carbon dioxide as final decomposition products. However, the applications of PHB are limited because of its stiffness and brittleness. Because cellulose nanocrystals (CNCs) have excellent intrinsic mechanical properties such as high specific strength and modulus, they may compensate for the insufficient physical properties of PHB by producing their nanocomposites. In this study, natural polyesters were extracted from *Cupriavidus necator* fermentation with CNCs, which were well-dispersed in nitrogen-limited liquid culture media. Fourier-transform infrared spectroscopy results revealed that the additional O–H peak originating from cellulose at 3500–3200 cm^{-1} was observed for PHB along with the C=O and –COO bands at 1720 cm^{-1}. This suggests that PHB–CNC nanocomposites could be readily obtained using *C. necator* fermented in well-dispersed CNC-supplemented culture media.

Keywords: polyhydroxybutyrate; natural polyester; cellulose nanocrystals; nanocomposites

1. Introduction

Synthetic plastics are ubiquitous in modern human life. Moreover, they have a wide range of applications owing to their excellent processability, mechanical properties, and low cost. Most petroleum-based synthetic polymers have accumulated consistently because of their high resistance to nature, resulting in serious environmental pollution. Therefore, researches to develop bio-based or oil-based biodegradable materials that can substitute conventional plastics have been actively conducted.

Polyhydroxybutyrate (PHB) is one of the representative natural polyesters of the polyhydroxyalkanoate (PHA) family and is an energy-storage product synthesized by various microorganisms in vivo [1–3]. PHB, a thermoplastic polymer that has the advantages of good biodegradability and biocompatibility, is evaluated as an alternative polymer with high commercial value for industrial materials, drug delivery systems, and pharmaceuticals [4,5]. However, owing to the stiff and brittle characteristics of the PHB materials, their applications are limited. For commercial applications and expanding the range of applications of PHB, several studies to improve the physical properties while maintaining the advantages of PHB by making it in a blend or composites are being conducted. PHB composites incorporated with Poly(ethylene glycol) and cellulose nanowhiskers showed a larger processing window and higher elongation at break than neat PHB [6]. Plasma-treated PHB/bacterial cellulose nanocomposites showed improved mechanical properties and antibacterial activity [7]. Blended polylactic acid (PLA) and PHB composites exhibited higher toughness than neat PHB [8].

Cupriavidus necator is a gram-negative β-proteobacterium that has a natural biosynthetic pathway to produce PHB [9,10]. Under nutrient limitation, *C. necator* can produce PHB in granules by redirecting carbon flux exceeding 70% of its dry cell weight for energy and carbon storage [11–13]. *C. necator* is a promising microorganism that can be used to produce PHB from various carbon sources, including waste or non-edible substances such as waste cooking oil or agricultural wastes [14,15]. Furthermore, *C. necator* is recognized as a useful strain for developing alternative materials for petroleum-based plastics and solving the CO_2 problem because it has the ability to produce PHB through CO_2 fixation [16,17].

Cellulose nanocrystals (CNCs) are rod-shaped colloidal particles with a diameter and length of 3–20 nm and 50–3000 nm, respectively, and are generally prepared from cellulose by hydrolysis with sulfuric acid [18,19]. CNCs have attracted significant attention as natural fillers because of their high aspect ratio, large specific surface area, high mechanical strength, and high elastic modulus. CNCs are used as composites of other polymers to enhance the mechanical strength or increase the interfacial adhesion between materials [20–22]. In addition, CNCs have several advantages such as low cost owing to the high abundance of cellulose on earth, excellent biocompatibility, and eco-friendliness. Various approaches for the modification of PHB with CNCs have been developed to enhance the thermal stability, strength, and stiffness through blending or preparation of composites [23–25].

In this study, PHB–CNC nanocomposites were obtained through one-pot biosynthesis and nanocomposite preparation. By dispersing the CNCs well in the culture medium, the conditions were optimized to fabricate the PHB–CNC nanocomposites during the extraction of PHB biosynthesized from *C. necator*. The effects of the CNCs on bacterial cell growth and chemical composition of the extracted polymer were investigated.

2. Materials and Methods

2.1. Materials

CNCs were purchased from the Process Development Center (PDC) at the University of Maine (Orono, ME, USA). All reagents for cell culture were purchased from BD Biosciences (Franklin Lakes, NJ, USA). All chemical reagents were purchased from Sigma-Aldrich (St. Louis, MO, USA).

2.2. Microorganism and Culture Media

In this study, *C. necator* (KCTC 22469) obtained from the Korean Collection for Type Cultures (Jeongeup, Korea) was used for the biosynthesis of PHB. To obtain the seed culture, *C. necator* was grown in an LB broth at 30 °C for 24 h under agitation at 250 rpm. For PHB production, the seed culture (0.5 mL) was transferred to a 250 mL flask containing 50 mL of culture media, which contained 1 g/L peptone, 1 g/L beef extract, 20 g/L glucose, and 14.5 g/L NaCl. The CNCs powder was added to the culture media at a concentration of 0.01–0.1 g/L, and then the CNCs were dispersed by ultrasonic treatment using a vibracell ultrasonic processor (Sonics and Materials Inc., Newtown, CT, USA), followed by transferring the seed culture.

2.3. Extraction of PHB

PHB was recovered from *C. necator* using sodium hypochlorite [26]. For PHB extraction, 50 mL of the culture broth was centrifuged at 4000 rpm for 10 min. The cell pellet was washed twice with 10 mM phosphate-buffered saline (pH 7.4). Following centrifugation, the cell pellet was resuspended in 5 mL of sodium hypochlorite solution (13% v/v) and incubated at 30 °C for 1 h at 250 rpm. After centrifugation and removal of the supernatant, the polymer was washed twice with deionized water and once with ethanol. The extracted PHB was dried in a vacuum oven.

2.4. Measurement of Cell Growth

The effect of CNCs on the growth of *C. necator* was observed using a ultraviolet-visible (UV-Vis) spectrophotometer (UV-2600, Shimadzu, Tokyo, Japan). Optical density was measured at 600 nm.

2.5. Biological Transmission Electron Microscopy

Intracellular PHB granules were observed by biological transmission electron microscopy (bio-TEM) using an FEI Tecnai G2 F20 TWIN TMP microscope (FEI, Hillsboro, OR, USA). After fermentation, the cells were fixed in a 4% glutaraldehyde solution at 4 °C for 4 h at 4 °C and then post-fixed in 1% osmium tetroxide.

2.6. Fourier-Transform Infrared Spectroscopy

To characterize PHB recovered from *C. necator* fermentation, Fourier-transform infrared (FTIR) spectroscopy was performed using Nicolet iS50 (Thermo Fisher Scientific, Waltham, MA, USA). The polymers were scanned 128 times at a resolution of 4 cm^{-1} in the wavelength range of 800–4000 cm^{-1}.

2.7. Mechanical Properties Measurement

The PHB films were prepared by solvent casting method. PHB or PHB-CNC nanocomposites extracted from *C. necator* were dissolved in chloroform and poured into the aluminum dishes at ambient temperature. The solution was dried for 5 days to evaporate chloroform, and then vacuum dried overnight to remove residual solvent. The prepared PHB films were cut into rectangular shapes (30 mm × 5 mm × 0.15 mm) for the tensile test. A universal testing machine (Model 5943, Instron, UK) was used to determine the mechanical properties of the films. Tensile properties were examined at 10 mm min^{-1} with a 50 N load cell.

2.8. Water Contact Angle Measurement

The water contact angle of PHB films was measured by dropping 1 µL of deionized water to the surface of the film using a contact angle analyzer DSA25 Basic (KRUSS, Hamburg, Germany).

2.9. Nuclear Magnetic Resonance Spectroscopy

^1H and ^{13}C nuclear magnetic resonance (NMR) spectroscopy were conducted using an AVANCE NEO 600 (Bruker, Billerica, MA, USA) at 600 MHz and 150 MHz, respectively.

3. Results and Discussion

3.1. Effect of CNCs on the Growth of C. necator

PHB biosynthesis by *C. necator* was conducted under nitrogen-limited conditions using media containing 20 g/L of glucose, 1 g/L of peptone, and 1 g/L of beef extract. After 24 h of fermentation, the cells were harvested, and intracellular PHB was observed using TEM. The TEM image in Figure 1a shows the formation of intracellular PHB granules under the given fermentation conditions.

The development of a system to fabricate PHB–CNC nanocomposites using *C. necator* and the effect of CNCs on bacterial growth was evaluated for different concentrations of CNCs. To determine the effect of CNCs on the growth of *C. necator*, bacterial cultivation was conducted by adding CNCs in the culture media at concentrations ranging from 0 to 0.1 g/L. As shown in Figure 1b, the bacterial growth was not significantly different after 24 h of cultivation with CNCs compared to the control. This result indicates that the CNCs did not inhibit the growth of *C. necator* at a concentration of 0.1 g/L.

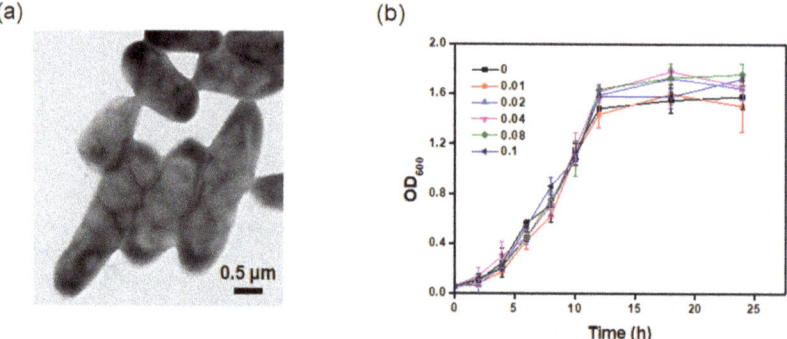

Figure 1. (a) TEM image of *C. necator* with PHB granules and (b) growth of *C. necator* after 24 h of incubation with different concentrations of CNCs (0–0.1 g/L).

3.2. Characterization of the PHB–CNC Nanocomposite

The effects of adding CNCs and NaCl to the culture media during PHB synthesis were characterized using FTIR spectroscopy. Figure 2 shows the FTIR spectra of PHB synthesized using *C. necator* under various conditions supplemented with CNCs and NaCl. The spectra of PHB with 0.1 g/L CNC and PHB with 14.5 g/L NaCl closely matched the spectrum of the control PHB. The peak at 2934–2977 cm^{-1} corresponds to –CH, and the band at 1720 cm^{-1} indicates the presence of C=O and –COO bonds [8]. The additional broad band at 3500–3200 cm^{-1} was observed from PHB synthesized in the culture media containing 0.1 g/L CNCs and 14.5 g/L NaCl. This transmittance peak indicates the presence of O–H stretching vibration in cellulose [27]. When NaCl was added, the surface charge of both bacteria and CNCs became less negative [28]. It is speculated that the CNCs could be well-dispersed and the bacterial cells could be surrounded by the CNCs owing to weaker electrical repulsion between bacteria and CNCs.

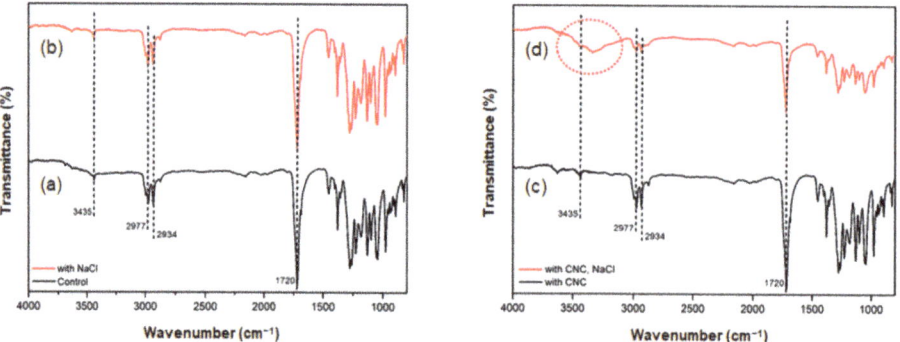

Figure 2. FTIR spectra of a PHB film obtained from *C. necator* incubated (a) without additives, (b) with 14.5 g/L NaCl, (c) with 0.1 g/L CNC, and (d) with 0.1 g/L CNC and 14.5 g/L NaCl.

To determine the effective concentration of CNCs for the preparation of PHB–CNC nanocomposite, PHB was produced supplemented with 0–0.1 g/L CNCs and 14.5 g/L NaCl. After cultivation, extracted PHB was characterized through FTIR spectroscopy. As shown in Figure 3, the PHB–CNC nanocomposites were obtained in the CNC suspension at a concentration of ≥0.04 g/L.

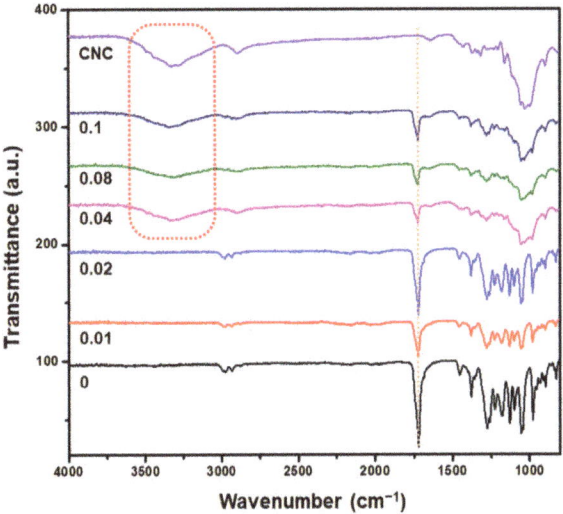

Figure 3. FTIR spectra of PHB synthesized by *C. necator* with different concentrations of CNCs.

Figure 4a displays the mechanical properties of PHB and PHB-CNC nanocomposite. The tensile strength and elongation at break of PHB were measured to be 22 MPa and 4%. The PHB-CNC nanocomposite fermented with 0.1 g/L CNC show the higher tensile strength (27 MPa) and elongation at break (5%) than neat PHB. Water contact angle measurements for PHB and the nanocomposite were conducted to investigate hydrophilicity on the surface. As shown in Figure 4b, the contact angle value was decreased from $84.5 \pm 2.9°$ (neat PHB) to $74.3 \pm 1.4°$ (PHB-CNC nanocomposite). It suggests that the incorporation of CNC increases the hydrophilicity of PHB due to the abundant hydroxyl groups of CNC. Moreover, ^{1}H NMR and ^{13}C NMR were used to determine the chemical structure of the polymer matrix. As shown in ^{1}H NMR and ^{13}C NMR spectra (Figures S1 and S2 in the Supplementary Materials), it can be seen there was no difference in the structure of polymer matrix between neat PHB and PHB-CNC nanocomposite. In other words, CNC supplement does not affect the chemical structure of PHB in the biosynthesis process.

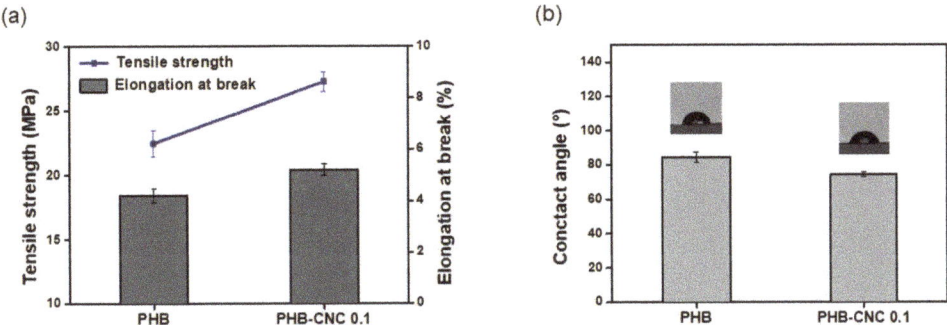

Figure 4. (**a**) Tensile strength and elongation at break, (**b**) water contact angle of PHB and PHB-CNC nanocomposite (0.1 g/L CNC).

3.3. Effects of CNCs Supplemented before and after Cultivation for the Nanocomposite Formation

Figure 5 shows the FTIR spectra of extracted PHBs under various conditions, that is, in the absence of CNCs, presence of 0.04 g/L CNCs, and 0.04 g/L CNCs added after fermentation. When the CNCs were added after cultivation, the FTIR spectrum of the extracted polymer showed a faint peak around 1720 cm^{-1}, representing carbonyl and ester bonds. This spectrum corresponds well with the results of CNCs. This FTIR result indicates that the polymer did not form CNC nanocomposites and that the polymers, which were extracted biasedly by the CNCs or PHB, were coated with CNCs during extraction.

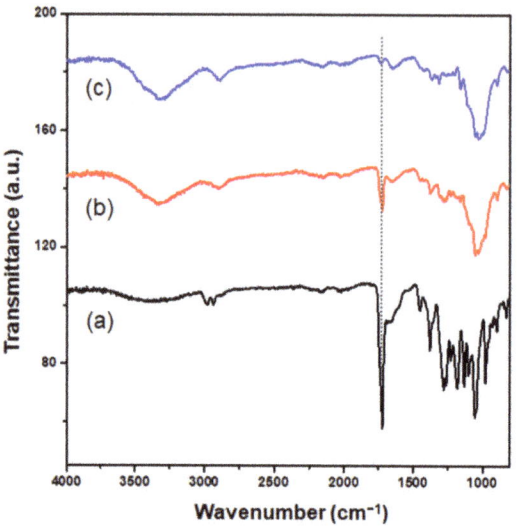

Figure 5. FTIR spectra of PHB produced from *C. necator* fermentation (**a**) without CNCs, (**b**) with 0.04 g/L CNCs, and (**c**) with 0.04 g/L CNCs added after cultivation.

Comparing the FTIR peaks of the CNC-supplemented polymers before and after fermentation, it can be concluded that the PHB–CNC nanocomposites can be obtained by adding CNCs in the initial state of cultivation. It was assumed that PHB was extracted with CNCs because of the colloidal stability of the CNCs in the suspension during bacterial growth and fermentation. During cultivation, the CNCs were well-dispersed around the cell. Consequently, the PHB–CNC nanocomposites can be obtained.

4. Conclusions

In this study, PHB–CNC nanocomposites were obtained using CNC-dispersed culture media and PHB-producing *C. necator*. The culture conditions were optimized by adding NaCl to reduce the electrical repulsion between the CNCs and bacterial cells, thereby resulting in a well-dispersed colloid suspension. By investigating the characteristics of the extracted polymer according to the concentration of CNCs, it was confirmed that the PHB–CNC nanocomposites could be obtained when 0.04 g/L or more CNCs were added. PHB-CNC nanocomposite obtained from *C. necator* through this fermentation process showed higher mechanical properties and hydrophilicity than neat PHB. In addition, when CNCs were added after PHB fermentation, the nanocomposites could not be obtained, which is expected because CNCs cause bacterial flocculation and thus the degree of dispersion between the CNCs and bacterial cells decreases. These results suggest that synthesis of PHB-CNC nanocomposite can be achieved by applying culture media supplemented various concentrations of CNC, expanding their applications as alternative biodegradable polyester materials.

Supplementary Materials: The following are available online at https://www.mdpi.com/article/10.3390/polym13162604/s1, Figure S1: ^{1}H NMR spectra of PHB (a) without CNC and (b) with 0.1 g/L CNC., Figure S2: ^{13}C NMR spectra of PHB (a) without CNC and (b) with 0.1 g/L CNC.

Author Contributions: Data curation was performed by G.S., J.P. and D.X.O. Formal analysis was conducted by D.-W.J., H.K., S.-A.P. and S.K. Methodology was designed by J.Y.L. Visualization was performed by G.S. and S.Y.H. Conceptualization by D.-W.J. and D.X.O. The original draft was written by G.S., and the draft was reviewed and edited by S.Y.H., J.P. and D.X.O. All authors have read and agreed to the published version of the manuscript.

Funding: This work was supported by the Bio-Industrial Technology Development Program (20008628) funded by the Ministry of Trade, Industry & Energy (MI, Korea) and the Korea Research Institute of Chemical Technology (KRICT) core project (SS2142-10).

Institutional Review Board Statement: Not applicable.

Informed Consent Statement: Not applicable.

Data Availability Statement: The data presented in this study are available on reasonable request from the corresponding author.

Conflicts of Interest: The authors declare no conflict of interest.

References

1. Li, Z.; Yang, J.; Loh, X.J. Polyhydroxyalkanoates: Opening doors for a sustainable future. *NPG Asia Mater.* **2016**, *8*, e265. [CrossRef]
2. Sudesh, K.; Abe, H.; Doi, Y. Synthesis, structure and properties of polyhydroxyalkanoates: Biological polyesters. *Prog. Polym. Sci.* **2000**, *25*, 1503–1555. [CrossRef]
3. Aragosa, A.; Specchia, V.; Frigione, M. Isolation of two bacterial species from argan soil in morocco associated with polyhydroxybutyrate (PHB) accumulation: Current potential and future prospects for the bio-based polymer production. *Polymers* **2021**, *13*, 1870. [CrossRef]
4. Elmowafy, E.; Abdal-Hay, A.; Skouras, A.; Tiboni, M.; Casettari, L.; Guarino, V. Polyhydroxyalkanoate (PHA): Applications in drug delivery and tissue engineering. *Expert Rev. Med. Devices* **2019**, *16*, 467–482. [CrossRef] [PubMed]
5. Bonartsev, A.P.; Bonartseva, G.A.; Reshetov, I.V.; Kirpichnikov, M.P.; Shaitan, K.V. Application of polyhydroxyalkanoates in medicine and the biological activity of natural Poly(3-hydroxybutyrate). *Acta Nat.* **2019**, *11*, 4–16. [CrossRef] [PubMed]
6. S. de O. Patrício, P.; Pereira, F.V.; dos Santos, M.C.; de Souza, P.P.; Roa, J.P.B.; Orefice, R.L. Increasing the elongation at break of polyhydroxybutyrate biopolymer: Effect of cellulose nanowhiskers on mechanical and thermal properties. *J. Appl. Polym. Sci.* **2013**, *127*, 3613–3621. [CrossRef]
7. Panaitescu, D.M.; Ionita, E.R.; Nicolae, C.A.; Gabor, A.R.; Ionita, M.D.; Trusca, R.; Lixandru, B.E.; Codita, I.; Dinescu, G. Poly(3-hydroxybutyrate) modified by nanocellulose and plasma treatment for packaging applications. *Polymers* **2018**, *10*, 1249. [CrossRef] [PubMed]
8. Zhang, M.; Thomas, N.L. Blending polylactic acid with polyhydroxybutyrate: The effect on thermal, mechanical, and biodegradation properties. *Adv. Polym. Technol.* **2011**, *30*, 67–79. [CrossRef]
9. Pradhan, S.; Dikshit, P.K.; Moholkar, V.S. Production, ultrasonic extraction, and characterization of Poly(3-hydroxybutyrate) (PHB) using *Bacillus megaterium* and *Cupriavidus necator*. *Polym. Adv. Technol.* **2018**, *29*, 2392–2400. [CrossRef]
10. Pavan, F.A.; Junqueira, T.L.; Watanabe, M.D.B.; Bonomi, A.; Quines, L.K.; Schmidell, W.; de Aragao, G.M.F. Economic analysis of polyhydroxybutyrate production by *Cupriavidus necator* using different routes for product recovery. *Biochem. Eng. J.* **2019**, *146*, 97–104. [CrossRef]
11. Panich, J.; Fong, B.; Singer, S.W. Metabolic engineering of *Cupriavidus necator* H16 for sustainable biofuels from CO_2. *Trends Biotechnol.* **2021**, *39*, 412–424. [CrossRef] [PubMed]
12. Ishizaki, A.; Tanaka, K.; Taga, N. Microbial production of Poly-D-3-hydroxybutyrate from CO_2. *Appl. Microbiol. Biotechnol.* **2001**, *57*, 6–12. [PubMed]
13. Tian, J.; Sinskey, A.J.; Stubbe, J. Kinetic studies of polyhydroxybutyrate granule formation in *Wautersia eutropha* H16 by transmission electron microscopy. *J. Bacteriol.* **2005**, *187*, 3814–3824. [CrossRef] [PubMed]
14. Martino, L.; Cruz, M.V.; Scoma, A.; Freitas, F.; Bertin, L.; Scandola, M.; Reis, M.A.M. Recovery of amorphous polyhydroxybutyrate granules from *Cupriavidus necator* cells grown on used cooking oil. *Int. J. Biol. Macromol.* **2014**, *71*, 117–123. [CrossRef]
15. Soto, L.R.; Byrne, E.; van Niel, E.W.J.; Sayed, M.; Villanueva, C.C.; Hatti-Kaul, R. Hydrogen and polyhydroxybutyrate production from wheat straw hydrolysate using *Caldicellulosiruptor* species and *Ralstonia eutropha* in a coupled process. *Bioresour. Technol.* **2019**, *272*, 259–266. [CrossRef]
16. Mozumder, M.S.I.; Garcia-Gonzalez, L.; De Wever, H.; Volcke, E.I. Poly(3-hydroxybutyrate) (PHB) production from CO_2: Model development and process optimization. *Biochem. Eng. J.* **2015**, *98*, 107–116. [CrossRef]
17. Park, I.; Jho, E.H.; Nam, K. Optimization of carbon dioxide and valeric acid utilization for polyhydroxyalkanoates synthesis by *Cupriavidus necator*. *J. Polym. Environ.* **2014**, *22*, 244–251. [CrossRef]

18. Elazzouzi-Hafraoui, S.; Nishiyama, Y.; Putaux, J.L.; Heux, L.; Dubreuil, F.; Rochas, C. The shape and size distribution of crystalline nanoparticles prepared by acid hydrolysis of native cellulose. *Biomacromolecules* **2008**, *9*, 57–65. [CrossRef]
19. Gong, J.; Mo, L.; Li, J. A comparative study on the preparation and characterization of cellulose nanocrystals with various polymorphs. *Carbohydr. Polym.* **2018**, *195*, 18–28. [CrossRef] [PubMed]
20. Kim, H.; Jeon, H.; Shin, G.; Lee, M.; Jegal, J.; Hwang, S.Y.; Oh, D.X.; Koo, J.M.; Eom, Y.; Park, J. Biodegradable nanocomposite of Poly(ester–*co*–carbonate) and cellulose nanocrystals for tough tear-resistant disposable bags. *Green Chem.* **2021**, *23*, 2293–2299. [CrossRef]
21. Park, S.-A.; Eom, Y.; Jeon, H.; Koo, J.M.; Lee, E.S.; Jegal, J.; Hwang, S.Y.; Oh, D.X.; Park, J. Preparation of synergistically reinforced transparent bio-polycarbonate nanocomposites with highly dispersed cellulose nanocrystals. *Green Chem.* **2019**, *21*, 5212–5221. [CrossRef]
22. Kim, T.; Jeon, H.; Jegal, J.; Kim, J.H.; Yang, H.; Park, J.; Oh, D.X.; Hwang, S.Y. Trans crystallization behavior and strong reinforcement effect of cellulose nanocrystals on reinforced Poly(butylene succinate) nanocomposites. *RSC Adv.* **2018**, *8*, 15389–15398. [CrossRef]
23. Seoane, I.T.; Manfredi, L.B.; Cyras, V.P.; Torre, L.; Fortunati, E.; Puglia, D. Effect of cellulose nanocrystals and bacterial cellulose on disintegrability in composting conditions of plasticized PHB nanocomposites. *Polymers* **2017**, *9*, 561. [CrossRef] [PubMed]
24. Seoane, I.T.; Fortunati, E.; Puglia, D.; Cyras, V.P.; Manfredi, L.B. Development and characterization of bionanocomposites based on Poly(3-hydroxybutyrate) and cellulose nanocrystals for packaging applications. *Polym. Int.* **2016**, *65*, 1046–1053. [CrossRef]
25. Zhang, B.; Huang, C.; Zhao, H.; Wang, J.; Yin, C.; Zhang, L.; Zhao, Y. Effects of cellulose nanocrystals and cellulose nanofibers on the structure and properties of polyhydroxybutyrate nanocomposites. *Polymers* **2019**, *11*, 2063. [CrossRef]
26. Heinrich, D.; Madkour, M.H.; Al-Ghamdi, M.A.; Shabbaj, I.I.; Steinbüchel, A. Large scale extraction of Poly(3-hydroxybutyrate) from *Ralstonia eutropha* H16 using sodium hypochlorite. *AMB Express* **2012**, *2*, 59. [CrossRef]
27. Mandal, A.; Chakrabarty, D. Isolation of nanocellulose from waste sugarcane bagasse (SCB) and its characterization. *Carbohydr. Polym.* **2011**, *86*, 1291–1299. [CrossRef]
28. Sun, X.; Shao, Y.; Boluk, Y.; Liu, Y. The impact of cellulose nanocrystals on the aggregation and initial adhesion to a solid surface of *Escherichia coli* K12: Role of solution chemistry. *Colloids Surf. B Biointerfaces* **2015**, *136*, 570–576. [CrossRef]

Article

Multiresponsive Cellulose Nanocrystal Cross-Linked Copolymer Hydrogels for the Controlled Release of Dyes and Drugs

Yuchen Jiang [1], Guihua Li [1,*], Chenyu Yang [1], Fangong Kong [2] and Zaiwu Yuan [1,2,*]

[1] Key Laboratory of Fine Chemicals in Universities of Shandong, School of Chemistry and Chemical Engineering, Qilu University of Technology (Shandong Academy of Sciences), Jinan 250353, China; 1043117140@stu.qlu.edu.cn (Y.J.); 1043118122@stu.qlu.edu.cn (C.Y.)

[2] State Key Laboratory of Biobased Material and Green Papermaking, Qilu University of Technology (Shandong Academy of Sciences), Jinan 250353, China; kfg@qlu.edu.cn

* Correspondence: ghli@qlu.edu.cn (G.L.); zyuan@qlu.edu.cn (Z.Y.)

Abstract: Multiresponsive hydrogels have attracted tremendous interest due to their promising applications in tissue engineering, wearable devices, and flexible electronics. In this work, we report a multiresponsive upper critical solution temperature (UCST) composite hydrogel based on poly (acrylic acid-co-acrylamide), PAAc-co-PAAm, sequentially cross-linked by acid-hydrolysis cellulose nanocrystals (CNCs). Scanning electron microscopy (SEM) observations demonstrated that the hydrogels are formed by densely cross-linked porous structures. The PAAc/PAAm/CNC hybrid hydrogels exhibit swelling and shrinking properties that can be induced by multiple stimuli, including temperature, pH, and salt concentration. The driving force of the volume transition is the formation and dissociation of hydrogen bonds in the hydrogels. A certain content of CNCs can greatly enhance the shrinkage capability and mechanical strength of the hybrid hydrogels, but an excess addition may impair the contractility of the hydrogel. Furthermore, the hydrogels can be used as a matrix to adsorb dyes, such as methylene blue (MB), for water purification. MB may be partly discharged from hydrogels by saline solutions, especially by those with high ionic strength. Notably, through temperature-controlled hydrogel swelling and shrinking, doxorubicin hydrochloride (DOX-HCl) can be controllably adsorbed and released from the prepared hydrogels.

Keywords: hydrogels; cellulose nanocrystals; UCST polymer; hydrogen bond; drug release

1. Introduction

Polymer hydrogels are three-dimensional polymer networks that can hold a large amount of water in the interspaces of their network [1–5]. Smart hydrogels are stimulus-responsive materials that can intelligently respond to environmental stimuli, such as the temperature and pH, along with the presence of electricity, a magnetic field, and light [6–10]. During the response process, smart hydrogels exhibit drastic volume changes and volume phase transitions; thus, they can be widely applied in wastewater treatment, drug delivery systems, and soft robotics [11–13]. Numerous polymers have been used to synthesize multiresponsive hydrogels, such as poly(N-isopropyl acrylamide) (PNIPAM) [14], polyacrylamide (PAAm) [15], poly (acrylic acid) (PAAc) [16], and poly(N, N-diethylacrylamide) [17]. It is well established that the volume change of hydrogels can be induced by individual or a combination of molecular interactions, including ionic interactions, hydrophobic interactions, hydrogen bonding, and van der Waals forces.

Lower critical solution temperature (LCST) and upper critical solution temperature (UCST) are two major categories of thermal transitions for thermoresponsive polymers. Poly(2-(dimethylamino)-ethyl methacrylate) (PDMAEMA) and PNIPAM are typical LCST-type polymers and have been widely reported [18]. Notably, most thermoresponsive hydrogels are based on polymers with LCSTs, and polymers with UCSTs are rarely reported in the fabrication of smart hydrogels [19–21]. Some researchers have found that nonresponsive polymers can be converted into thermoresponsive polymers in water by controlling the strength of polymer–polymer interactions. For example, thermoresponsive hydrogels were constructed with the incorporation of PAAm, which is not a thermoresponsive polymer. When polymer–polymer hydrogen bonds are stronger than polymer–water bonds, UCST polymers in water can be obtained. Dai and coworkers designed smart hydrogels with UCST characteristics by constructing an interpenetrating network (IPN) of PAAm and PAAc with chlorophyllin incorporated as the chromophore [16].

Cellulose is one of the most abundant natural biopolymers, composed of glucose, and is responsible for the structural scaffolding of cells in all green plants [22]. Because of their intrinsic biocompatibility and biodegradability, cellulose-based functional materials have attracted great attention in scientific research. Needle-like cellulose nanocrystals (CNCs), derived from acid hydrolysis, have been widely used to reinforce nanocomposite hydrogels [23]. Chang et al. reported a nanocomposite network of poly(acrylic acid-co-acrylamide) (PAAAM) sequentially cross-linked by quaternized tunicate cellulose nanocrystals (Q-TCNCs) and Fe^{3+} [24]. Q-TCNCs contain many hydroxyl groups, the surfaces of which are positively charged. In the nanocomposite hydrogels, Q-TCNCs act as both interfacial compatible reinforcers and cross-linkers. Nanocomposite hydrogels reinforced by poly(N-vinylpyrrolidone)-grafted cellulose nanocrystals (CNCs-g-PVP) were developed by a dual physical cross-linking strategy [25]. The obtained hydrogels exhibited high tensile strength, remarkable toughness, rapid self-recovery, and favorable fatigue resistance. The introduction of CNCs into polymer networks not only enriches raw material choices for finite components but also facilitates the formation of multiple hydrogen bonds, the formation and dissociation of which are important for understanding the swelling and shrinking properties of polymer hydrogels.

Herein, a series of CNC-reinforced nanocomposite hydrogels were fabricated by incorporating CNCs into the polymer network of PAAc-co-PAAm through hydrogen bonds, denoted as PAAc/PAAm/CNC (Scheme 1). The carboxylic acid (–COOH) groups of PAAc and amide (–$CONH_2$) groups of PAAm form intra- and intermolecular hydrogen bonds only at low temperature, while they dissociate at a high temperature. The so-called "zipper effect" also exists in the PAAc/PAAm/CNC hydrogels. Deprotonation of the carboxy groups of PAAc and the amide groups of PAAm at high pH weakens the hydrogen bonds, which induces swelling of the prepared hydrogels. According to the "Hofmeister series" effect, the sulfate ester (–SO_3H) groups on the CNC surface help to precipitate polymers, which drives hydrogels to shrink. The cross-sectional scanning electron microscopy (SEM) results show that interconnected porous structures were formed in the PAAc/PAAm/CNC hydrogels. The shrinkage capability and toughness of the PAAc/PAAm/CNC hydrogels are significantly improved compared to those of the PAAm-co-PAAc hydrogel. Based on these properties, the prepared hydrogels were used as a matrix to efficiently adsorb toxic dyes. Additionally, through hydrogel swelling and shrinking, doxorubicin hydrochloride (DOX-HCl) that is loaded in the hydrogels can be reversibly and controllably released and adsorbed in solution.

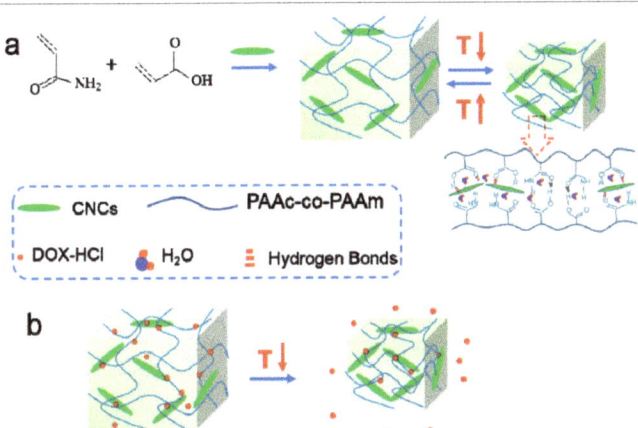

Scheme 1. (**a**) Synthesis route of PAAc/PAAm/CNC hydrogels; the shrinkage property induced by the temperature is mainly derived from the formation of hydrogen bonds between the polymer chains and CNCs; (**b**) Schematic illustration of the release of doxorubicin hydrochloride (DOX-HCl) from the hydrogel.

2. Materials and Methods

2.1. Materials

The monomers acrylamide (AAm, 98%) and acrylic acid (AAc, AR grade) and the initiator potassium persulfate (KPS, AR grade) were obtained from Sinopharm Chemical Reagent Co., Ltd. (Shanghai, China), H_2SO_4, HCl, NaOH, $MgCl_2$, and NaCl were purchased from Tianjin Damao Chemical Reagent Factory (Tianjin, China), N,N'-Methylenebis (acrylamide) (MBA) was supplied by Shanghai Mackin Biochemical Co., Ltd. (Shanghai, China), doxorubicin hydrochloride (DOX-HCl) was purchased from Dalian MeiLun Biotechnology Co., Ltd. (Dalian, China), and methylene blue (MB) and methyl orange (MO) were purchased from Wenzhou Red Flag Auxiliary Factory (Wenzhou, China). Milli-Q water was used in all experiments.

2.2. Preparation of the Cellulose Nanocrystals (CNCs)

Milled cellulose pulp was hydrolyzed by 64 wt% H_2SO_4 (8.5 mL solution/1 g pulp) at 45 °C and stirred for 30 min. Then, the reaction was terminated by adding 1000 mL of cold water. After undergoing sedimentation overnight, the bottom CNC suspension was removed, then alternately centrifuged and washed 3 times with ultrapure water. The obtained concentrated CNC suspension was placed inside dialysis membrane bags (with an 8000–14,000 molecular weight cutoff) for dialysis for ~7 days until the pH reached 6. The dialyzed suspension was sonicated for 10 min at 90 W and then evaporated to obtain a 3 wt% CNC dispersion. For the desulfation of CNCs, a 3 wt% CNC dispersion (270 mL) was mixed with NaOH (10.8 g), and the mixture was stirred at 60 °C for 5 h. The obtained suspension was purified by dialysis against Milli-Q water until the effluent remained at neutral pH, resulting in a desulfated CNC (DCNC) suspension.

2.3. Desulfation Procedure of the Cellulose Nanocrystals (CNCs)

The sulfate ester groups were removed as previously described by Grey et al. [26]. A 3.68% w/w CNC suspension (250 mL) was mixed with NaOH (10 g) to give a concentration of 1 M NaOH, and the mixture was stirred at 60 °C. The mixture turned slightly yellowish, and the nanoparticles started to aggregate and settle. After 5 h the reaction was stopped, and the suspension was purified by dialysis against deionized water until the effluent remained at neutral pH. The weight percentage of the resulting CNC suspension was

2.73%, and the yield of the reaction was 75%. The sulphur contents of the CNCs before and after desulfation were measured to be 0.81 ± 0.04 wt% and 0.27 ± 0.03 wt%, respectively, by a conductometric titration method [27].

2.4. Preparation of the PAAc/PAAm/CNC Composite Hydrogels

Quantities of 0.13 g of AAc, 0.13 g of AAm (n_{AAc}/n_{AAm} molar ratio of approximately unity), 0.0075 g of KPS, and 0.0065 g of MBA (2.5 wt% of the total weight of the monomers) were added to 5 g of dispersions respectively containing 1 wt%, 2 wt%, or 3 wt% CNCs. These solutions were heated at 75 °C for 6 h, resulting in the PAAc/PAAm/CNC composite hydrogels, which were denoted as PAAc/PAAm/CNC$_x$ (x = 1, 2, or 3, representing the 1 wt%, 2 wt%, or 3 wt% CNC dispersion, respectively).

2.5. Shrinking/Swelling Measurement of the Hydrogels

The shrinking/swelling properties of the PAAc/PAAm/CNC hydrogels at various temperatures and pH values and in various saline solutions were evaluated by weighing them after keeping them at equilibrium for an adequate amount of time. For example, to observe the swelling response to varying pH, the hydrogels were cut into pieces of the same weight (~0.5 g) and then immersed in 10 mL of solutions at various pH values (such as 2, 6, and 12) and kept at equilibrium for 24 h before measurement. For examination of the response to alternating temperatures between 2 °C and 37 °C, the pH was fixed at 2 or 4. In addition, two representative salt solutions, NaCl and MgCl$_2$, were used to detect the swelling–shrinking response to salt solutions at various ionic strengths.

2.6. Dye/Drug Loading and Release

First, the maximum uptakes of MB and DOX-HCl in the PAAc/PAAm/CNC$_2$ hydrogel were determined. A PAAc/PAAm/CNC$_2$ hydrogel of 0.50 g was immersed into a 10 mL solution of 0.1 mg·mL^{-1} MB for 3 days at 25 °C for adequate adsorption. The maximum uptake was measured to be 16.3 milligrams of MB per gram of hydrogel, and the resulting hydrogel was denoted MB@PAAc/PAAm/CNC$_2$. Under the same conditions, a 0.53 g PAAc/PAAm/CNC$_2$ hydrogel was immersed in 10 mL aqueous solution of 0.85 mg·mL^{-1} DOX-HCl, resulting in a DOX-HCl@PAAc/PAAm/CNC$_2$ hydrogel with a maximum uptake of 13.3 milligrams of DOX-HCl per gram of hydrogel. Thereafter, release experiments were carried out by immersing the above as-prepared MB@PAAc/PAAm/CNC$_2$ and DOX-HCl@PAAc/PAAm/CNC$_2$ hydrogels in 10 mL of an aqueous phase at a fixed temperature and pH. The concentrations of MB and DOX-HCl in the media were monitored by taking 500 µL aliquots at specific time points. Every time, the media needed to be restored to 10 mL by water at the same pH value. The release or resorption concentrations in the media were determined at 664 nm (MB) and 484 nm (DOX-HCl) using a UV spectrophotometer. Every concentration point was repeated three times. Calibration curves of MB and DOX-HCl were made in advance using standard solutions at known concentrations.

2.7. Characterizations

Fourier transform infrared (FT-IR) spectroscopy was carried out on an FT-IR spectrometer (Thermo Fisher Scientific, 5225 Verona Rd, Fitchburg, WI, USA) in the wavenumber range of 4000–400 cm^{-1}. X-ray diffraction (XRD) patterns for the dried hydrogels were obtained using a D8-ADVANCE diffractometer (Bruker, Germany) equipped with a Cu Kα X-ray source (λ = 0.15418 nm) and a graphite monochromator. SEM images were recorded on a field-emission microscope (Regulus 8220) at an accelerating voltage of 5 kV. The hydrogel specimens needed to be freeze-dried and fractured for SEM observation of the fractured surface. Thermal gravimetric analysis (TGA) was performed from 50 °C to 950 °C using a thermogravimetric analyzer (TGA Q50, Micromeritics Instrument Corporation) at 10 °C/min in nitrogen. Differential scanning calorimetry (DSC) experiments were carried out using a Q2000 DSC (TA instrument, New Castle, DE, USA) at 10 °C/min in nitrogen. Note that the samples for the FT-IR and XRD experiments were freeze-dried under vacuum before use. For DSC characterization, hydrogels were used without any treatment. The rheology of the PAAc/PAAm/CNC$_x$ hydrogels was systematically measured on a HAAKE Rheostress 6000 with a Thermo Scientific heating system at two temperatures (4 °C and 25 °C), with all the specimens thermally equilibrated for 5 min before measurement. Hydrogel films with a 20 mm diameter were cut into 0.7 mm thick pieces and placed into a parallel plate geometry 45.0 mm in diameter. To determine the linear viscoelastic region of each hydrogel sample, the stress sweep was checked at a fixed frequency (1.0 Hz). In the drug loading, release, and resorption experiments, an ultraviolet–visible (UV–vis) spectrometer (UV-2600) was used to determine the UV–vis absorption spectra of the MB and DOX-HCl aqueous solutions.

3. Results and Discussion

3.1. Preparation and Characterization of the PAAc/PAAm/CNC Hydrogels

A series of PAAc/PAAm/CNC hydrogels were synthesized by thermal polymerization at 75 °C. Photographs of typical PAAc/PAAm/CNC hydrogels kept at a high temperature (52 °C) are shown in Figure S1. Unlike the transparent PAAc-co-PAAm hydrogel, the PAAc/PAAm/CNC hydrogels exhibited an opaque appearance, and their transparency decreased with increasing CNC concentration. The FT-IR spectra of the freeze-dried as-prepared hydrogels (Figure 1a) were obtained to investigate the combination and hydrogen bonds between components. For pure CNCs (blue line), the surface of which has abundant –OH groups, the –OH stretching peak appeared at approximately 3650–3200 cm^{-1} [28]. A broad band from 3455 to 3100 cm^{-1} was observed for the PAAc-co-PAAm (black line) and PAAc/PAAm/CNC$_2$ (red line) hydrogels, which is characteristic of hydrogen bonds between the –COOH groups of PAAc, –CONH$_2$ groups of PAAm, and –OH groups of CNCs [28,29]. Two peaks appeared at 897 cm^{-1} and 1056 cm^{-1} for both the CNCs and the PAAc/PAAm/CNC$_2$ hydrogel, which are characteristic of the asymmetrical stretching vibration peak of C–O–C and the bending vibration peak of C–O at the β-(1–4)-glycosidic linkage. In addition, the peak at 1209 cm^{-1} is attributed to the stretching vibration of S=O in the –OSO$_3$H groups, which also exists in both CNCs and the PAAc/PAAm/CNC$_2$ hydrogels. The stretching vibration of carbonyl at 1667 cm^{-1} and 1727 cm^{-1} and the deformation vibration of –NH$_2$ at 1612 cm^{-1} were weakened for the PAAc/PAAm/CNC$_2$ hydrogel, further indicating the formation of hydrogen bonds [30–32].

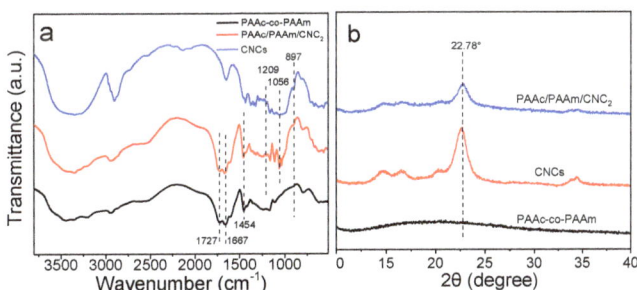

Figure 1. (a) FTIR spectra and (b) XRD patterns of CNCs and the PAAc-co-PAAm and PAAc/PAAm/CNC$_2$ hydrogels.

To detect the structural changes between pure CNCs and the hybrid hydrogels, the XRD patterns are shown in Figure 1b. For pure CNCs, typical cellulose I crystalline peaks at 14.9°, 16.4°, 22.8°, and 34.4° were observed [33–35]. Compared to the PAAc-co-PAAm hydrogel, the PAAc/PAAm/CNC$_2$ hydrogel exhibited the characteristic diffraction peak of CNCs at 22.8°, although the strength of the peak was slightly lower. At the same time, the PAAc/PAAm/CNC$_2$ hydrogel displayed a stronger amorphous nature than pure CNCs, as evidenced by its elevated broad peak. The diffraction peaks of the PAAc/PAAm/CNC$_2$

hydrogels were assumed to be the superimposition of those of CNCs and the PAAc-co-PAAm hydrogel.

The thermal properties of the PAAc/PAAm/CNC$_2$ hydrogels were examined by thermogravimetric analysis (TGA) and differential scanning calorimetry (DSC). From the TGA curves shown in Figure 2a, less weight loss was recorded for the PAAc/PAAm/CNC$_2$ hydrogel than for the hydrogel without CNCs in the range of 500–800 °C. The presence of CNCs helps to improve the thermal stability of the hybrid PAAc/PAAm/CNC hydrogels. The DSC curve is a useful tool to assess the miscibility and combination in polymer blends [16,36]. The single CNC component showed a prominent endothermic peak at 245 °C (Figure 2b and Figure S2), which may be attributed to the decomposition temperature of CNCs according to the TGA data. The individual dried PAAc-co-PAAm hydrogel showed one single glass transition temperature at 203 °C. In the case of the PAAc/PAAm/CNC$_2$ hydrogel, it can be observed that the curve exhibits a single glass transition temperature located at 210 °C, which is approximately between those of the two individual components of CNCs and the PAAc-co-PAAm polymer. Dai et al. [16] reported that the glass transition Tg value of the PAAm–PAAc IPN was 125 °C. This suggests that the hydrogels formed in this work exhibited higher thermal stability than did the PAAm–PAAc hydrogels. Comparatively, for the physical mixture of CNCs and the PAAc-co-PAAm hydrogel (with a dry weight ratio of 1:1), two separate glass transition temperatures occurred. These results demonstrate the good combination and interaction between PAAc-co-PAAm and CNCs.

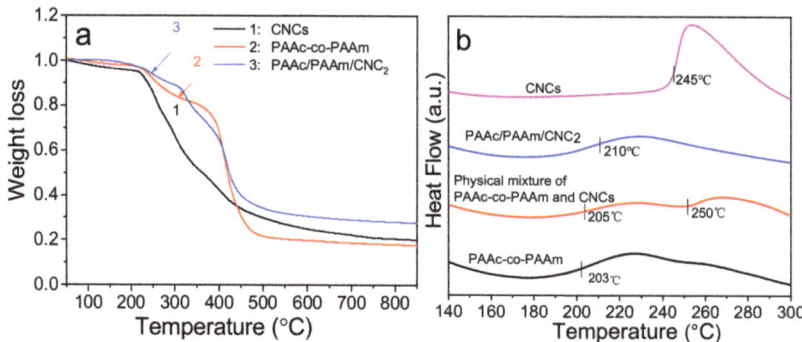

Figure 2. (**a**) TGA curves and (**b**) DSC thermograms of CNCs and the PAAc-co-PAAm and PAAc/PAAm/CNC$_2$ hydrogels. For comparison, the DSC thermogram of the physical mixture of PAAc-co-PAAm and CNCs is shown.

3.2. Shrinkage and Swelling Properties of the PAAc/PAAm/CNC Hydrogels

Due to temperature-dependent intermolecular interactions between PAAc and PAAm, the composite hydrogels composed of PAAc and PAAm exhibited remarkable shrinking–swelling properties with varying temperature. Specifically, at high temperatures, the PAAc and PAAm molecules in the hydrogels maintain a high dissociation state, while at a relatively low temperature, zipper-like cooperative hydrogen bonds can form between PAAc and PAAm, and water molecules are expelled from the hydrogel matrix [15,37,38]. Therefore, the IPN hydrogel composed of PAAc and PAAm possesses outstanding shrinking–swelling properties that are dependent on temperature. However, despite consisting of the same PAAc and PAAm segments, their copolymer, i.e., the PAAc-co-PAAm hydrogels, exhibited comparatively weak shrinking–swelling capability in response to various temperatures. Thus, the introduction of a small amount of CNCs to PAAc-co-PAAm can greatly improve the shrinking–swelling capability of the hydrogel.

3.2.1. Temperature Effects

Similar to the PAAc–PAAm IPN hydrogels [8,39], the PAAc-co-PAAm hydrogels have an upper critical solution temperature (UCST), below which the hydrogel will shrink from a swollen volume. When the temperature was reduced from the original preparation temperature (75 °C) to 2 °C, the PAAc-co-PAAm hydrogel (n_{AAc}/n_{AAm} = 1:1) showed a volume shrinkage accompanied by water expulsion from the hydrogel matrix (Figure 3a). The shrinkage ratio, R, is calculated via Equation (1):

$$R = (m_{52} - m_t)/m_{52} \tag{1}$$

where m_{52} is the weight of the hydrogel at 52 °C and m_t is that at t °C. It should be noted that the weight and volume of the hydrogels at 52 °C remained the same as those at 75 °C. From Figure 3b, the shrinkage of the PAAc-co-PAAm hydrogel was significantly low, merely 18.5%. It is well known that in the PAAc-co-PAAm hydrogels, the network is interwoven by the direct covalent linkages of polymer chains. These linkages enhance the rigidity of the hydrogel network to a certain extent but may simultaneously prevent cross-linked polymer chains from migrating and further associating through H-bonds (Figure 3c). Thus, only a small number of H-bonds can form between the PAAc and PAAm segments when the temperature is reduced. This leads to the PAAc-co-PAAm hydrogels having a relatively weaker shrinkage capability than the PAAc-PAAm IPN hydrogels [16,40].

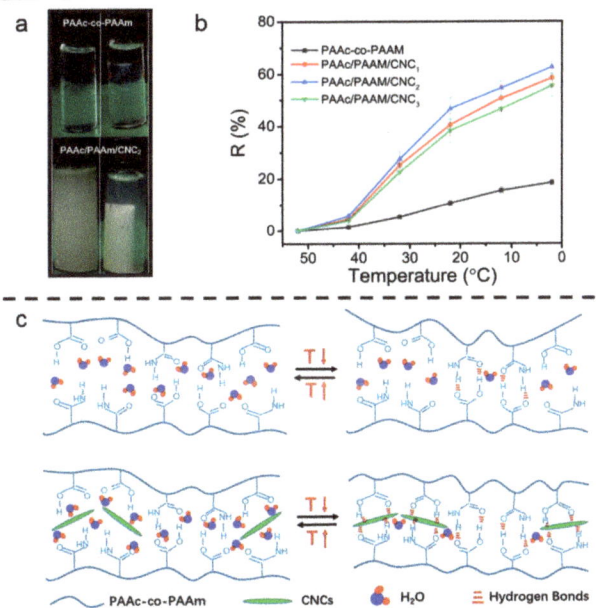

Figure 3. (a) Photographs of the PAAc-co-PAAm and PAAc/PAAm/CNC$_2$ hydrogels at 52 °C (**left**) and 2 °C (**right**); (b) shrinkage ratio curves of the typical hydrogels at varying CNC concentrations and different temperatures; and (c) structural representations of both the copolymer hydrogels without and with CNCs, indicating their structural changes after temperature variation.

As CNCs are the nanocrystalline form of cellulose, there are rich hydroxyl groups on their surface. When CNCs are incorporated and dispersed uniformly in the PAAc-co-PAAm matrix, many more H-bonding associations tend to form, such as the complex between –COOH and –CONH$_2$ and the additional complexes between –OH and –COOH or –CONH$_2$ during the cooling process (refer to Scheme 2). This prompts the polymer chains to be much more curled and compressed, greatly improving the shrinkage property of the hydrogel network after a decrease in the temperature. For the PAAc/PAAm/CNC$_1$ sample, R reached 55.5%, while it was 58.4% for the PAAc/PAAm/CNC$_2$ sample when both were reduced to an equilibrium temperature of 2 °C (Figure 3b). Figure 3c shows the structural representations of both the copolymer hydrogels without and with CNCs. These results indicate that as the temperature increases, the polymer–polymer or polymer–CNC complexes can be dissociated, and water enters due to the swollen hydrogel network; as the temperature decreases, these complexes can be reconstructed, and many water molecules are expelled.

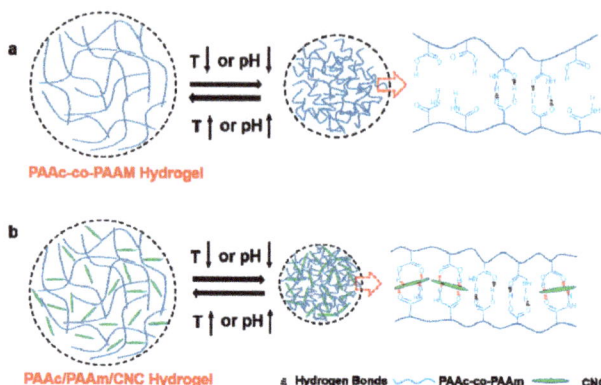

Scheme 2. Schematic drawings of the microstructures in the PAAc-co-PAAm (**a**) and PAAc/PAAm/CNC hydrogels (**b**), indicating that the larger shrinkage of the latter, which is induced by the temperature/pH, is mainly due to the formation of hydrogen bonds between the polymer chains and CNCs.

During acid hydrolysis in the preparation of CNCs, sulfate ester groups (–OSO$_3$H) are produced on the CNC surfaces. The sulfur content of the CNCs was evaluated to be 0.82 ± 0.04 wt% by a conductometric titration method [27]. According to the "Hofmeister series" [41–43], where –OSO$_3$H is on the left of the Hofmeister series, CNCs are expected to tend to precipitate polymers through interactions with surface –OSO$_3$H groups. To verify whether this was the case, the –OSO$_3$H groups were removed by heating CNCs at 60 °C in 1 M NaOH for 5 h, and the resultant CNCs were named DCNCs. The sulphur content of the DCNCs was measured to be 0.27 ± 0.03 wt%. Instead of CNCs, an equal amount of DCNCs was used to prepare a PAAc/PAAm/DCNC$_2$ hybrid hydrogel, where the corresponding components were identical to those in the PAAc/PAAm/CNC$_2$ hydrogel. Then, the shrinkage behaviors of the two hydrogels were observed. Interestingly, from the photographs in Figure S3, the PAAc/PAAm/DCNC$_2$ hydrogel demonstrated a weaker shrinkage ability than the PAAc/PAAm/CNC$_2$ hydrogel at 2 °C, with a value of 30%. It is obvious that the –OSO$_3$H groups play an important role in the shrinkage capability of the composite hydrogels. We speculate that the –OSO$_3$H groups may promote the formation of complexes both between polymer and polymer and between the polymers and CNCs, which drives the hydrogels to shrink.

In addition to the amount of CNCs and their surface groups, the influence of the mole ratio of AAc to AAm (n_{AAc}/n_{AAm}) on hydrogel shrinkage was also considered. From Figure S4a, the PAAc/PAAm/CNC hydrogel with n_{AAc}/n_{AAm} = 1 displayed a larger shrinkage capability than did those with n_{AAc}/n_{AAm} values of 1/2, 1/3, and 2/1, and this result is consistent with the reports by Dai et al. and Ilmain et al. [30,37]. The shrinkage ratios corresponding to n_{AAc}/n_{AAm} values of 1/2, 1/3, and 2/1 were approximately 18.1%, 31.4%, and 44.5%, respectively (Figure 4a). This result demonstrates that both the interactions between the PAAc and PAAm segments and those between the polymers and CNCs may contribute to the volume transition of hydrogels. In view of the optimal shrinkage capability, the PAAc/PAAm/CNC$_2$ hydrogel with n_{AAc}/n_{AAm} = 1 and 2 wt% CNCs was chosen as a typical sample for subsequent studies on volume transition and controlled release properties.

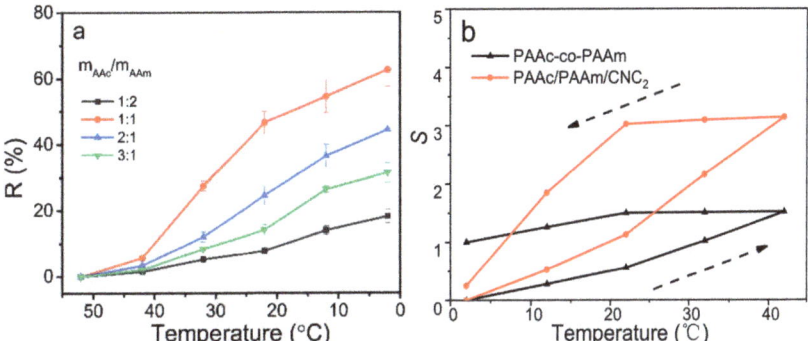

Figure 4. (a) Shrinkage ratio curves of the hydrogels at various mole ratios of AAc to AAm and varying temperatures, and (b) the temperature dependence of the equilibrium swelling ratio of the PAAc-co-PAAm and PAAc/PAAm/CNC$_2$ hydrogels.

As discussed above, H-bonds are formed at low temperatures, while a high temperature causes the dissociation of polymer complexes in hydrogels, resulting from the breakage of H-bonds. This mechanism may further induce the inner network volume to expand and allow large amounts of water to enter the hydrogel phase from an ambient aqueous phase. Thus, a swelling ratio (S) was defined to evaluate the swelling property:

$$S = (m_t - m_2)/m_2 \qquad (2)$$

where m_2 is the weight of the hydrogel at 2 °C and m_t is that at t °C. The S value of the PAAc/PAAm/CNC$_2$ hydrogel gradually increased to 3.2 at 42 °C. From Figure 4b, one can see that when the temperature was reduced back to 2 °C, S was restored to 0.25, which is close to 0, indicating its good scalability and reversibility under temperature variation. Comparatively, the S value of the PAAc-co-PAAm hydrogel was merely 1.5 at 42 °C and restored to 1, reflecting its poor reversibility with low extensibility and retractility. Interestingly, from the images in Figure S4b, after a heating–cooling process, the PAAc-co-PAAm hydrogel was seriously crinkled, while the PAAc/PAAm/CNC$_2$ hydrogel maintains its original 3D shape, i.e., a regular cylinder with a smooth surface, demonstrating the excellent free-standing property of the hydrogel when combined with CNCs.

The microstructures of the hybrid hydrogels at different temperatures were characterized by scanning electron microscopy (SEM), as shown in Figure 5. The interconnected porous morphology was viewed from the cross-sectional SEM images of all the hydrogels. For the PAAc-co-PAAm hydrogel, the average pore size only seemed to slightly decrease from 17.4 to 14.6 µm when the temperature changed from 42 °C to 2 °C (Figure 5a,b), which is consistent with the small shrinkage in hydrogel volume. As shown in Figure 5d,e, with the same temperature variation, the pore size of the PAAc/PAAm/CNC$_2$ hydrogel was greatly reduced from 14.0 µm to 3.3 µm. Undoubtedly, the hydrogel network became tightly packed due to the large shrinkage effect. It was reported that the pore size of a hydrogel network is influenced by the cross-link density and the amount of water absorbed inside the hydrogel [8]. We may consider that the addition of CNCs provides multiple anchor sites for the PAAc-co-PAAm chains, which is responsible for the increase in cross-link density. As a result, less water can be held in the hydrogel. In addition, from the magnified SEM image, one can see that the pore wall and anchor sites of the PAAc-co-PAAm hydrogel (Figure 5c) are relatively smooth. In the case of the PAAc/PAAm/CNC$_2$ hydrogel, many

CNC nanorods are inserted into the pore wall and anchor sites and are arranged regularly, which makes the pore wall relatively rough (Figure 5f).

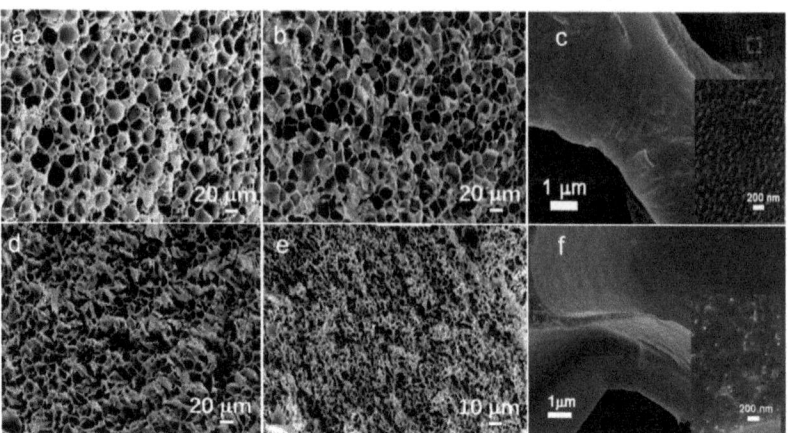

Figure 5. Cross-sectional SEM images. Hydrogels of PAAc-co-PAAm at (**a**) 42 °C and (**b**) 2 °C. Hydrogels of PAAc/PAAm/CNC$_2$ at (**d**) 42 °C and (**e**) 2 °C. Morphologies of the pore wall of (**c**) the PAAc-co-PAAm and (**f**) PAAc/PAAm/CNC$_2$ hydrogels at 2 °C; the inset images are magnified SEM images of the areas outlined by red frames.

3.2.2. pH Effects

As mentioned above, theoretically, a significant portion of H-bonds are formed between the –COOH groups and –CONH$_2$ groups. These bonds are important for the gelation and mechanical performance of hydrogels. The formation and dissociation of H-bonds, as well as the swelling and shrinking behaviors, can be tuned by altering the pH of the ambient solution. As shown in Figure 6a, at room temperature, a piece of hydrogel was immersed in water at different pH values, and reaching dynamic equilibrium usually took 4 h. As the pH changed from 2 to 6 and then to 12, the hydrogel expanded continuously, which is ascribed to the dissociation of both the PAAc–PAAm complex and H-bonds between polymers and CNCs. At high pH, –CONH$_2$ and –COOH groups are deprotonated; thus, both polymer hydrophilicity and dissociation of the polymer complex are greatly enhanced [44,45]. As a result, the three-dimensional polymer network greatly expanded, and water molecules were absorbed into the hydrogel matrix. From the SEM image (Figure 6b,c), one can see that as the pH varied from 2 to 6, the pore size became significantly larger (from 18.3 to 122.7 μm), and the cross-linkage and pore density decreased correspondingly. At pH = 12, the rigidity of the hydrogel network was largely destroyed, with the pores seriously stretched and ruptured by excessive swelling (Figure 6d). It was determined that a pH above 8 causes permanent destruction of the PAAc/PAAm/CNC$_2$ hydrogel, that is, the swollen hydrogel will no longer contract when it is replaced in a low-pH or low-temperature environment. For the swollen hydrogel at pH = 6 (<8), the volume decreased and the hydrogel became much more opaque when it was placed in water at a low pH or low temperature. Comparatively, in a low-pH environment (such as pH = 2), the PAAc/PAAm/CNC$_2$ hydrogel exhibited reversible swelling–shrinking behavior upon switching of the temperature between 37 and 2 °C (Figure S5). Heating-induced swelling and cooling-induced shrinkage could be repeated at least 10 times without fatigue (Figure S5b).

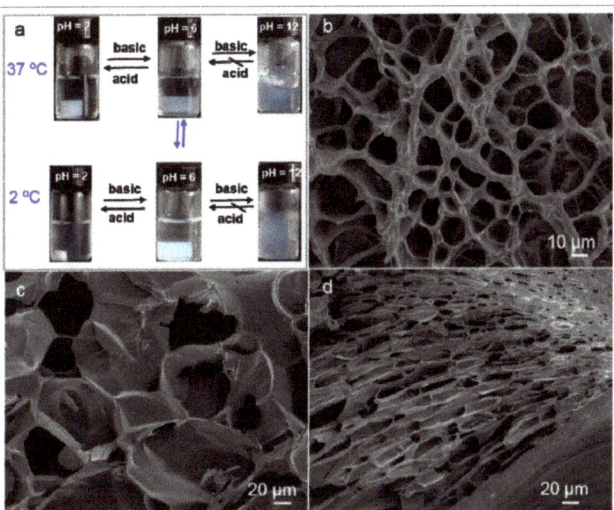

Figure 6. (a) Photographs of the swelling–shrinking behaviors of the PAAc/PAAm/CNC$_2$ hydrogels at different pH values and at 37 and 2 °C. (**b**–**d**) Cross-sectional SEM images of the PAAc/PAAm/CNC$_2$ hydrogels at varying pH values: pH = 2 (**b**), 6 (**c**), and 12 (**d**).

3.3. Rheological Properties of PAAc/PAAm/CNC Hydrogels

The influence of CNC concentration and temperature on the mechanical strength of the PAAc/PAAm/CNC composite hydrogels was investigated, as shown in Figure 7. To determine the linear viscoelastic region of each hydrogel sample, the stress sweep was checked at a fixed frequency (1.0 Hz). As is well known, in the linear viscoelastic region, the elastic modulus (G′) is independent of the yield stress [46–48]. Additionally, the solid-like network structures of gels are suddenly broken when the yield stress is above a critical value (τ^*). Typically, the critical value τ^* reflects the strength of the network structures. Figure 7a illustrates that, at 2 °C, the τ^* values of PAAc–PAAm, PAAc/PAAm/CNC$_1$, PAAc/PAAm/CNC$_2$, and PAAc/PAAm/CNC$_3$ were about 50, 90, 200, and 500 Pa, respectively. This means that for hydrogels with a higher CNC concentration, the linear viscoelastic region became wider. With increasing CNC concentration, the elastic modulus (G′) also increased to reach a maximum value (1000 Pa) at 3% CNCs, which can be ascribed to more tightly entangled networks. At 25 °C (Figure 7b), the variation of the rheological properties of the composite hydrogels influenced by CNC concentration was similar to that at 2 °C. For pure PAAc–PAAm hydrogel, both the τ^* and G′ values showed slight change with varying temperature, of about 20 and 30 Pa, respectively. For the PAAc/PAAm/CNC$_3$ hydrogel, when the temperature changed from 2 to 25 °C, the τ^* value showed nearly no change, while the G′ value decreased from 1000 to 200 Pa. One should note that for all hydrogels formed by PAAc/PAAm/CNC, the loss modulus (G″) was lower than G′ in the linear viscoelastic region; this elastic dominant behavior indicates typical reversible cross-linked networks in the hydrogels. As a general phenomenon, at the same temperature, with increasing CNC concentration, G″ increased to reach a highest value at 3% CNCs.

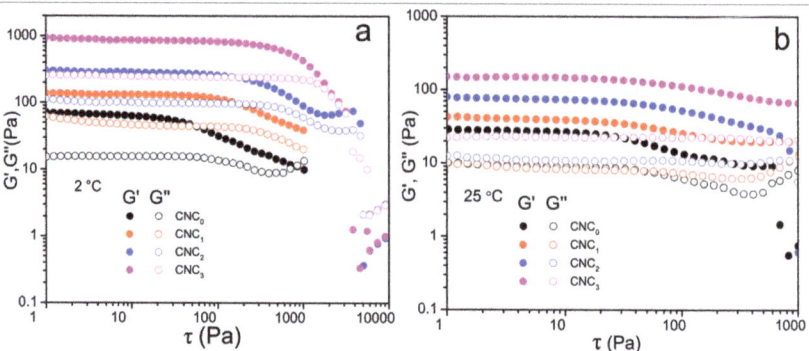

Figure 7. Stress sweep of the hydrogels with varying CNC concentrations at (**a**) 2 °C and (**b**) 25 °C.

3.4. Adsorption and Controlled Release of Dyes or Drugs

Stimulation-responsive intelligent hydrogels have been widely used to adsorb different types of toxic dyes from industrial wastewater [49–53]. In this work, two typical water-soluble dyes, negatively charged methyl orange (MO) and positively charged methylene blue (MB), were selected as simple models. Typically, 0.5 g of polymer hydrogel was immersed into 10 mL of dye solution at a concentration of 100 mg/L and then left undisturbed. Combined with the calibration curves of MB and MO (Figure S6a,b), the dye concentration after adsorption was monitored by UV–vis spectroscopy. The positively charged MB molecules were efficiently trapped in the hydrogels within two days, and the dark blue solution became clear, as shown in Figure S6c. The maximum uptake was measured to be 16.3 milligrams of MB per gram of hydrogel. However, for the negatively charged MO, the solution retained its dark-yellow color (Figure S6d), indicating the poor adsorption of MO in the hydrogel. Two reasons may explain the selective adsorption of MB. First, the surface of the composite hydrogel is negatively charged because of the sulfate ester groups on the CNCs, and strong electrostatic attraction can thus be formed between the hydrogel matrix and positively charged MB. Second, the pH of the MB solution was 7.5, and the hydrogel was expanded under this condition, which may have contributed to the adsorption of MB molecules.

It was reported that the solubility of polymers in water can be controlled by salts, which may influence the swelling–shrinking performance of the composite hydrogel, thus further controlling the release of adsorbed molecules. Aqueous solutions of two representative salts, NaCl and $MgCl_2$, at different concentrations were used to trigger the shrinkage of the the $PAAc/PAAm/CNC_2$ hydrogel at room temperature. From the photographs (Figure S7), the volumes of the $PAAc/PAAm/CNC_2$ hydrogels were remarkably compressed by these salt solutions. First, large salt concentrations clearly led to high shrinkage ratios (R) for both the NaCl and $MgCl_2$ solutions. Second, the triggered shrinkage ratios of the hydrogels gradually increased in the order of 0.2 mol/L NaCl, 0.1 mol/L $MgCl_2$, 0.4 mol/L NaCl, and 0.2 mol/L $MgCl_2$, which is exactly consistent with the increase in ionic strength.

The release of MB from the as-prepared $MB@PAAc/PAAm/CNC_2$ hydrogel (refer to Section 2.6) in pure water and different salt solutions was monitored over time. One can see from the release curves in Figure 8a that the initial release rate was comparatively rapid and then reached equilibrium after several hours. The addition of NaCl and $MgCl_2$ enhanced the release of MB, since the equilibrium concentrations of MB were greatly improved. Furthermore, the released equilibrium concentrations also corresponded to the ionic strengths of the different salt solutions, implying that the release of MB can be simply tuned by the use of various salt solutions. The release of MB can also be tuned by changing the temperature and pH value. As shown in Figure 8b, at pH = 4, MB was released in ambient water at a concentration of ~3 mg/L at 37 °C. When the system was stored at 2 °C, MB was partially resorbed. At pH = 2, the release–resorption cycle at 37 °C and 2 °C was similar to that at pH = 4, but the equilibrium concentrations were much higher.

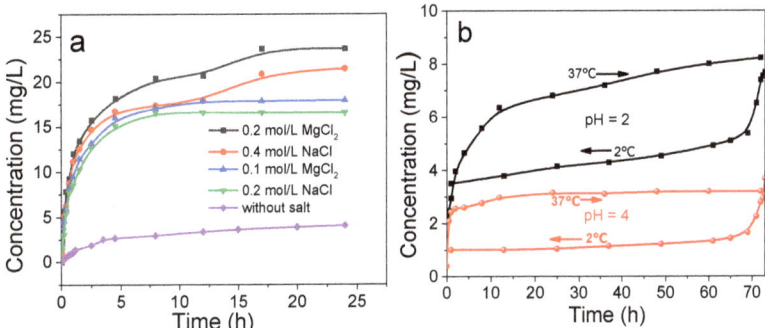

Figure 8. (a) Concentrations of methylene blue (MB) released from the as-prepared MB@PAAc/PAAm/CNC$_2$ hydrogels in 10 mL of different salt solutions at room temperature as a function of time. (b) Release–resorption cycles of the MB@PAAc/PAAm/CNC$_2$ hydrogel in 10 mL of water at two pH values and at 37 °C and 2 °C. The released concentration of MB was consecutively detected over time.

The above results inspired us to assess the controlled release of water-soluble ionic drugs through in vitro hydrogel shrinking and swelling tests. Doxorubicin hydrochloride (DOX-HCl), widely used as an anticancer drug [54], was selected in this work. An experiment to determine DOX-HCl release from the as-prepared DOX-HCl@PAAc/PAAm/CNC$_2$ hydrogel in 10 mL of pure water was performed. The calibration curves of DOX-HCl are shown in Figure S8, and the release profile was obtained by detecting the time-dependent UV–vis spectra of the release media. As shown in Figure 9, the in vitro release studies with the PAAc/PAAm/CNC$_2$ hydrogel revealed that the release behavior of DOX-HCl exhibited a trend similar to that of MB. At pH = 4, the released percentage of DOX-HCl at equilibrium in the aqueous phase was relatively low, i.e., 0.17% at 2 °C and 2.1% at 37 °C. Therefore, most of the DOX-HCl remained in the hydrogel phase, especially at a low temperature. When the pH value of the aqueous phase was 2, the released percentage of DOX-HCl in the aqueous phase increased, reaching 5.8% at 2 °C and 24% at 37 °C. Therefore, a high pH and low temperature are beneficial to the adsorption and storage of DOX-HCl, while in an acidic stomach environment (~37 °C and pH = 2), the DOX-HCl@PAAc/PAAm/CNC$_2$ hydrogel can provide a high and stable release rate of DOX-HCl along with its gradual consumption. In addition, the release–adsorption behavior of the DOX-HCl@PAAc/PAAm/CNC$_2$ hydrogel showed good reversibility: the cooling-induced adsorption and heating-induced release could be repeated at least 10 times without any decrease in the drug loading efficiency. The SEM image in Figure S9 shows that their structures were nearly unchanged. Based on the above results, the release–adsorption property makes this hydrogel a promising new candidate for targeted drug delivery and controlled release.

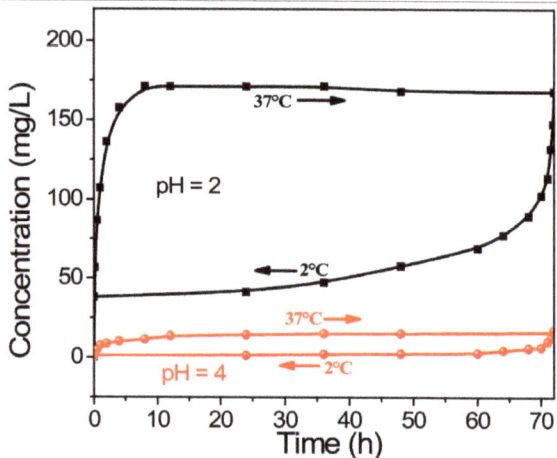

Figure 9. Release–resorption cycles of DOX-HCl from an as-prepared DOX-HCl@PAAc/PAAm/CNC$_2$ hydrogel in 10 mL of water at pH = 2 or pH = 4 and at 37 °C and 2 °C. The released percentage of DOX-HCl was consecutively detected.

4. Conclusions

In conclusion, we successfully developed a multiresponsive UCST composite hydrogel based on PAAC-co-PAAM sequentially cross-linked with CNCs through hydrogen bonds. The microstructures of the PAAc-co-PAAm/CNC hybrid hydrogels were densely cross-linked porous structures. The pore size varied with the volume transition of the hydrogels. With tuning of the environmental temperatures, pH values, and salt concentrations, the PAAc-co-PAAm/CNC hybrid hydrogels exhibited swelling and shrinking. Particularly, at pH = 2, cooling-induced shrinking and heating-induced swelling could be repeated at least 10 times without damaging the structure and properties of the hydrogel. Compared to the PAAc-co-PAAm hydrogel, the shrinkage performance and toughness of the PAAc/PAAm/CNC hydrogels were significantly improved. These excellent properties of the composite hydrogel inspired us to utilize the shrinkable polymer hydrogel as an intelligent carrier for the efficient adsorption of ionic dyes and the controlled release of drugs. A high pH and low temperature were beneficial to the adsorption and storage of drugs in the PAAc/PAAm/CNC hydrogels, while a low pH and high temperature were helpful for the release of drugs from the hydrogels. These hybrid hydrogels are expected to be promising candidates for drug delivery and controlled release.

Supplementary Materials: The following are available online at https://www.mdpi.com/article/10.3390/polym13081219/s1, Figure S1: Photographs of the typical hydrogels at 52 °C. From left to right: hydrogels of PAAc-co-PAAm, PAAc/PAAm/CNC$_1$, PAAc/PAAm/CNC$_2$, and PAAc/PAAm/CNC$_3$. Figure S2: DSC curves of CNCs and the PAAc-co-PAAm and PAAc/PAAm/CNC$_2$ hydrogels in the whole temperature range. Figure S3: Photographs of the PAAc-co-PAAm/DCNC$_2$ hydrogels at 52 °C and 2 °C. Figure S4: Photographs of (a) PAAc-co-PAAm/CNC hydrogels with varying mole ratios of AAc to AAm at a fixed CNC concentration of 2%, and (b) PAAc-co-PAAm and PAAc/PAAm/CNC$_2$ in the initial state and after a heating–cooling process in the final state. Figure S5: (a) Reversibility of the PAAc/PAAm/CNC$_2$ hydrogel after 10 cycles at 37 and 2 °C; (b) SEM image of PAAc/PAAm/CNC$_2$ hydrogels after 10 cycles of cooling and heating. Figure S6: Calibration curves of (a) MB and (b) MO in aqueous solutions, and the MB (c) and MO (d) solutions before (left) and after (right) adsorption. Figure S7: (a) Photographs of PAAc/PAAm/CNC$_2$ hydrogels in water (top) and in different salt solutions (bottom). (b) Shrinkage ratios of PAAc/PAAm/CNC$_2$ hydrogels in different salt solutions to those in pure water. All systems were equilibrated at room temperature for 2 days. Figure S8: Calibration curve of DOX-HCl in aqueous solution. Figure S9. Cross-sectional SEM image of the dried DOX-HCl@PAAc/PAAm/CNC$_2$ hydrogel after 10 cycles of cooling-induced adsorption and heating-induced release by alternating the temperature between 2 °C and 37 °C.

Author Contributions: Conceptualization, G.L. and Z.Y.; methodology, Y.J. and C.Y.; software, Y.J. and C.Y.; Supervision, G.L. and F.K.; formal analysis, Y.J.; investigation, Y.J. and C.Y.; data curation, F.K.; writing—original draft preparation, G.L. and Z.Y.; writing—review and editing, G.L. and Z.Y.; funding acquisition, G.L. and Z.Y.; All authors have read and agreed to the published version of the manuscript.

Funding: This work was financially supported by the NSFC (Grant No. 21703111 and 31570570), the Program for Scientific Research Innovation Team in Colleges and Universities of Jinan (No. 2018GXRC006), the Foundation (No. ZZ20190114) of State Key Laboratory of Biobased Material and Green Papermaking, Qilu University of Technology, Shandong Academy of Sciences, and the Program for Longcheng Yingcai Engineering of Zhucheng (No. LCYC2018-010).

Institutional Review Board Statement: Not applicable.

Informed Consent Statement: Not applicable.

Data Availability Statement: The data presented in this study are available on request from the corresponding author.

Conflicts of Interest: The authors declare no conflict of interest.

References

1. Weng, G.; Thanneeru, S.; He, J. Dynamic coordination of Eu-iminodiacetate to control fluorochromic response of polymer hydrogels to multistimuli. *Adv. Mater.* **2018**, *30*, 1706526. [CrossRef] [PubMed]
2. Morelle, X.P.; Illeperuma, W.R.; Tian, K.; Bai, R.; Suo, Z.; Vlassak, J.J. Highly stretchable and tough hydrogels below water freezing temperature. *Adv. Mater.* **2018**, *30*, 1801541. [CrossRef] [PubMed]
3. Li, Z.; Meng, X.; Xu, W.; Zhang, S.; Ouyang, J.; Zhang, Z.; Liu, Y.; Niu, Y.; Ma, S.; Xue, Z.; et al. Single network double cross-linker (SNDCL) hydrogels with excellent stretchability, self-recovery, adhesion strength, and conductivity for human motion monitoring. *Soft Matter* **2020**, *16*, 7323–7331. [CrossRef]
4. Rong, Q.; Lei, W.; Chen, L.; Yin, Y.; Zhou, J.; Liu, M. Anti-freezing, conductive self-healing organohydrogels with stable strain-sensitivity at subzero temperatures. *Angew. Chem. Int. Ed.* **2017**, *56*, 14347–14351. [CrossRef]
5. Huang, S.; Kong, X.; Xiong, Y.; Zhang, X.; Chen, H.; Jiang, W.; Niu, Y.; Xu, W.; Ren, C. An overview of dynamic covalent bonds in polymer material and their applications. *Eur. Polym. J.* **2020**, *141*, 110094. [CrossRef]
6. Li, M.; Wang, H.; Hu, J.; Hu, J.; Zhang, S.; Yang, Z.; Li, Y.; Cheng, Y. Smart hydrogels with antibacterial properties built from all natural building blocks. *Chem. Mater.* **2019**, *31*, 7678–7685. [CrossRef]
7. Veleva, V.R.; Cue, B.W.; Todorova, S. Benchmarking green chemistry adoption by the global pharmaceutical supply chain. *ACS Sustain. Chem. Eng.* **2018**, *6*, 2–14. [CrossRef]
8. Hua, L.; Xie, M.; Jian, Y.; Wu, B.; Chen, C.; Zhao, C. Multiple-responsive and amphibious hydrogel actuator based on asymmetric UCST-type volume phase transition. *ACS Appl. Mater. Interfaces* **2019**, *11*, 43641–43648. [CrossRef]
9. Downs, F.G.; Lunn, D.J.; Booth, M.J.; Sauer, J.B.; Ramsay, W.J.; Klemperer, R.G.; Hawker, C.J.; Bayley, H. Multi-responsive hydrogel structures from patterned droplet networks. *Nat. Chem.* **2020**, *12*, 363–371. [CrossRef]
10. Zarzar, L.D.; Kim, P.; Aizenberg, J. Bio-inspired design of submerged hydrogel-actuated polymer microstructures operating in response to pH. *Adv. Mater.* **2011**, *23*, 1442–1446. [CrossRef]
11. Deng, Z.; Guo, Y.; Zhao, X.; Ma, P.X.; Guo, B. Multifunctional stimuli-responsive hydrogels with self-healing, high conductivity, and rapid recovery through Host–Guest interactions. *Chem. Mater.* **2018**, *30*, 1729–1742. [CrossRef]
12. Fernandes, R.; Gracias, D.H. Self-folding polymeric containers for encapsulation and delivery of drugs. *Adv. Drug Deliv. Rev.* **2012**, *64*, 1579–1589. [CrossRef]
13. Rus, D.; Tolley, M.T. Design, fabrication and control of soft robots. *Nature* **2015**, *521*, 467–475. [CrossRef] [PubMed]
14. Kim, J.; Cho, Y.; Kim, S.; Lee, J. 3D cocontinuous composites of hydrophilic and hydrophobic soft materials: High modulus and fast actuation time. *ACS Macro Lett.* **2017**, *6*, 1119–1123. [CrossRef]
15. Yang, M.; Liu, C.; Li, Z.; Gao, G.; Liu, F. Temperature-responsive properties of poly(acrylic acid-co-acrylamide) hydrophobic association hydrogels with high mechanical strength. *Macromolecules* **2010**, *43*, 10645–10651. [CrossRef]
16. Xu, Y.; Ghag, O.; Reimann, M.; Sitterle, P.; Chatterjee, P.; Nofen, E.; Yu, H.; Jiang, H.; Dai, L.L. Development of visible-light responsive and mechanically enhanced "Smart" UCST interpenetrating network hydrogels. *Soft Matter* **2018**, *14*, 151–160. [CrossRef] [PubMed]
17. Tuan, H.N.A.; Nhu, V.T.T. Synthesis and properties of pH-thermo dual responsive semi-iPN hydrogels based on N, N′-diethylacrylamide and itaconamic acid. *Polymers* **2020**, *12*, 1139. [CrossRef]

18. Jana, S.; Biswas, Y.; Anas, M.; Saha, A.; Mandal, T.K. Poly [oligo(2-ethyl-2-oxazoline)acrylate]-based poly(ionic liquid) random copolymers with coexistent and tunable lower critical solution temperature- and upper critical solution temperature-type phase transitions. *Langmuir* **2018**, *34*, 12653–12663. [CrossRef] [PubMed]
19. Ding, Y.; Yan, Y.; Peng, Q.; Wang, B.; Xing, Y.; Hua, Z.; Wang, Z. Multiple stimuli-responsive cellulose hydrogels with tunable LCST and UCST as smart windows. *ACS Appl. Polym. Mater.* **2020**, *2*, 3259–3266. [CrossRef]
20. Zheng, J.; Xiao, P.; Le, X.; Lu, W.; Théato, P.; Ma, C.; Du, B.; Zhang, J.; Huang, Y.; Chen, T. Mimosa inspired bilayer hydrogel actuator functioning in multi-environments. *J. Mater. Chem. C* **2018**, *6*, 1320–1327. [CrossRef]
21. Liu, F.; Jiang, S.; Ionov, L.; Agarwal, S. Thermophilic films and fibers from photo cross-linkable UCST-type polymers. *Polym. Chem.* **2015**, *6*, 2769–2776. [CrossRef]
22. Isogai, A. Emerging nanocellulose technologies: Recent developments. *Adv. Mater.* **2020**, *2020*, 2000630. [CrossRef]
23. Yue, Y.; Wang, X.; Han, J.; Yu, L.; Chen, J.; Wu, Q.; Jiang, J. Effects of nanocellulose on sodium alginate/polyacrylamide hydrogel: Mechanical properties and adsorption-desorption capacities. *Carbohydr. Polym.* **2019**, *206*, 289–301. [CrossRef]
24. Zhang, T.; Zuo, T.; Hu, D.; Chang, C. Dual physically cross-linked nanocomposite hydrogels reinforced by tunicate cellulose nanocrystals with high toughness and good self-recoverability. *ACS Appl. Mater. Interfaces* **2017**, *9*, 24230–24237. [CrossRef]
25. Li, B.; Han, Y.; Zhang, Y.; Cao, X.; Luo, Z. Dual physically crosslinked nanocomposite hydrogels reinforced by poly (N-vinylpyrrolidone) grafted cellulose nanocrystal with high strength, toughness, and rapid self-recovery. *Cellulose* **2020**, *27*, 9913–9925. [CrossRef]
26. Kloser, E.; Gray, D. Surface grafting of cellulose nanocrystals with poly (ethylene oxide) in aqueous media. *Langmuir* **2010**, *26*, 13450. [CrossRef]
27. Abitbol, T.; Kloser, E.; Gray, D. Estimation of the surface sulfur content of cellulose nanocrystals prepared by sulfuric acid hydrolysis. *Cellulose* **2013**, *20*, 785. [CrossRef]
28. Zhou, C.; Wu, Q.; Yue, Y.; Zhang, Q. Application of rod-shaped cellulose nanocrystals in polyacrylamide hydrogels. *J. Colloid Interface Sci.* **2011**, *353*, 116–123. [CrossRef] [PubMed]
29. Shao, C.; Wang, M.; Meng, L.; Chang, H.; Wang, B.; Xu, F.; Yang, J.; Wan, P. Mussel-inspired cellulose nanocomposite tough hydrogels with synergistic self-healing, adhesive, and strain-sensitive properties. *Chem. Mater.* **2018**, *30*, 3110–3121. [CrossRef]
30. Dai, H.; Chen, Q.; Qin, H.; Guan, Y.; Shen, D.; Hua, Y.; Tang, Y.; Xu, J. A temperature-responsive copolymer hydrogel in controlled drug delivery. *Macromolecules* **2006**, *39*, 6584–6589. [CrossRef]
31. Jeong, D.; Kim, C.; Kim, Y.; Jung, S. Dual crosslinked carboxymethyl cellulose/polyacrylamide interpenetrating hydrogels with highly enhanced mechanical strength and superabsorbent properties. *Eur. Polym. J.* **2020**, *127*, 109586. [CrossRef]
32. Ryu, J.H.; Han, N.K.; Lee, J.S.; Jeong, Y.G. Microstructure, thermal and mechanical properties of composite films based on carboxymethylated nanocellulose and polyacrylamide. *Carbohydr. Polym.* **2019**, *211*, 84–90. [CrossRef]
33. Zhang, Y.; Tian, Z.; Fu, Y.; Wang, Z.; Qin, M.; Yuan, Z. Responsive and patterned cellulose nanocrystal films modified by N-methylmorpholine-N-oxide. *Carbohydr. Polym.* **2020**, *228*, 115387. [CrossRef]
34. Yang, N.; Ji, X.; Sun, J.; Zhang, Y.; Xu, Q.; Fu, Y.; Li, H.; Qin, M.; Yuan, Z. Photonic actuators with predefined shapes. *Nanoscale* **2019**, *11*, 10088–10096. [CrossRef] [PubMed]
35. Sun, J.; Ji, X.; Li, G.; Zhang, Y.; Liu, N.; Li, H.; Qin, M.; Yuan, Z. Chiral nematic latex-GO composite films with synchronous response of color and actuation. *J. Mater. Chem. C* **2019**, *7*, 104–110. [CrossRef]
36. Gómez Ribelles, J.L.; Pradas, M.M.; Dueñas, J.M.M.; Cabanilles, C.T. Glass transition in homogeneous and heterogeneous interpenetrating polymer networks and its relation to concentration fluctuations. *J. Non-Cryst. Solids.* **2002**, *307–310*, 731–737. [CrossRef]
37. Ilmain, F.; Tanaka, T.; Kokufuta, E. Volume transition in a gel driven by hydrogen bonding. *Nature* **1991**, *349*, 400–401. [CrossRef]
38. Zhang, L.Q.; Chen, L.W.; Zhong, M.; Shi, F.K.; Liu, X.Y.; Xie, X.M. Phase transition temperature controllable poly(acrylamide-co-acrylic acid) nanocomposite physical hydrogels with high strength. *Chin. J. Polym Sci.* **2016**, *34*, 1261–1269. [CrossRef]
39. Zhang, H.; Guo, S.; Fan, W.; Zhao, Y. Ultrasensitive pH-induced water solubility switch using UCST polymers. *Macromolecules* **2016**, *49*, 1424–1433. [CrossRef]
40. Yue, Y.F.; Haque, M.A.; Kurokawa, T.; Nakajima, T.; Gong, J.P. Lamellar hydrogels with high toughness and ternary tunable photonic stop-band. *Adv. Mater.* **2013**, *25*, 3106–3110. [CrossRef]
41. Zhang, Y.; Furyk, S.; Bergbreiter, D.E.; Cremer, P.S. Specific ion effects on the water solubility of macromolecules: PNIPAM and the Hofmeister series. *J. Am. Chem. Soc.* **2005**, *127*, 14505–14510. [CrossRef]
42. Salis, A.; Ninham, B.W. Models and mechanisms of Hofmeister effects in electrolyte solutions, and colloid and protein systems revisited. *Chem. Soc. Rev.* **2014**, *43*, 7358–7377. [CrossRef]
43. Sun, J.; Liu, Y.; Jin, L.; Chen, T.; Yin, B. Coordination-induced gelation of an L-glutamic acid Schiff base derivative: The anion effect and cyanide-specific selectivity. *Chem. Commun.* **2016**, *52*, 768–771. [CrossRef] [PubMed]
44. Han, Z.; Wang, P.; Mao, G.; Yin, T.; Zhong, D.; Yiming, B.; Hu, X.; Jia, Z.; Nian, G.; Qu, S.; et al. Dual pH-responsive hydrogel actuator for lipophilic drug delivery. *ACS Appl. Mater. Interfaces* **2020**, *12*, 12010–12017. [CrossRef] [PubMed]
45. Dai, L.; Ma, M.; Xu, J.; Si, C.; Wang, X.; Liu, Z.; Ni, Y. All-lignin-based hydrogel with fast pH-stimuli responsiveness for mechanical switching and actuation. *Chem. Mater.* **2020**, *32*, 4324–4330. [CrossRef]
46. Li, G.; Liu, M.; Song, C.; Yuan, Z. Printable and conductive supramolecular hydrogels facilitated by peptides and group 1B metal ions. *Appl. Surf. Sci.* **2019**, *493*, 94–104. [CrossRef]

47. Wang, H.; Xu, W.; Song, S.; Feng, L.; Song, A.; Hao, J. Hydrogels facilitated by monovalent cations and their use as efficient dye adsorbents. *J. Phys. Chem. B* **2014**, *118*, 4693. [CrossRef] [PubMed]
48. Rohani Rad, E.; Vahabi, H.; Formela, K.; Saeb, M.R.; Thomas, S. Injectable poloxamer/graphene oxide hydrogels with well-controlled mechanical and rheological properties. *Polym. Advan. Technol.* **2019**, *30*, 2250–2260. [CrossRef]
49. Yang, K.; Li, X.; Cui, J.; Zhang, M.; Wang, Y.; Lou, Z.; Shan, W.; Xiong, Y. Facile synthesis of novel porous graphene-like carbon hydrogel for highly efficient recovery of precious metal and removal of organic dye. *Appl. Surf. Sci.* **2020**, *528*, 146928. [CrossRef]
50. Adhikari, B.; Palui, G.; Banerjee, A. Self-assembling tripeptide based hydrogels and their use in removal of dyes from waste-water. *Soft Matter* **2009**, *5*, 3452–3460. [CrossRef]
51. Basak, S.; Nandi, N.; Paul, S.; Hamley, I.W.; Banerjee, A. A tripeptide-based self-shrinking hydrogel for waste-water treatment: Removal of toxic organic dyes and lead (Pb^{2+}) ions. *Chem. Commun.* **2017**, *53*, 5910–5913. [CrossRef] [PubMed]
52. Chatterjee, S.; Hui, P.C.; Kan, C.; Wang, W. Dual-responsive (pH/temperature) Pluronic F-127 hydrogel drug delivery system for textile-based transdermal therapy. *Sci. Rep.* **2019**, *9*, 11658. [CrossRef] [PubMed]
53. Pasparakis, G.; Tsitsilianis, C. LCST polymers: Thermoresponsive nanostructured assemblies towards bioapplications. *Polymer* **2020**, *211*, 123146. [CrossRef]
54. Xu, W.; Hong, Y.; Song, A.; Hao, J. Peptide-assembled hydrogels for pH-controllable drug release. *Colloids Surf. B* **2020**, *185*, 110567. [CrossRef] [PubMed]

Communication

Glycerin/NaOH Aqueous Solution as a Green Solvent System for Dissolution of Cellulose

Ke Li [1,†], Huiyu Yang [1,2,†], Lang Jiang [1], Xin Liu [1], Peng Lang [3], Bo Deng [1,*], Na Li [4,*] and Weilin Xu [1,*]

1. State Key Laboratory of New Textile Materials and Advanced Processing Technologies, Wuhan Textile University, Wuhan 430200, China; dabing_ke@163.com (K.L.); hy-yang_wtu@hotmail.com (H.Y.); 15629129772@163.com (L.J.); xinliu_wtu@163.com (X.L.)
2. College of Material Science and Engineering, Wuhan Institute of Technology, Wuhan 430073, China
3. College of Materials Science and Engineering, Qingdao University, Qingdao 266071, China; 17852156975@163.com
4. School of Environmental Engineering and Chemistry, Luoyang Institute of Science and Technology, Luoyang 471023, China
* Correspondence: dengjianguo88@outlook.com (B.D.); weini.baobao@163.com (N.L.); weilin_xu@hotmail.com (W.X.)
† Ke Li and Huiyu Yang contributed equally to this work.

Received: 13 January 2020; Accepted: 1 August 2020; Published: 3 August 2020

Abstract: Dissolving cellulose in water-based green solvent systems is highly desired for further industrial applications. The green solvent glycerin—which contains hydrogen-bonding acceptors—was used together with NaOH and water to dissolve cellulose. This mixed aqueous solution of NaOH and glycerin was employed as the new green solvent system for three celluloses with different degree of polymerization. FTIR (Fourier-transform infrared), XRD (X-ray diffractometer) and TGA (thermogravimetric analysis) were used to characterize the difference between cellulose before and after regenerated by HCl. A UbbeloHde viscometer was used to measure the molecule weight of three different kinds of cellulose with the polymerization degree of 550, 600 and 1120. This solvent system is useful to dissolve cellulose with averaged molecule weight up to 2.08×10^5 g/mol.

Keywords: cellulose; glycerin; dissolution; green solvent; aqueous solution

1. Introduction

Cellulose has been put under the spotlight in the preparation of novel polymers and materials as one of the most affluent biopolymer sources in the world [1,2]. However, giant intra- and inter-molecule hydrogen bonds in the natural structure of cellulose result in its insolubility in both water and normal organic solvents, which greatly limits its application in industry [3].

Some solvent systems such as ammonium thiocyanate [4], calcium thiocyanate, sodium thiocyanate [5], lithium chloride/N, N-dimethylacetamide (LiCl/DMAc) [6,7] and NH_3/NH_4SCN [8], have been successfully applied to dissolve cellulose in the last century. However, resulting environmental pollution and high cost have confined these solvent systems as merely applicable at the lab scale.

Some green solvent systems including N-methylmorpholine-N-oxide (NMMO) [9], ionic liquid [10], water-based solvent systems [11] and mixed solvent systems—including amino acid ionic liquid/dimethyl sulfoxide (DMSO), tetra(n-butyl) ammonium hydroxide (TBAH) aqueous solution [12] and deep eutectic solvents (DESs) [13]—have been successively developed to dissolve cellulose. Although the mechanisms of cellulose dissolution varies with solvents, most researchers believe that—regardless of the molecule weight and crystallinity of cellulose [14]—the destruction of inter-molecule and intra-molecule hydrogen bonds in the complex structure of cellulose is a prerequisite to dissolve cellulose. Moreover, the interaction between the hydroxyl protons of D-dehydrated pyran glucose unit and dissociated solvent anions is the main driving force for cellulose dissolution [15].

Yuan et al. [16] first revealed that cellulose could be dissolved in a NaOH/urea aqueous solution after freezing the suspension into an ice-state, following a thawing process at room temperature under rigid agitation. This report opens the new window to dissolve cellulose in water-based solvent systems. To better understand the necessary of precooling procedure in the dissolving of cellulose, solid-state ^{13}C-NMR [17], low temperature DSC [18], small-angle X-ray scattering [19,20] and synchrotron radiation micro-diffraction have been extensively applied. The results reveal that Na-cellulose complexes and hydrated alkali ions were two key factors that account mainly for the dissolving mechanism. The destruction of hydrogen bonds can be achieved by the formation of new hydrogen bonds between Na$^+$ and hydroxyl groups of cellulose [21]. Moreover, the cellulose dissolution has been confined into an 8–9-wt% NaOH concentration region and became more remarkable at the temperature of or under four degrees Celsius [21].

Furthermore, Yuan et al. found that cellulose could be quickly dissolved in a precooled aqueous solution of LiOH/urea [18], NaOH/urea [16] or NaOH/thiourea [22] by generating a stable cellulose solution. In this, urea and thiourea acted as the hydrogen-bonding acceptor associated with the hydroxyl inside cellulose. These inter- and intra-chains associations stopped the regeneration of cellulose and ensures the solvation of cellulose.

Poly(ethylene glycol) (PEG)—another molecule which is possible candidate to stable the cellulose solution has also been successfully used by Yan to dissolve cellulose [23]. Instead, the oxygen atoms in the PEG chain are the hydrogen-bonding acceptor which stabilize the cellulose solution. The obtained cellulose solution in PEG/NaOH solvent system could be stable even for 30 days' storage at room temperature at the cellulose concentration up to 13 wt%.

This is to say, any molecules with hydrogen-bonding acceptors are possible candidate for dissolving cellulose. Hydroxyl groups in a typical environmentally friendly molecule such as glycerin, could be alternative for the urea or thiourea to stable the cellulose solution [24].

In this study, a mixed aqueous solution of NaOH and glycerin was employed as a new green solvent system for three celluloses with different degree of polymerization (DP). Glycerin acts as a hydrogen bond acceptor which could prevent the ressociation of cellulose hydroxyl groups to form a stable and uniform solution. The proposed green solvent system could dissolve cellulose with a number average molecule weight of up to 2.08×10^5 g/mol which was much higher than the reported 1.3×10^5 g/mol [23].

2. Materials and Methods

2.1. Materials

Three kinds of celluloses (cotton linter pulps) with an α-cellulose content more than 95% were purchased from Shanghai Hengxin Chemical Reagent Co., Ltd (Shanghai, China).

Their DPs were 550 (short fiber cotton linter pulps, referred to as SF-C), 600 (long fiber cotton linter pulps, referred to as LF-C), 1120 (high mechanical cotton linter pulps, referred to as HM-C), respectively. Corresponding physical properties of SF-C, LF-C and HM-C are shown in Table 1.

Table 1. Physical properties of cellulose with different degrees of polymerization (DPs).

	DP	α-Cellulose /%	Viscosity/mPa·s	Ash/%	Fe/ppm	Alkali Absorption Value/%	H$_2$O/%	Whiteness/%	Small Dust/mm^2/kg
SF-C	550	95.1	9.2	0.1	11	513	13.1	78	154
LF-C	600	96.0	11.4	0.07	8	699	13.2	80	79
HM-C	1120	98.6	29.9	0.06	11	582	8.7	82	88

The cellulose was dried in vacuum at 35 °C overnight to remove the water content prior to use. Deionized water used in all solutions was taken from a Milli-Q Plus 185 water purification system (Millipore, Bedford, MA) and had a resistivity of 10–16 MΩ·cm at 25 °C. Deuterium water (D_2O, 99.9 atom% D) was purchased from Aladdin Reagent (Shanghai, China) Co., Ltd. Sodium hydroxide (NaOH), hydrochloric acid (HCl) and urea were purchased from Sinopharm Chemical Reagent Co., Ltd. (Shanghai, China).

2.2. Dissolution of Cellulose in Glycerin/NaOH Aqueous Solution

One gram of glycerin and 9.0 g of NaOH were added into 90 mL of deionized water to prepare the mixed aqueous solution of glycerin/NaOH. Then, 4.0 g of three kinds of cellulose were added into the mixture, respectively and allowed to swell for 6 h at room temperature. Then, the suspension was precooled down to -20 °C and held at that temperature overnight to make a solid frozen mass. The frozen solid was then stirred strongly under the action of a homogenizer (S10, Scientz company, Ningbo, China) at 20,000 rpm until the cellulose solution was completely thawed. Three kinds of homogeneous cellulose solutions were then finally obtained.

2.3. Preparation of Regenerated Cellulose

Around 2 mL 3-M HCl was added dropwise into 1 mL uniform cellulose solution and allowed to stand for 10 min. After the cellulose aggregates being completely collected by discarding the supernatant, the precipitate was washed three times by deionized water and then oven-dried at 40 °C for 20 h. Finally, pure and dry regenerated cellulose could be obtained.

2.4. Characterization

Photographs of dissolved and regenerated cellulose were taken with an Apple X mobile phone.

The microscopic morphology of cellulose before and after dissolution and further regenerated by 3-M hydrochloric acid (HCl) was analyzed with a scanning electron microscope (SEM, JSM-6510LV, JEOL Co., Ltd., Tokyo, Japan) using an accelerating voltage of 10 kV. The original cellulose and dried regenerated cellulose were cut into 5-mm × 5-mm samples and directly attached to the conductive adhesive for testing. Fifty microliters dissolved cellulose solution was directly dropped onto the surface of a 5 mm × 5 mm silicon wafer. The silicon wafer was cleaned with ethanol prior to solution deposition. Then, the silicon wafer was oven-dried at 45 °C for 1 hour and attached to a conductive adhesive on an aluminum sample holder for electron microscopy scanning.

^{13}C-NMR spectrum of the cellulose solution was measured on a Bruker spectrometer (^{13}C-NMR, AVANCE 400, BRUKER Co., Ltd., Karlsruhe, Germany). The number of scans was 1024; the time of each scan was 14.48 s. The solvent in this study was prepared by substituting D_2O for H_2O and the cellulose was dissolved to obtain a 4-wt% cellulose solution and 4 mL of this solution was measured for ^{13}C-NMR spectrum in a nuclear magnetic tube with diameter of 5 mm and length of 20–25 cm.

An X-ray diffractometer (XRD, Bruke D8 Advance, Karlsruhe, Germany) was used to analyze the crystalline structure of the cellulose and the regenerated cellulose.

Thermogravimetric analysis (TGA) was carried out on a TA Instruments (TGA5500, New Castle DE, USA). A five-milligram sample was heated from 30 to 800 °C under nitrogen with 25-mL/min flow rate at a constant heating rate of 10 °C/min.

Fourier-transform infrared (FTIR, Nicolet NEXUS 670, Wisconsin, USA) spectroscopic analyses of all samples were done with a resolution of 2 cm^{-1} by averaging 64 scans in the range of 4000–400 cm^{-1}. The FTIR spectra of dried original, dried regenerated cellulose and glycerin were taken under an attenuated total reflection (ATR) mode using corresponding accessory.

The viscosity of the cellulose in 4.6-wt% NaOH/ 15-wt% urea aqueous solution was measured at 25 ± 0.1 °C with a clean Ubbelohde viscometer (Ubbelohde viscometer, Youlaibo Technology Co., Ltd., Beijing, China). The used cellulose concentrations are 0.5×10^{-3} g/mL, 1.0×10^{-3} g/mL, 2.0×10^{-3} g/mL, respectively.

3. Result and Discussion

Figure 1 shows the photos of three kinds of cellulose aqueous solutions in 1.0-wt% glycerin/9.0-wt% NaOH before and after regeneration. Clear cellulose aqueous solutions without any aggregate in 1.0-wt% glycerin/9.0-wt% NaOH confirms that the good solubility of cellulose in glycerin/NaOH solvent system. After the addition of 3-M HCl, delamination appears by forming white precipitations at the bottom of solution which implies the regeneration of cellulose occurs.

Figure 1. Photos of three kinds of cellulose aqueous solutions in 1.0-wt% glycerin/9.0-wt% NaOH and regenerated cellulose by 3-M HCl.

The dissolving status of cellulose in the new solvent system could be detected by SEM. As shown in Figure 2, SF-C, LF-C and HM-C show clear fiber-like morphology before dissolution. After completely dissolved in glycerin/NaOH solvent system, fiber-like morphologies disappeared, and a final homogeneous morphology could be obtained. This convinced the good dissolution ability of glycerin/NaOH solvent system to the cellulose. After further adding of HCl, the apparently strip morphology with much smaller size (compared with the cellulose before dissolution) may attribute to the reduced crystallinity which accompanied by increased amorphous area after cellulose regeneration [25].

To further study the dissolution status of cellulose macromolecule in glycerin/NaOH solvent system. D_2O was used to replace the deionized water to dissolve cellulose, SF-C, LF-C and HM-C in glycerin/NaOH solvent system. The concentration of cellulose was fixed to 4 wt% for three kind of celluloses. Figure 3 shows the ^{13}C-NMR spectra of three different celluloses. Peaks located at 104.1 ppm, 79.5 ppm, 74.3 ppm and 61 ppm were ascribed to the carbon atom of C1, C4, C2, C6 as inserts in Figure 3. Twin peak located at 75.7 ppm was ascribed to the carbon atom of C3,5 [26]. Compared with the reported cellulose I obtained from the NaOH/urea solvent system, a higher magnetic field shifting from 79.2 ppm [27] to 79.5 ppm implies the destruction of intra-molecule hydrogen bonding, which is similar with that reported dissolution of wood pulp in LiCl/DMAc [28].

Peaks located at 73.1 ppm and 63.6 ppm were ascribed to the carbon in glycerin. Thus, we can conclude that three cellulose with different DPs could dissolve well in glycerin/NaOH. Moreover, the 1.0-wt% glycerin/9.0-wt% NaOH is the direct solvent of cellulose instead of a derivation aqueous solution.

Figure 4 shows the XRD patterns of pristine and regenerated cellulose from glycerin/NaOH aqueous solution. Before regeneration, celluloses with different DPs showed characteristic 2θ peak at 14.9 ° and 16.4 ° which are correspond to (1–10), (110) lattice plane. Peak at 26.7° of SF-C and HM-C is in accordance with (200) lattice plane. However, the peak at 23.7° of LF-C is characteristic signal for (101) lattice plane. All these lattice planes are identical signals of cellulose I [29]. After regeneration, all celluloses showed peaks at 20.0 ° and 22.1 °, which are identical signals of (110) and (020) lattice planes of cellulose II, respectively [30]. These cellulose I to cellulose II transition convinced the successful regeneration of cellulose [31].

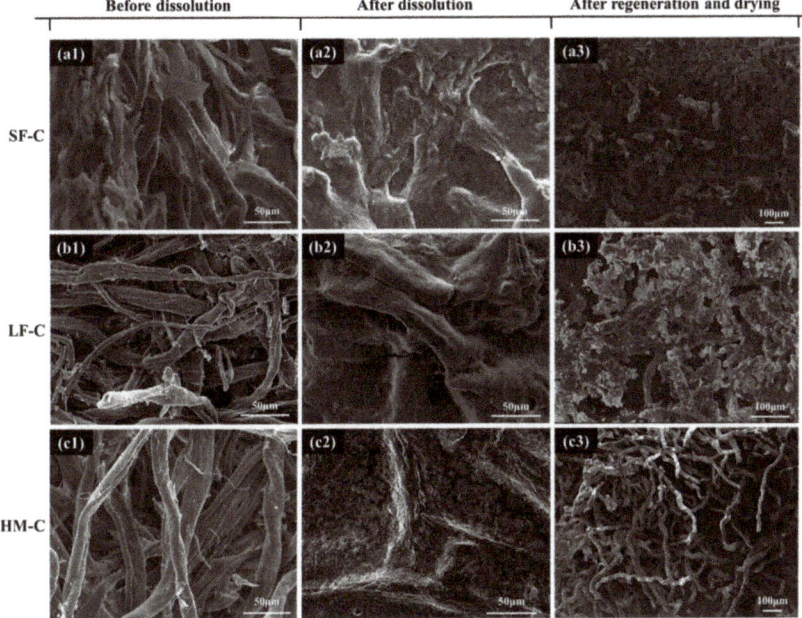

Figure 2. SEM images of (**a**) SF-C, (**b**) LF-P and (**c**) HM-C (a1, b1, c1) before dissolving the (a2, b2, c2) as-prepared cellulose solution under strong stir and (a3, b3, c3) the regenerated cellulose particles after the evaporation of water.

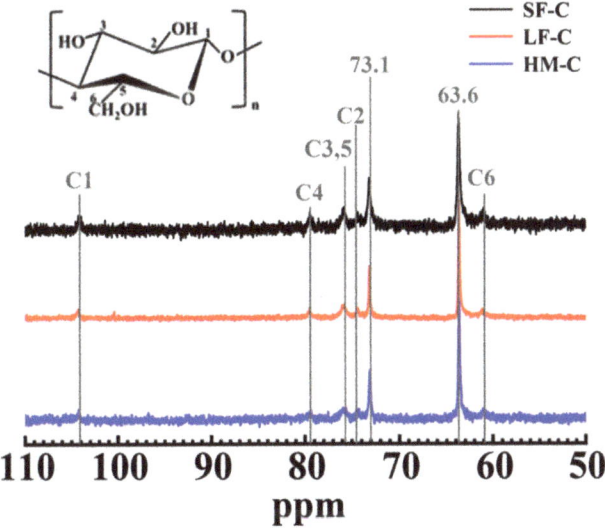

Figure 3. ^{13}C-NMR spectra of three kinds of 4.0-wt% cellulose in 1.0-wt% glycerin /9.0-wt% NaOH /D$_2$O aqueous solution.

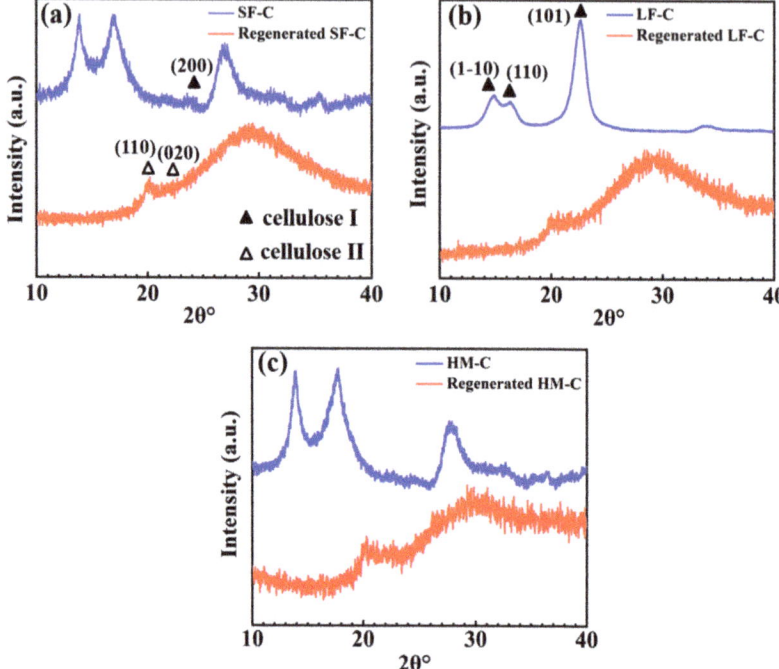

Figure 4. XRD patterns of the cellulose and regenerated cellulose from its glycerin/NaOH aqueous solution: (**a**) SF-C; (**b**) LF-C; (**c**) HM-C.

In addition, we calculated the crystallinity of celluloses with different DPs using Rietveld method [32]. The results show that the crystallinity of cellulose with different DPs in glycerol/NaOH aqueous solution changes significantly, and the destruction of molecular chain structure leads to a sharp decrease in crystallinity, as shown in Table 2. The sharply decreased crystallinity of cellulose after regeneration accounts mainly the excellent solubility of cellulose in NaOH/glycerin.

Table 2. Crystallinity of cellulose with different DPs before and after regeneration.

Sample	Before Regeneration (%)	After Regeneration (%)
SF-C	47.3±1.7	1.6±0.2
LF-C	57.7±0.6	4.1±0.3
HM-C	61.6±1.7	14.9±0.9

The regenerated cellulose could be obtained by adding diluted HCl into cellulose aqueous solution and followed by rinsing and drying. Figure 5 reveals the FTIR spectra of the cellulose, glycerin and regenerated for SF-C, LF-C and HM-C.

The bathochromic shift of hydroxyl from 3270 cm^{-1} to 3320 cm^{-1} after regeneration is due to the weakening effect of glycerin to the inter- and intra-molecule hydrogen bonding [33]. Peak around 2890 cm^{-1} is the stretching vibration of CH while that of 1427 cm^{-1} is the bending vibration of CH_2 in pristine cellulose. The transition of CH_2 around 1427 cm^{-1} (in pristine cellulose) to CH around 1369 cm^{-1} (in regenerated cellulose) after regenerating indicates the rotational isomer variation from C3–O3 and C6–O6. This further convinces the transition from cellulose I to cellulose II [34].

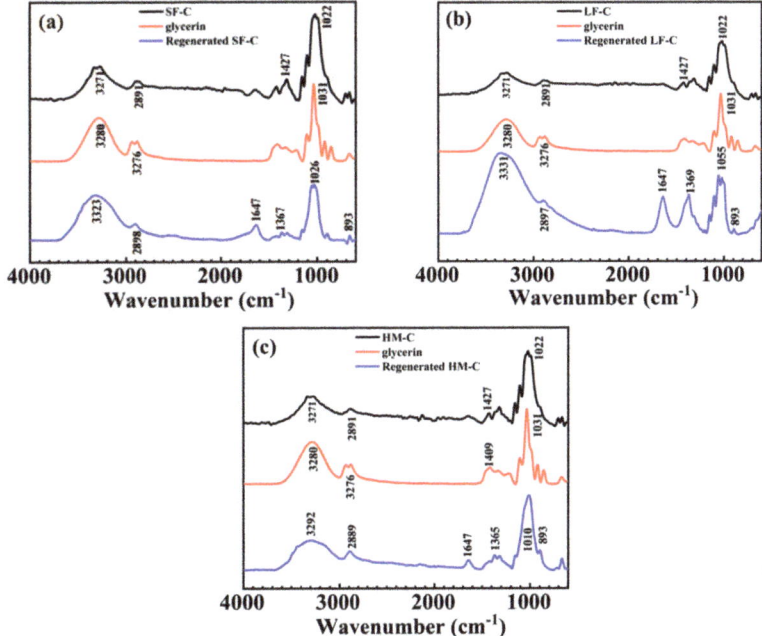

Figure 5. FTIR spectra of the cellulose, glycerin and regenerated cellulose from its glycerin/NaOH aqueous solution: (**a**) SF-C; (**b**) LF-C; (**c**) HM-C.

The absorption peak around 893 cm^{-1} is the outward stretching vibration of asymmetric rings which corresponds to the vibration band of C5 and C6. The existence of this peak means the cellulose is well dissolved in the solvent [35].

Compared with spectrum of glycerin, peak at 1647 cm^{-1} in regenerated cellulose indicates the trace amount of glycerin residue [36].

Figure 6 shows the contrastive TGA curves of three kinds of cellulose and corresponding regenerated cellulose. The decomposition temperature of all pristine celluloses starts from 290 °C while that for regenerated cellulose shows a sharply decreased temperature of 230 °C. This is because a large number of regular hydrogen bonds are destroyed during the regeneration in crystalline region which lowers the thermal stability of cellulose [37]. Char residual weight percentages of pristine and regenerated celluloses are summarized in Figure 6d. Much higher amount of char from regenerated cellulose is due to the existed bigger amorphous zone which is favorable for the forming of pyrolytic char [38].

The dependence of intrinsic viscosity on the concentration of three kinds of cellulose in 4.6-wt% NaOH/15-wt% urea aqueous solution at 25°C was plotted in Figure 7.

Clear intersection points in all Huggins–Kraemer curves demonstrate that the glycerin/NaOH system is good solvent for cellulose with different DPs. Intercepts which indicate the intrinsic viscosity for SF-C, LF-C and HM-C are 238 mL/g, 253 mL/g and 464 mL/g, respectively.

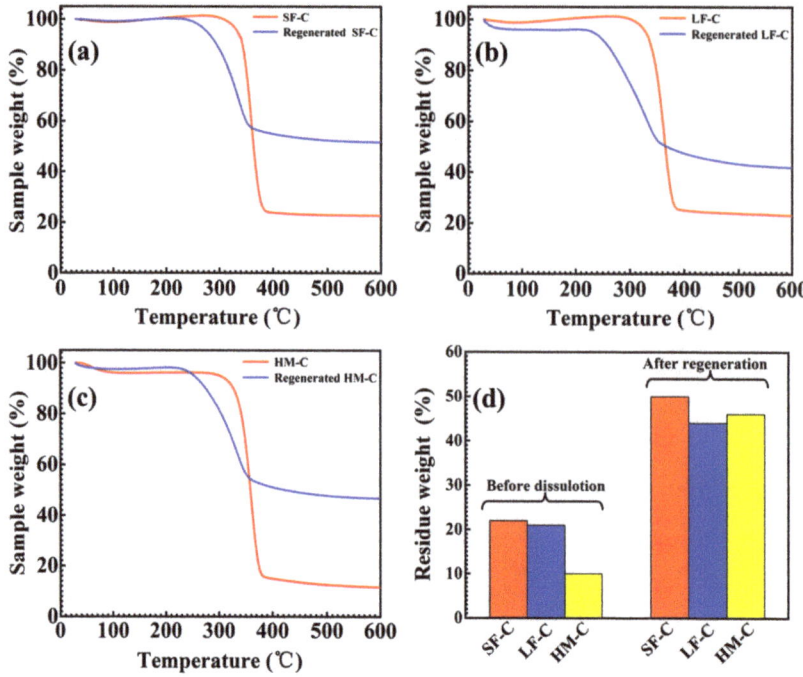

Figure 6. Thermal analysis of the cellulose and regenerated cellulose from its glycerin/NaOH aqueous solution: (**a**) SF-C; (**b**) LF-C; (**c**) HM-C; (**d**) comparison of TGA of different state cellulose.

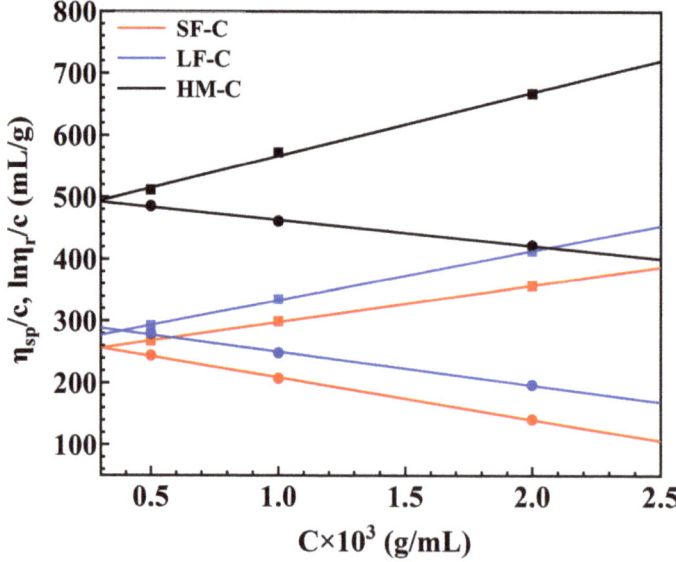

Figure 7. Dependence of intrinsic viscosity ([η]) on the concentration of three kinds of cellulose in 4.6-wt% NaOH/15-wt% urea aqueous solution at 25 °C.

According to the formula (1), the calculated number averaged molecular weights of SF-C and HM-C are 8.77×10^4 g/mol, 9.49×10^4 g/mol, 2.08×10^5 g/mol, respectively. According to the literature [39], the K and α value used here are 3.72×10^{-2} cm^3/g and 0.77, respectively.

$$[\eta] = KM^\alpha \tag{1}$$

All these results indicate the glycerin/NaOH is a novel good solvent system to dissolve cellulose with the molecular weight up to 2.08×10^5 g/mol. Zhang et al. [22] reported that the NaOH/thiourea system can dissolve cellulose with a viscosity average molecular weight of 2.0×10^5 while introduced thiourea will bring about secondary pollution.

4. Conclusions

In this communication, we developed a novel green solvent system, glycerin/NaOH, to dissolve cellulose. The aqueous solution of 1.0-wt% glycerin/9.0-wt% NaOH could dissolve the cellulose well and form a homogeneous solution. Glycerin acts as the hydrogen-bonding acceptor which could stop the reassociation of hydroxyl groups of cellulose to form homogeneous solution. Moreover, also this method is applicable to dissolve cellulose with the number averaged molecular weight up to 2.08×10^5 g/mol.

Author Contributions: K.L. and H.Y. performed the experiments, analysis, or interpretation of data for the work, and drafting the work. L.J. and P.L. collected data for the work. B.D. wrote the paper, designed all the work, and revising it critically for important intellectual content. X.L., N.L. and W.X. supervised the entire research progress. All authors have read and agreed to the published version of the manuscript.

Funding: This work was supported by the National Natural Science Foundation of China (Grant 51773158); Excellent Young Science and Technology Innovation Team of Hubei High School (Grant 201,707); and the Key Laboratory of Textile Fiber & Product (Wuhan Textile University) (Grant FZXW2017013), Ministry of Education.

Conflicts of Interest: The authors declare no conflicts of interest.

References

1. Parsamanesh, M.; Tehrani, A.D. Synthesize of new fluorescent polymeric nanoparticle using modified cellulose nanowhisker through click reaction. *Carbohyd. Polym.* **2016**, *136*, 1323–1331. [CrossRef] [PubMed]
2. Hui, W.; Gabriela, G.; Rogers, R.D. Ionic liquid processing of cellulose. *Chem. Soc. Rev.* **2012**, *41*, 1519–1537.
3. Meng, H.; Yanteng, Z.; Jiangjiang, D.; Zhenggang, W.; Yun, C.; Lina, Z. Fast contact of solid-liquid interface created high strength multi-layered cellulose hydrogels with controllable size. *ACS Appl. Mater. Interfaces* **2014**, *6*, 1872–1878.
4. Liu, C.K.; Cuculo, J.A.; Smith, B. Coagulation studies for cellulose in the ammonia/ammonium thiocyanate (NH$_3$/NH$_4$SCN) direct solvent system. *J. Polym. Sci. Pol. Phys.* **1989**, *27*, 2493–2511. [CrossRef]
5. Hattori, M.; Koga, T.; Shimaya, Y.; Saito, M. Aqueous calcium thiocyanate solution as a cellulose solvent. structure and interactions with cellulose. *Polym. J.* **1998**, *30*, 43–48. [CrossRef]
6. Ramos, L.A.; Assaf, J.M.; Seoud, O.A.E.; Frollini, E. Influence of the supramolecular structure and physicochemical properties of cellulose on its dissolution in a lithium chloride/N,N-dimethylacetamide solvent system. *Biomacromolecules* **2005**, *6*, 2638–2647. [CrossRef]
7. Mccormick, C.L.; Callais, P.A.; Hutchinson, B.H. Solution studies of cellulose in lithium chloride and N,N-dimethylacetamide. *Macromolecules* **1985**, *18*, 2394–2401. [CrossRef]
8. Cuculo, J.A.; Smith, C.B.; Sangwatanaroj, U.; Stejskal, E.O.; Sankar, S.S. Study on the mechanism of dissolution of the cellulose/NH$_3$/NH$_4$SCN system. I. *J. Polym. Sci. Pol. Chem.* **2010**, *32*, 229–239. [CrossRef]
9. Rosenau, T.; Hofinger, A.; Potthast, A.; Kosma, P. On the conformation of the cellulose solvent N-methylmorpholine-N-oxide (NMMO) in solution. *Polymer* **2003**, *44*, 6153–6158. [CrossRef]
10. Doherty, T.V.; Morapale, M.; Foley, S.E.; Linhardt, R.J.; Dordick, J.S. Ionic liquid solvent properties as predictors of lignocellulose pretreatment efficacy. *Green Chem.* **2010**, *12*, 1967–1975. [CrossRef]

11. Ghalami-Choobar, B.; Shekofteh-Gohari, M.; Sayyadi-Nodehi, F. Thermodynamic study of ternary electrolyte KCl + 1-ProH + water system based on pitzer and pitzer-simonson-clegg models using potentiometric measurements. *J. Mol. Liq.* **2013**, *188*, 49–54. [CrossRef]
12. Mitsuru, A.; Yukinobu, F.; Hiroyuki, O. Fast and facile dissolution of cellulose with tetrabutylphosphonium hydroxide containing 40 wt% water. *Chem. Commun.* **2012**, *48*, 1808–1810.
13. Hongwei, R.; Chunmao, C.; Shaohui, G.; Dishun, Z.; Qinghong, W. Synthesis of a Novel Allyl-Functionalized Deep Eutectic Solvent to Promote Dissolution of Cellulose. *BioResources* **2016**, *11*, 8457–8469.
14. Hu, Y.; Thalangamaarachchige, V.D.; Acharya, S.; Abidi, N. Role of low-concentration acetic acid in promoting cellulose dissolution. *Cellulose* **2018**, *25*, 4389–4405. [CrossRef]
15. Medronho, B.; Lindman, B. Competing forces during cellulose dissolution: From solvents to mechanisms. *Curr. Opin. Colloid Interface Sci.* **2014**, *19*, 32–40. [CrossRef]
16. Yuan, M.; Zhang, L.; Jie, C.; Zhou, J.; Kondo, T. Effects of coagulation conditions on properties of multifilament fibers based on dissolution of cellulose in NaOH/Urea aqueous solution. *Ind. Eng. Chem. Res.* **2008**, *47*, 8676–8683.
17. Bi, X.; Zhao, P.; Ping, C.; Zhang, L.; Kai, H.; Cheng, G. NMR spectroscopic studies on the mechanism of cellulose dissolution in alkali solutions. *Cellulose* **2013**, *20*, 613–621.
18. Jie, C.; Lina, Z. Rapid dissolution of cellulose in LiOH/urea and NaOH/urea aqueous solutions. *Macromol. Biosci.* **2010**, *5*, 539–548.
19. Endo, T.; Hosomi, S.; Fujii, S.; Ninomiya, K.; Takahashi, K. Nano-Structural Investigation on Cellulose Highly Dissolved in Ionic Liquid: A Small Angle X-ray Scattering Study. *Molecules* **2017**, *22*, 178. [CrossRef]
20. Yu, O.; Hidaka, H.; Kimura, S.; Kim, U.J.; Kuga, S.; Wada, M. Formation and stability of cellulose-copper-NaOH crystalline complex. *Cellulose* **2014**, *21*, 999–1006.
21. Lindman, B.; Karlström, G.; Stigsson, L. On the mechanism of dissolution of cellulose. *J. Mol. Liq.* **2010**, *156*, 76–81. [CrossRef]
22. Zhang, L.; Ruan, D.; Gao, S. Dissolution and regeneration of cellulose in NaOH/thiourea aqueous solution. *J. Polym. Sci. Part B Polym. Phys.* **2002**, *40*, 1521–1529. [CrossRef]
23. Yan, L.; Gao, Z. Dissolving of cellulose in PEG/NaOH aqueous solution. *Cellulose* **2008**, *15*, 789–796. [CrossRef]
24. He, Z.; Liu, J.; Li, L.; Lan, D.; Zhang, J. Absorption properties and spectroscopic Studies of dilute sulfur dioxide in aqueous glycerol solutions. *Ind. Eng. Chem. Res.* **2012**, *51*, 13882–13890. [CrossRef]
25. Moniruzzaman, M.; Mahmood, H.; Ibrahim, M.F.; Yusup, S.; Uemura, Y. Effects of pressure and temperature on the dissolution of cellulose in ionic liquids. *Adv. Mater. Res.* **2016**, *1133*, 588–592. [CrossRef]
26. Jiang, Z.; Fang, Y.; Ma, Y.; Liu, M.; Liu, R.; Guo, H.; Lu, A.; Zhang, L. Dissolution and metastable solution of cellulose in NaOH/thiourea at 8 °C for construction of nanofibers. *J. Phy. Chem. B* **2017**, *121*, 1793–1801. [CrossRef]
27. Zhang, L.; Ruan, D.; Zhou, J. Structure and Properties of Regenerated Cellulose Films Prepared from Cotton Linters in NaOH/Urea Aqueous Solution. *Ind. Eng. Chem. Res.* **2001**, *40*, 5923–5928. [CrossRef]
28. Berthold, F.; Gustafsson, K.; Berggren, R.; Sjöholm, E.; Lindström, M. Dissolution of softwood kraft pulps by direct derivatization in lithium chloride/N,N-dimethylacetamide. *J. Appl. Polym. Sci.* **2010**, *94*, 424–431. [CrossRef]
29. Chen, X.; Chen, X.; Cai, X.M.; Huang, S.; Wang, F. Cellulose dissolution in a mixed solvent of tetra(n-butyl)ammonium hydroxide/dimethyl sulfoxide via radical reactions. *ACS Sustain. Chem. Eng.* **2018**, *6*, 2898–2904. [CrossRef]
30. Nomura, S.; Kugo, Y.; Erata, T. ^{13}C NMR and XRD studies on the enhancement of cellulose II crystallinity with low concentration NaOH post-treatments. *Cellulose* **2020**, 10570-020-03036-6. [CrossRef]
31. Tang, S.; Baker, G.A.; Ravula, S.; Jones, J.E.; Zhao, H. PEG-functionalized ionic liquids for cellulose dissolution and saccharification. *Green Chem.* **2012**, *14*, 2922–2932. [CrossRef]
32. Oudiani, A.E.; Chaabouni, Y.; Msahli, S.; Sakli, F. Crystal transition from cellulose I to cellulose II in NaOH treated Agave americana L. fibre. *Carbohyd Polym.* **2011**, *86*, 1221–1229. [CrossRef]
33. Gibril, M.E.; Li, H.F.; Li, X.D.; Li, H.; Zhong, X.; Zhang, Y.; Han, K.Q.; Yu, M.H. Application of twin screw eextruder in cellulose dissolution with ionic liquid. *Appl. Mec. Mater.* **2013**, *268*, 605–609.
34. Mohan, M.; Banerjee, T.; Goud, V.V. Effect of protic and aprotic solvents on the mechanism of cellulose dissolution in Ionic liquids: A combined molecular dynamics and experimental insight. *Chemistryselect* **2016**, *1*, 4823–4832. [CrossRef]

35. Wu, R.N.; Zhu, H.; He, B.H. Dissolution of absorbent cotton in ionic liquid and characterization of the regenerated cellulose. *Adv. Mater. Res.* **2014**, *830*, 163–166. [CrossRef]
36. Gómez-Siurana, A.; Marcilla, A.; Beltrán, M.; Berenguer, D.; Martínez-Castellanos, I.; Menargues, S. TGA/FTIR study of tobacco and glycerol-tobacco mixtures. *Thermochim. Acta* **2013**, *573*, 146–157. [CrossRef]
37. Lethesh, K.C.; Wilfred, C.D.; Shah, S.N.; Uemura, Y.; Mutalib, M.I.A. Synthesis and characterization of nitrile-functionalized azepanium ionic liquids for the dissolution of cellulose. *Proc. Eng.* **2016**, *148*, 385–391. [CrossRef]
38. Mettler, M.S.; Mushrif, S.H.; Paulsen, A.D.; Javadekar, A.D.; Vlachos, D.G.; Dauenhauer, P.J. Revealing pyrolysis chemistry for biofuels production: Conversion of cellulose to furans and small oxygenates. *Energy Environ. Sci.* **2012**, *5*, 5414–5424. [CrossRef]
39. Swart, H.C.; Du, P.J.; De Villiers, M.M.; Lotter, A.P.; Liebenberg, W. Correlation between in vitro release from topical delivery vehicles and microbicidal activity of triclosan. *Pharmazie Die* **2006**, *61*, 35–40.

© 2020 by the authors. Licensee MDPI, Basel, Switzerland. This article is an open access article distributed under the terms and conditions of the Creative Commons Attribution (CC BY) license (http://creativecommons.org/licenses/by/4.0/).

Article

Low Molecular Weight and Polymeric Modifiers as Toughening Agents in Poly(3-Hydroxybutyrate) Films

Adriana Nicoleta Frone [1,*], Cristian Andi Nicolae [1], Mihaela Carmen Eremia [2], Vlad Tofan [3], Marius Ghiurea [1], Ioana Chiulan [1], Elena Radu [1], Celina Maria Damian [4] and Denis Mihaela Panaitescu [1,*]

1. Polymer Department, National Institute for R&D in Chemistry and Petrochemistry ICECHIM, 202 Splaiul Independentei, 060021 Bucharest, Romania; ca_nicolae@yahoo.com (C.A.N.); ghiurea@gmail.com (M.G.); ioana.chiulan@icechim.ro (I.C.); nina.radu58@yahoo.ro (E.R.)
2. National Institute for Chemical Pharmaceutical Research and Development ICCF, 112 Calea Vitan, 031299 Bucharest, Romania; mihaelaceremia@yahoo.com
3. Cantacuzino National Institute of R&D for Microbiology and Immunology, 103 Splaiul Independentei, 050096 Bucharest, Romania; tofan.vlad@gmail.com
4. Advanced Polymer Materials Group, Faculty of Applied Chemistry and Materials Science, University Politehnica of Bucharest, 1-7 Gheorghe Polizu, 011061 Bucharest, Romania; celina.damian@yahoo.com
* Correspondence: ciucu_adriana@yahoo.com (A.N.F.); panaitescu@icechim.ro (D.M.P.)

Received: 5 October 2020; Accepted: 20 October 2020; Published: 22 October 2020

Abstract: The inherent brittleness of poly(3-hydroxybutyrate) (PHB) prevents its use as a substitute of petroleum-based polymers. Low molecular weight plasticizers, such as tributyl 2-acetyl citrate (TAC), cannot properly solve this issue. Herein, PHB films were obtained using a biosynthesized poly(3-hydroxyoctanoate) (PHO) and a commercially available TAC as toughening agents. The use of TAC strongly decreased the PHB thermal stability up to 200 °C due to the loss of low boiling point plasticizer, while minor weight loss was noticed at this temperature for the PHB-PHO blend. Both agents shifted the glass transition temperature of PHB to a lower temperature, the effect being more pronounced for TAC. The elongation at break of PHB increased by 700% after PHO addition and by only 185% in the case of TAC; this demonstrates an important toughening effect of the polymeric modifier. Migration of TAC to the upper surface of the films and no sign of migration in the case of PHO were highlighted by X-ray photoelectron spectroscopy (XPS) and atomic force microscopy (AFM) results. In vitro biocompatibility tests showed that all the PHB films are non-toxic towards L929 cells and have no proinflammatory immune response. The use of PHO as a toughening agent in PHB represents an attractive solution to its brittleness in the case of packaging and biomedical applications while conserving its biodegradability and biocompatibility.

Keywords: poly(3-hydroxyoctanoate); polyhydroxybutyrate; bio-based modifiers; toughening; biocompatibility; thermal properties

1. Introduction

A huge percentage close to 90% of the globally produced plastics uses virgin fossil fuel feedstocks [1]. It is estimated that plastics' share will grow to around 20% of the global oil consumption by 2050 if this trend continues [1]. In this context, the plastics market has turned to polymers derived from renewable resources as eco-friendly alternatives to petroleum-based polymers. Poly(3-hydroxybutyrate) (PHB) belongs to the large family of polyhydroxyalkanoates (PHAs), which are fully biodegradable biopolyesters of hydroxycarbonic acids produced by either chemical or bacterial synthesis from biorenewable and biowaste resources [2,3]. Due to its biodegradability, biocompatibility, and competitive physical properties, PHB is the most well studied PHA for biomedical and food packaging materials [4–6]. However, its high production cost and advanced brittleness narrow its application as a substitute for common synthetic polymers [4–7]. Plasticizers may increase PHB flexibility and toughness, thus solving the problem related to its high brittleness [4,7–15]. Tributyl 2-acetylcitrate (TAC) is obtained from naturally occurring citric acid. It may be used as a plasticizer for applications starting from food packaging to biomedical applications as it is environmentally friendly and, in small amounts, does not raise safety concerns for humans [14,16–18]. Nevertheless, a major issue when using TAC or other citrate esters as plasticizers is related to their relatively low boiling point, which results in considerable weight loss at temperatures from 160 to 200 °C, therefore within the processing temperature range of PHB [19,20]. Moreover, TAC addition could trigger the migration at the surface of the film [13], thus deteriorating the mechanical properties. In particular, Corrêa et al. [20] explained the decrease in the thermal stability of poly(3-hydroxybutyrate-co-3-hydroxyvalerate)-TAC/organo-modified montmorillonite nanocomposites through the low degradation temperature of neat plasticizer (around 130 °C). Similarly, the addition of only 5 wt% TAC lowered the temperature to 5% weight loss of PHB and also influenced the maximum degradation peak [17]. Maiza et al. [16] observed that triethyl citrate and TAC, used as plasticizers in PLA, migrate out of the matrix, the weight loss being directly proportional to the temperature (100 or 135 °C) and plasticizer concentration. In general, low molecular weight plasticizers are liquids which are not chemically bonded to the polymer matrix and, therefore, at room or elevated temperature, they leach out from the polymer matrix [21]. When plasticized PHB is used for scaffolds and other medical devices or for food packaging, the plasticizer will leach out into the surrounding medium (human body or foods), raising health problems overtime or damaging food quality. In this context, the use of polymeric toughening agents could be a better solution to the inherent brittleness of PHB. Therefore, elastomeric medium chain length PHAs may be a better alternative to TAC and other citrate esters provided they successfully increase the flexibility and other properties of PHB while preventing migration.

Polyhydroxyoctanoate (PHO) belongs to the same PHA family as PHB, but it has a higher side-chain length, much lower crystallinity and melting point, and it exhibits elastomeric properties [22,23]. Therefore, it can provide increased ductility to PHB, reduced migration, and no leakage when processed at high temperatures. However, PHO biopolymer is not compatible with PHB, forming a biphasic system over the whole composition range [24]. Previous studies showed a noticeable increase in the elongation at break only from a high amount of PHO in the blends [22,24]. Thus, a doubling of the elongation at break was observed at 20 [24] or 15 wt% [24] PHO in PHB/PHO blends, depending on PHO type and preparation conditions, together with a strong decrease in tensile strength and modulus [22,24]. In biocomposites with a low amount of PHO (5 wt%), the supplementary addition of bacterial cellulose restored the Young's modulus [25]. Therefore, the domain of low PHO concentrations is more interesting from both a scientific and economical point of view and this was thoroughly investigated in the case of PHB/PHO blends obtained by solution casting.

In this work, we synthesized a PHO homopolymer (95 mol% 3-hydroxyoctanoate units) and we comparatively studied the toughening effect of PHO and low molecular weight TAC in neat PHB. PHB films with different concentrations of the two bio-based modifiers were characterized to obtain information on the surface morphology, crystalline structure, biocompatibility, and thermal and mechanical properties, useful for the intended application in packaging and biomedicine.

2. Materials and Methods

2.1. Materials

PHB powder with density of 1.24 g/cm^3 was acquired from Biomer(Schwalbach am Taunus, Germany) and was used as received. Tributyl 2-acetylcitrate ≥98% was purchased from Sigma-Aldrich.

Chloroform used for polymer dissolution was supplied by a local company (Chimopar SRL, Bucharest, Romania).

2.2. Biosynthesis of PHO

Pseudomonas fluorescens ICCF 392 strain from ICCF Culture Collection of Microorganisms was used for the biosynthesis of PHO. The strain *Pseudomonas fluorescens* was isolated from 1 g of rotten beech wood powder, homogenized in 20 mL sterile broth. The pre-inoculum culture (10%, w/v) was maintained on a specific medium consisting of (w/v) yeast extract 1.0%, peptone 1.0%, glycerol 5.0%, and agar 2%. The composition of culture medium for inoculum (IPS medium) was glucose 1 g/100 mL, corn extract 1.5 g/100 mL, KH_2PO_4 1 g/100 mL, NaCl 1 g/100 mL, and $MgSO_4$ 0.05 g/100 mL. Inoculum culture was developed in Erlenmeyer flasks of 500 mL under stirring at around 30 °C, for 24 h. The fermentation medium was inoculated with 10% (w/v) of inoculum. Besides the carbon source (sodium octanoate 2.0 g/L), the culture medium used in the fermentation process contained $NaNH_4HPO_4 \cdot 4H_2O$ 3.5 g/L, K_2HPO_4 7.5 g/L, KH_2PO_4 3.7 g/L, citric acid 20 g/L, 0.1mL/100mL of trace element solution 1, and 0.1mL/100mL of solution 2. Trace element solution 1 consisted of (per liter 1M HCl) $MgSO_4 \cdot 7H_2O$, 120.0 g. Trace element solution 2 consisted of (per liter 1M HCl) $FeSO_4 \cdot 7H_2O$, 2.78 g, $CaCl_2 \cdot 2H_2O$, 1.47 g, $MnCl_2 \cdot 4H_2O$, 1.98 g, $CoSO_4 \cdot 7H_2O$, 2.81 g, $CuCl_2 \cdot 2H_2O$, 0.17 g, and $ZnSO_4 \cdot 7H_2O$, 0.29 g. Bioprocesses were carried out in 500 mL flasks containing 100 mL of culture broth, which were maintained on a rotary shaker at 220 min^{-1}, 29 ± 1 °C for 48 h. The cultivation was performed by nutrient addition (sodium octanoate stock solution, 83.33 g/L) with the sequences of 3 mL each at 0, 24, and 30h, respectively. A total biomass of 1.975 g dry cells/L of fermentation medium was obtained and the yield of PHO was 21.10 g/100 g dry biomass.

2.3. Preparation of PHB/PHO and PHB/TAC

The PHB/PHO films were prepared using a solvent-casting method. Dried PHO membrane was dissolved in chloroform and stirred at room temperature until complete dissolution. PHB powder was subsequently added, so as to obtain different PHB:PHO weight ratios: 100:0 (PHB), 95:5 (PHB/5PHO), 90:10 (PHB/10PHO), 85:15 (PHB/15PHO), and 80:20 (PHB/20PHO). The concentration of the total amount of polymers in the chloroform was kept constant at 4 wt%/v. The resulting mixture was stirred at room temperature for 10 min until the polymers were well dispersed and then heated at 50 °C for 1 h to ensure complete dissolution of both components. PHB/PHO films with a thickness of around 10 µm were cast from the resulting solutions onto glass slides. They were initially dried at room temperature for several hours and then in a vacuum oven at 45 °C for 48 h to remove any residual solvent. PHB/TAC films with the same weight ratios as for PHB/PHO films were prepared following the same procedure and denoted as PHB/5TAC, PHB/10TAC, PHB/15TAC, and PHB/20TAC, while PHB film served as a reference.

2.4. Characterization

2.4.1. Gas Chromatography–Mass Spectrometry (GC-MS) Analysis of PHO

Triple Quad GC/MS system (Agilent Technology) was used for both qualitative and quantitative analysis of the PHO monomers. First, 0.5 µL of PHO polyester sample was injected using the split mode and the injector temperature was set to 250 °C. The analysis was performed using a HP-FFAP DB-WAX column (30 m × 0.25 mm, with a 0.25 µm film thickness) with a mobile phase of helium at a constant flow rate of 1 mL min^{-1}. The GC oven temperature was held for 1 min at 90 °C, then raised by 7 °C/min up to 240 °C and held for 12 min. MS transfer temperature was set to 280 °C, MS ion source to 230 °C. The electron ionization (EI) source of the instruments was operated at 70 eV in the scan range (m/z) = 40–400. The fragmentation pattern of the obtained mass spectra was analyzed by NIST 98 Mass Spectral Database, Gaithersburg, MD, USA.

2.4.2. Thermogravimetric Analysis (TGA)

Thermogravimetric analysis (TGA) of the blends was carried out on a TA Q5000 analyzer (TA Instruments Inc., New Castle, DE, USA), from 25 to 700 °C, with 10 °C/min, using nitrogen as a purge gas (40 mL/min).

2.4.3. Differential Scanning Calorimetry (DSC)

Thermal parameters of the blends, including glass transition temperature (T_g), melting temperature (T_m), crystallization temperature (T_c), their specific enthalpies and crystalline ratio (X_c), were determined with a DSC Q2000 calorimeter (TA Instruments Inc., New Castle, DE, USA) under helium flow (25 mL/min). A heat–cool modulation program from −65 to 190 °C, at an average heating/cooling rate of 10 ± 0.80 °C/min with a period of 30 s, was used. The crystallinity degree of PHB in the films was estimated from the melting enthalpy values (ΔH_m) of the samples and the melting enthalpy of 100% crystalline PHB (ΔH_m^0, 146 J/g) [25] using Equation (1).

$$X_c[\%] = \frac{\Delta H_m}{\Delta H_m^0 \times w_i} \times 100 \quad (1)$$

where w_i is the weight fraction of PHB in the blends.

2.4.4. X-ray Diffraction (XRD)

The crystalline phase of neat PHB and PHB blends was analyzed using an X-ray diffractometer (Rigaku Corporation, Japan) with a Cu Kα (λ = 0.1541 nm) source. Scanning was performed at 45 kV and 200 mA at the 2θ scanning angle, between 5° and 40°, with a scanning step of 0.04 min^{-1}. The degree of crystallinity (CI) of the samples was calculated using Equation (2), where A_C is the sum of the areas under the crystalline peaks, and A_a is the area of the amorphous halo.

$$CI\ (\%) = \frac{A_C}{A_c + A_a} \times 100 \quad (2)$$

The inter-planar distances (d) were calculated using Bragg's law [26] and the crystallite sizes (D) at the main peaks using the Scherrer equation [27]. Fityk software was used for XRD data processing and nonlinear curve fitting.

2.4.5. Tensile Tests

Tensile tests were conducted according to ISO 527, Part 3, which is applicable to films, at room temperature using an Instron 3382 universal testing machine with a load cell of 1 kN. For each sample, at least 5 specimens of type 5A (10 μm thickness) were tested with a crosshead speed of 2 mm min^{-1}. The average values and the standard deviations for Young's modulus, tensile strength, and elongation at break were calculated using the Bluehill 2 Software.

2.4.6. Atomic Force Microscopy (AFM)

The surface morphology and the root mean square roughness of solvent-cast PHB films surfaces were determined using a MultiMode 8 atomic force microscope (Bruker, Santa Barbara, CA, USA). The characterization of each sample was performed in Peak Force (PF) Quantitative Nanomechanical Mapping (QNM) mode, in air, using silicon nitride tips at a scan rate of 1 Hz and a scan angle of 90°. The image processing and data analysis were conducted with NanoScope software version 1.20.

2.4.7. X-ray Photoelectron Spectroscopy (XPS)

The chemical composition at the surface of neat PHB and plasticized films was analyzed using a fully integrated K-Alpha system (Thermo Scientific, Waltham, MA, USA) equipped with a monochromated AlK$_\alpha$ source (1486.6 eV). Both survey (0–1200 eV) and high-resolution spectra were recorded for PHB blends. Charging effects were compensated by a flood gun and binding energies were calibrated by placing the C1s peak at 284.8 eV as internal standard. The pass energy for the survey spectra was set to 200 eV, and for the high-resolution spectra, it was 20 eV.

2.4.8. Contact Angle Measurements (CA)

CA measurements were carried out using a CAM 200 instrument (Biolin Scientific, Gothenburg, Sweden) equipped with a high-resolution camera (Basler A602f) and an auto-dispenser. CA was measured in air, at room temperature and ambient humidity, 2 s after the drop contacted the surface of the films. Drops of 6 µL deionized water were dispensed on each sample, and the value of the reported CA was the average of seven measurements. The images of the droplets were acquired with the high-resolution camera using CAM software.

2.4.9. Biocompatibility Test

Biocompatibility of PHB blends was tested using fluorescence microscopy as described elsewhere [25]. Film samples were sterilized by incubation in 70% ethanol solution overnight and then washed with sterile phosphate buffer saline (PBS) to discard residual ethanol. Afterwards, 0.32 cm^2 sterilized discs were carved out of each film and deposited in an ultra-low attachment flat-bottom 96-well plate (Sigma-Aldrich) to prevent cell adhesion to surfaces other than those tested. L929 murine fibroblast cells were seeded at a concentration of 1×10^4 cells/well on top of each sample and allowed to adhere for two different time intervals (1 and 9 days). After each incubation period, samples were gently washed to discard non-adherent cells. Further, samples were fixed with 4% paraformaldehyde (Sigma-Aldrich) solution, followed by staining with staining solution: 0.05% Triton X-100, SYBR-Green I, Texas-Red-X Phalloidin (Molecular Probes, Thermo Fisher Scientific). Images were recorded with an Eclipse TE2000 inverted fluorescence microscope (Nikon, Austria) and processed with Huygens software (SVI, Hilversum, The Netherlands).

2.4.10. Evaluation of Pro-Inflammatory Effect

Nearly equal amounts of each film were sterilized by overnight incubation in 70% ethanol solution followed by washing steps with PBS to completely remove ethanol. Samples were incubated at 37 °C in PBS for 20 days to allow elution of their constituents. A sample of PBS alone, incubated in a similar fashion, served as a control. Following incubation, elution samples were collected and tested for endotoxin activity with LAL QCL-1000 kit (Lonza, Verviers Belgium). Further, differentiated macrophage-like cells were incubated in the presence of collected elution samples (diluted 1/10 in complete culture medium) for 24 h. Cells incubated with lipopolysaccharide (LPS 100 ng/mL) or without stimuli were used as positive and negative controls. Supernatants were collected and stored at −80 °C until use. The influence of elution samples on differentiated macrophages was evaluated by measuring tumor necrosis factor-α (TNF-α) concentrations in culture supernatants using ELISA (DuoSet kits from R&D Systems Inc., Minneapolis, MN, USA).

3. Results and Discussion

3.1. PHO Characterization

PHO was characterized in terms of the monomer structure and composition by GC-MS analysis through its conversion to volatile carboxylic acids (Figure 1). The gas chromatogram showed a major peak with the retention time of 12.19 min and two minor peaks at 14.73 and 17.33 min (Figure 1a). The major peak corresponds to 3-octenoic acid, the first minor peak corresponds to trans-2-hexenoic acid, while the second minor peak was identified as 3-hydroxy-dodecanoic acid by comparing molecules in the GC database. Therefore, the biosynthesized PHO was mostly composed of 3-hydroxyoctanoate (3HO—C8), which represents 95.02 mol% of the total monomer content and very small amounts of

3-hydroxyhexanoate (3HHx—C6, 3.32 mol%) and 3-hydroxydodecanoate (3HDD—C12, 1.66 mol%) (Figure 1b–d). All products were detected in the form of single-type monomers only.

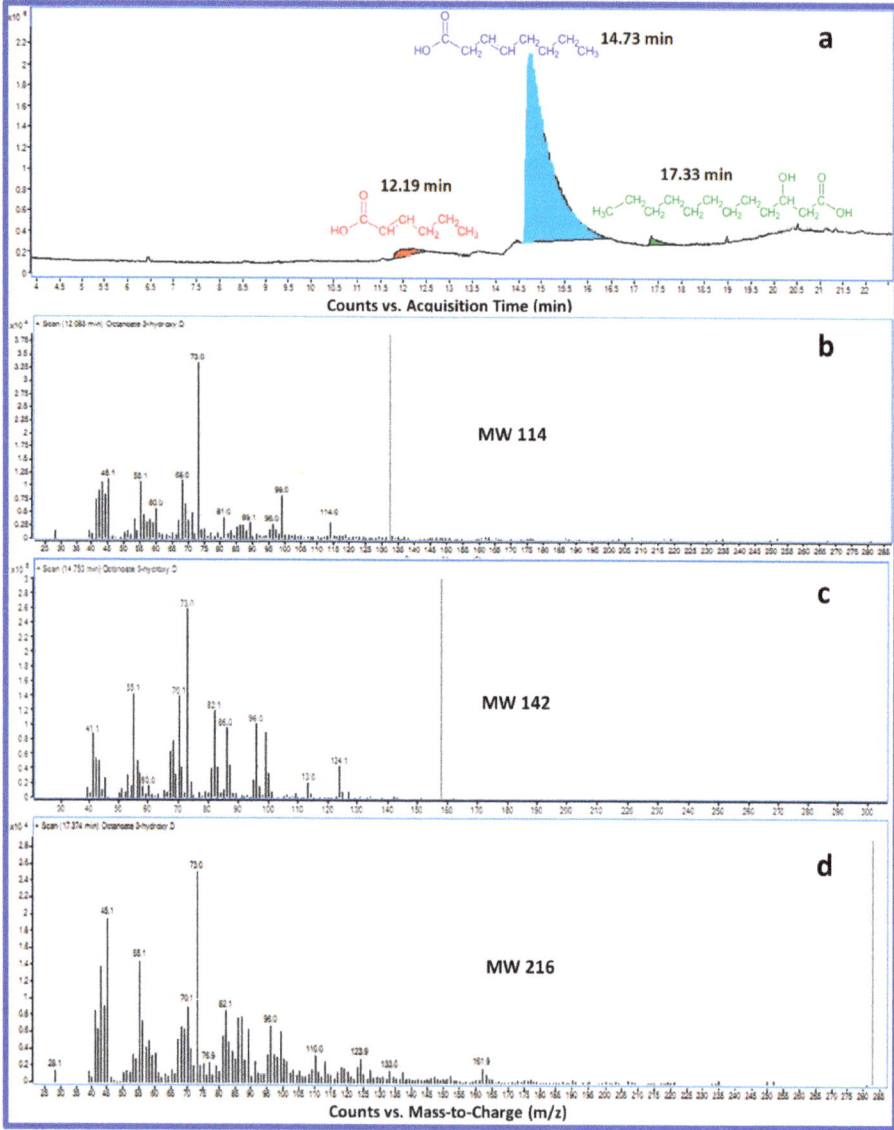

Figure 1. GC-MS chromatogram of PHO synthesized from *Pseudomonas fluorescens* showing monomeric composition of PHO (**a**) and mass spectra of trans-2-hexenoic acid (**b**), 3-octenoic acid (**c**), and 3-hydroxy-dodecanoic acid (**d**).

In addition, DSC analysis of PHO displayed a T_g at −35 °C and a double melting event with a major melting peak at around 43 °C and a shoulder at 51 °C. The total enthalpy of fusion was 16.22 J/g, suggesting a low degree of crystallinity of only 11% if the melting fusion of 100% crystalline PHB was used.

3.2. Thermal Stability of PHB Blends

The thermal degradation of neat PHB and PHB blends was investigated by TGA and the thermograms are plotted in Figure 2a–d, while the main parameters resulting from TGA and derivative curves (DTG) are summarized in Table 1. The thermal degradation of neat PHB and PHB/PHO films took place in a single stage, whereas that of PHB/TAC films showed two separated weight loss steps (Figure 2a–d). The temperature at which PHB/PHO films lost 5% of their mass ($T_{5\%}$) exhibited a downshift of 10–15 °C for up to 15 wt% PHO and of 33 °C for PHB/20PHO.

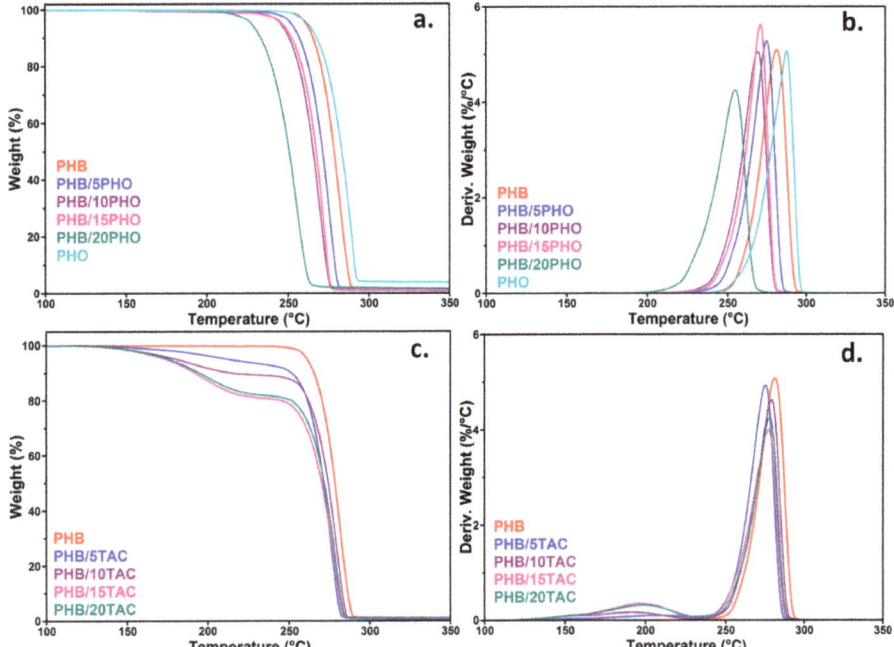

Figure 2. TGA and DTG curves for neat PHB, PHO (**a**,**b**) and TAC (**c**,**d**) plasticized PHB films.

Table 1. TGA parameters for neat PHB and plasticized PHB films.

Sample	$T_{5\%}$ (°C)	WL_{220} (%)	T_{d1} (°C)	T_{d2} (°C)	R_{500} * (%)
PHB	261.3	0.3	-	281.4	0.7
PHB/5PHO	252.2	0.4	-	275.0	0.7
PHB/10PHO	245.6	0.5	-	269.3	0.5
PHB/15PHO	246.6	1.1	-	271.0	0.7
PHB/20PHO	227.5	2.2	-	255.0	1.1
PHO	262.3	0.2	-	287.8	2.8
PHB/5TAC	213.1	5.0	202.5	275.5	0.4
PHB/10TAC	180.8	7.0	191.1	279.5	0.1
PHB/15TAC	174.6	17.0	195.9	277.4	0.3
PHB/20TAC	176.1	17.0	198.5	278.2	0.3

* R_{500} is the residue at 500 °C.

A similar behavior was observed in the case of the temperature of the maximum decomposition rate (T_d), which decreased by less than 10 °C for up to 15 wt% PHO and by around 26 °C at maximum PHO concentration (Table 1). This major degradation step may be ascribed to the random chain scission of PHB ester bonds by intramolecular cis-elimination, where degradation products like crotonic acid, linear oligomers with a crotonate end-group, and dimers and trimers of crotonic acid were formed [17,28].

The thermal stability of PHO is similar to that of PHB or better (Figure 2a). Therefore, the degradation of PHB/PHO may be determined by the presence of impurities, knowing that sodium and other metal salts are common elements in PHO biosynthesis and the thermal stability of PHB is very sensitive to metal traces [29,30]. Thus, it was reported that the residual metal compounds, derived from the fermentation process, catalyzed the depolymerization of PHB, resulting in a depression of the thermal stability [29,30]. Moreover, the difference observed between the thermal stability of PHB with up to 15 wt% PHO and that with 20 wt% PHO could indicate increased incompatibility and phase segregation at the highest concentration of PHO.

As for PHB/TAC films, the first degradation step was observed at around 220 °C and was attributed to the vaporization of most of the TAC plasticizer, while the second step, from 230 to 300°C, corresponds to PHB degradation [14]. Indeed, the weight loss at 220 °C (WL_{200}), around 5, 7, 17, and 17%, was close to the corresponding proportion of TAC in plasticized PHB (5, 10, 15, and 20%) (Table 1). The differences may come from some vaporization of TAC during the melt processing step. Moreover, much lower $T_{5\%}$ values compared to those of PHB/PHO films were noticed, the differences between the corresponding compositions ranging from 39 to 72 °C (Figure 2c vs. Figure 2a). Compared to PHB, the decrease in $T_{5\%}$ was higher, with 85 °C for the maximum TAC concentration (Table 1). The differences come from the low boiling point of TAC plasticizer [16,19,20], the decrease in the thermal stability of PHAs due to the vaporization of low molecular weight plasticizers being previously reported [9,10]. In addition, the T_d values, corresponding to the thermal degradation of PHB, remained close to each other and to that of neat PHB (Table 1, Figure 2d) due to the evaporation of most of the TAC at up to 220 °C.

3.3. DSC Analysis of Plasticized PHB

The influence of PHO and TAC on the melting and crystallization behaviors of PHB films was investigated by DSC. The values for the main thermal events, namely the glass transition temperature (T_g), melting (T_m) and crystallization (T_c) temperatures, their enthalpies, and the crystallinity degree (X_c), were reported in Table 2. Neat and plasticized PHB films displayed a well-defined melting peak with a shoulder at a lower temperature (Figure 3a,b). Double or multiple melting behaviors have been previously reported for PHB and were related either to the process of partial melting, recrystallization, and remelting or to the melting of crystals with different crystalline structures, perfection, or thickness [9,31]. Thus, the shoulder at low temperature (T_{m1}) could be associated with the fusion of the crystals with low perfection and thinner lamella and the main fusion peak at a higher temperature (T_{m2}) to the fusion of more perfect (recrystallized) crystals [32,33]. Another assumption related to the double melting peak is the presence of β crystals that melt at a lower temperature than the common α form of PHB [34].

PHB/PHO films showed similar thermal behavior with that of neat PHB (Table 2), suggesting a low influence of PHO on the mobility of PHB chains in the melting region due to the incompatibility of PHB and PHO in this temperature range. On the contrary, TAC addition induced a systematic and clear depression of T_{m1} values, with 5–13 °C, along with its increasing content in PHB. A less important shift, between 2 and 5 °C, was noticed for the main melting peak (Table 2). These observations indicate that the presence of TAC favors the segmental motion of the PHB chains, especially of those with lower molecular weight. It must be noted that the area of the main peak (T_{m2}) decreased notably in the case of PHB/PHO films and to a lesser extent for PHB/TAC ones (Table 2). This can be an indication of smaller population of more perfect and larger crystals in PHB films containing PHO as against PHB/TAC films.

Table 2. DSC results determined from the heating and cooling cycles for neat and plasticized PHB films.

Sample	Heating			Cooling		
	T_{m1}/T_{m2} (°C)	$\Delta H_1/\Delta H_2$ (J/g)	X_c (%)	T_g (°C)	T_c (°C)	ΔH_c (J/g)
PHB	163.3/173.6	17.8/59.3	53	7.1	88.9	59.0
PHB/5PHO	164.2/173.7	21.1/52.0	53	4.7	82.4	52.4
PHB/10PHO	163.03/173.3	16.7/52.4	53	6.6	84.4	50.6
PHB/20PHO	163.9/173.0	18.9/41.9	52	4.0	75.6	40.9
PHB/5TAC	158.3/172.1	14.5/58.4	53	4.5	83.6	51.6
PHB/10TAC	153.8/169.9	11.9/58.7	54	−1.5	76.5	49.5
PHB/20TAC	150.0/169.2	7.6/61.9	60	−1.5	77.0	51.2

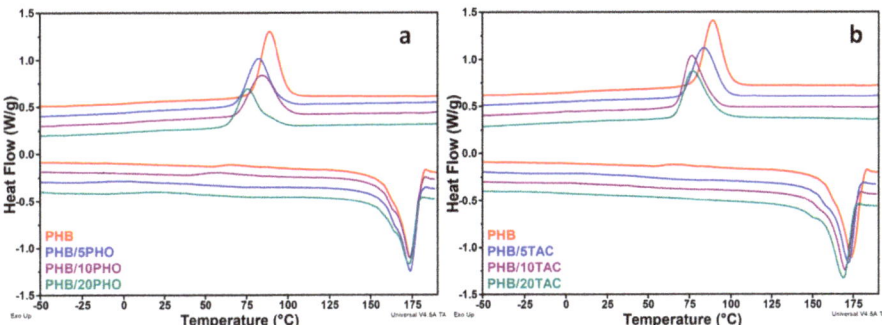

Figure 3. DSC first heating and cooling curves for neat PHB and PHB containing PHO (**a**) and TAC (**b**).

The T_g values were determined from the first cooling scan (Figure S1, Supplementary Materials). A decrease in the glass transition temperature of PHB was noticed in all the blends and it depended on both the concentration and the type of modifier. T_g was shifted from 7.1 °C for neat PHB to 4.0 °C for PHB/20PHO and to −1.5 °C for PHB/20TAC (Table 2). The smaller decrease in T_g in the case of PHB/PHO compared to PHB/TAC may be due to the lack of miscibility and the segregation of PHB and PHO domains following their melting during the previous heating cycle. A similar decreasing trend in T_g values was reported for PHB containing different plasticizers [12,35].

A very broad exothermic event, from around −25 to 75 °C, depending on the modifier type and concentration, was noticed for most of the films (Figure 3a,b) and may be related to a cold crystallization event. It was reported that the crystallization behavior of plasticized PHAs composites is a complex process due to several overlapping phenomena [20]. In this context, we assume that the cold crystallization process can overlap with the glass transition, which occurs in the same temperature range, remelting, or branching. However, TAC induced a less pronounced event in PHB films compared to PHO, probably due to the important increase in polymer chain mobility. This is obvious at maximum TAC content, when no such broad exothermic event occurred (Table 2, Figure 3b). No change in crystallinity was noticed in PHB blends compared to neat PHB except for a slight increase in crystallinity for the PHB/20TAC film (Table 2).

During cooling, a downshift of T_c was induced by the presence of both modifiers in PHB (Table 2, Figure 3a,b). For the highest amount of PHO and TAC, the depression of the T_c values was significant, of 13.3 and 11.9 °C, respectively (Table 2). This may be attributed to increased mobility up to a lower temperature. Further, with increasing PHO content in PHB films, T_c peaks became broader, pointing out the presence of a large range of crystallite sizes and possible increased compositional heterogeneity (Figure 3a,b). No major differences were noticed during the second heating cycle as compared to the first one (Figure S2, Supplementary Materials).

Thermal analyses showed different influences of TAC and PHO upon the crystallization of PHB. Thus, due to its low molecular weight and good miscibility with PHB, along with its high volatility,

TAC increased the chain mobility of PHB in the amorphous phase through a lubricity effect and leached out from the matrix above 200 °C. Contrarily, PHO was not miscible with PHB, especially in the domain of the melting temperatures when segregation occurred; however, it remained in the composition of PHB films, hindering PHB crystallization.

3.4. X-ray Diffraction (XRD)

The XRD spectra of neat PHB and PHB blends with the highest proportion of TAC and PHO are presented in Figure 4.

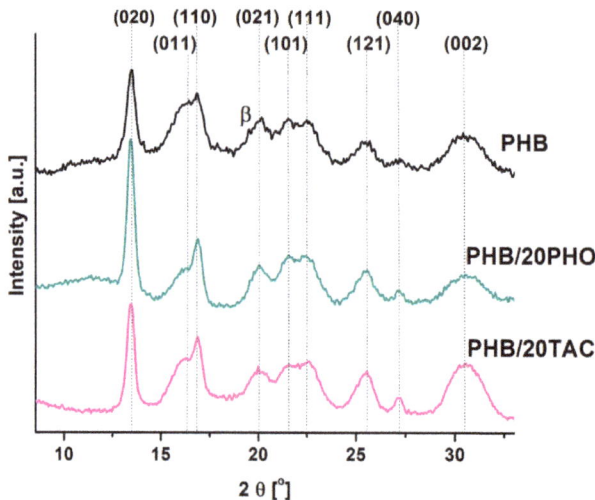

Figure 4. XRD spectra of neat PHB and PHB films containing 20 wt% PHO and TAC.

The diffractograms of neat PHB and PHB blends revealed a similar crystalline profile corresponding to orthorhombic crystal planes [36,37]. Two strong crystalline peaks were detected at 2θ 13.5° and 16.9°, which were assigned to the (020) and (110) planes of the orthorhombic unit cell and several weaker reflections, as shown in Figure 4. All the samples contain also a small amount of orthorhombic β-form crystal with zig-zag conformation, as revealed by the reflection of the (021) plane located at around 2θ = 20.1°. The occurrence of the β crystalline form may contribute to the presence of a shoulder before the melting peak, signaled by DSC analysis. Increased intensity of the peak corresponding to (020) crystal plane was observed in the blends (Figure 4), especially in PHB/20PHO, suggesting the presence of a preferred orientation and increased crystallinity. Indeed, higher CI values were calculated from the XRD data of PHB/PHO (76%) and PHB/TAC (71%) compared to neat PHB (57%). Therefore, PHO contributed to a larger extent to the formation of an ordered crystalline structure in PHB. The differences between the crystallinity values determined by XRD and DSC may come from the difference between the methods, XRD emphasizing the surface crystallinity and DSC that of the bulk material and from the influence of the temperature in DSC heating on the segregation of PHO and PHB domains.

The interplanar distances (d_{hkl}) and the apparent crystallite sizes perpendicular to the hkl (where h, k, and l are the Miller indices) plane (D_{hkl}) were estimated using Bragg's law and the Scherrer equation and the results are summarized in Table 3.

Table 3. XRD interplanar distances and crystallite sizes for neat PHB and PHB films with 20 wt% modifier.

Sample	d_{020} (nm)	d_{110} (nm)	d_{021} (nm)	d_{101} (nm)	d_{111} (nm)	d_{121} (nm)	d_{040} (nm)	d_{002} (nm)	D_{020} (nm)	D_{110} (nm)	D_{040} (nm)
PHB	0.6565	0.5259	0.4426	0.4140	0.3948	0.3497	0.3282	0.2928	14.8	7.5	8.8
PHB/20PHO	0.6580	0.5253	0.4428	0.4123	0.3927	0.3498	0.3279	0.2958	17.4	13.3	14.5
PHB/20TAC	0.6565	0.5240	0.4426	0.4118	0.3922	0.3597	0.3279	0.2923	15.9	13.3	20.4

Higher crystal sizes perpendicular to the (020), (110), and (040) planes were observed for the PHB films containing PHO and TAC, similar to other observations [38]. This increase in crystal size may be due to the greater order of the crystal structure after the addition of modifiers.

3.5. Mechanical Characterization

The evolution of the mechanical properties of neat and modified PHB films is shown in Figure 5. Both modifiers increased the flexibility of PHB films in correlation with their amount in the blends but the effect was much more significant in the case of PHO, which emerges as an efficient toughening agent. Thus, at maximum PHO and TAC content, the elongation at break was higher, with 700 and 185%, than that of neat PHB (Figure 5a). This means that PHO induced four times higher ductility. On the other hand, both modifiers led to a similar decrease in the elastic modulus down to close values, 944 and 938 MPa, for PHB films containing 20 wt% PHO and TAC (Figure 5b). Interestingly, at low PHO and TAC content (5 wt%), the strength of PHB blends was improved by around 22 and 27% as against neat PHB. However, when the content of modifiers exceeded 10%, the tensile strength of PHB/PHO films underwent a gradual reduction, while that of PHB/TAC remained in the same range (Figure 5c). Lower strength and elastic modulus values were reported for PHB modified by different plasticizers and toughening agents [12,39]. This may be due to the different preparation conditions and PHO composition, the PHO used in this study being almost a homopolymer, with 95 mol% 3-hydroxyoctanoate units. The eight times increase in the elongation at break is very important given that a small increase in PHB elongation was reported, even for high concentrations of low molecular weight plasticizers [9,10,17]. On the other hand, TAC behaves as a common plasticizer, acting as a spacer between PHB molecules and allowing limited flexibility. The excellent effect of PHO on the ductility of PHB may enable the design of new PHB-based materials with a wider range of mechanical properties, thereby increasing the potential of their application in soft tissue engineering, packaging, or other applications where the balance of ductility and stiffness is a must.

3.6. Surface Morphology

The morphology of neat PHB and PHB films containing 5 and 20wt% PHO and TAC, on their surface exposed to air, was investigated by AFM, and representative topographic images at different scan areas are shown in Figure 6a–c. Neat PHB film displayed a well-known spherulitic morphology consisting of an organized fibrous structure corresponding to the lamellar stacks (Figure 6a) [17,37,40]. All the analyzed samples showed a clearly edge-on lamellar growth. The addition of 5wt% PHO or TAC resulted in small changes; however, the crystalline structure was clearly noticed (Figure 6b,d). No significant changes were observed at a high concentration of PHO (PHB/20PHO) compared to neat PHB; however, compared to PHB/5PHO, the crystal structure was less well organized. Indeed, 20 wt% PHO may cause some restrictions in the PHB spherulites' growth compared to PHB/5PHO due to the limitations in chain mobility. On the contrary, a high concentration of TAC plasticizer led to a disordered structure and a different rearrangement on the surface of PHB (Figure 6e). A high degree of disorder is observed in the crystalline structure of PHB/20TAC (Figure 6d) and the blurred images suggest the migration of TAC to the surface of the film, forming a pellicle. These morphological changes can be better observed in the AFM images taken at a lower scan area of 5μm in Figure 6e. The PHB surface roughness was quantitatively characterized by the root mean square roughness (Rq) on the basis of four unprocessed topographic AFM images of 15 × 15 µm². An Rq value of 160 ± 1.2 nm was found for neat PHB film. After PHO addition, the Rq increased to 213 ± 9.8 nm for PHB/5PHO and to 168 ± 4.9 nm for PHB/20PHO. The increased Rq value found in PHB film with low PHO content may be attributed to the spherulites formed on the surface [41]. Since the rate of nucleation and crystal growth depends on the supersaturation of the employed solvent and obtaining conditions, either nucleation or growth may be dominant over the other, and crystals of different sizes and shapes may be obtained [42]. Contrarily, TAC alters the surface roughness of PHB; lower Rq values were obtained with increasing TAC content, 145 ± 8.5 nm and 124 ± 0.7 nm for PHB/5TAC and PHB/20TAC films. This may be due to the migration of small molecules of TAC to the upper surface of the PHB film, which was further investigated by XPS analysis.

3.7. Surface Properties by X-ray Photoelectron Spectroscopy

XPS analyses show the surface modification and composition of PHB after the addition of modifiers. XPS survey spectra of neat PHB and PHB blends are presented in Figure 7. The results show the presence of O and C as main elements and also some impurities consisting of Si and metal traces (less than 2%). Therefore, the lower thermal stability of PHB-PHO blends (Figure 1) may be due to these impurities, which catalyze the depolymerization reaction of PHB by β-elimination [30]. The corresponding elementary content and carbon to oxygen ratios are given in Table 4.

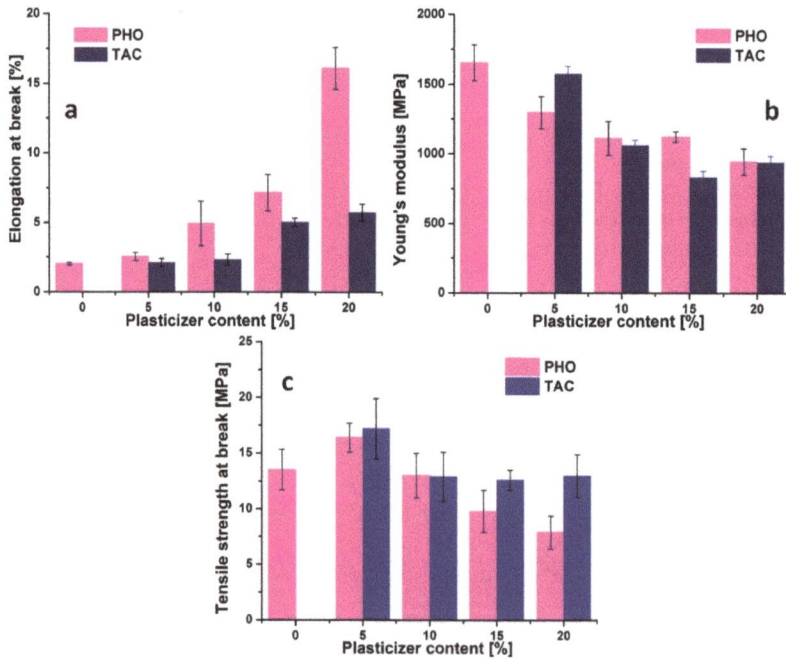

Figure 5. Mechanical properties determined from the tensile tests: elongation at break (**a**), Young's modulus (**b**), and tensile strength (**c**).

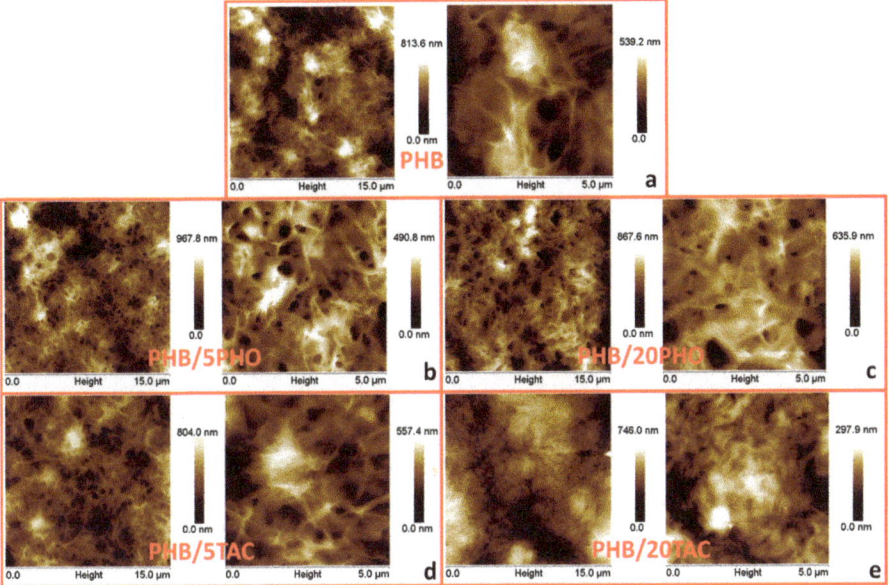

Figure 6. AFM topographic images (height) of neat (**a**) and modified (**b**–**e**) PHB films at 15 (left) and 5 μm (right) scan areas.

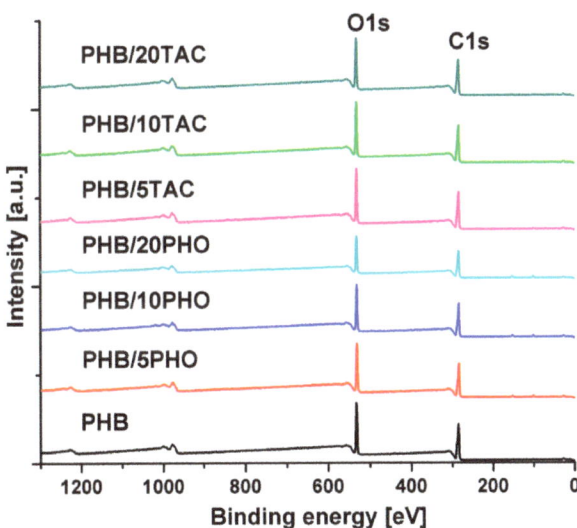

Figure 7. XPS survey spectra of neat and plasticized PHB films.

Table 4. Atomic concentrations of C and O elements and relative proportion of C species from C 1s fitting on the surface of neat PHB and plasticized PHB films acquired from XPS scans.

Sample	Atomic Percentage (%)					C1s Peak Fit Atomic Percentage (%)		
	C1s	O1s	C/O Experimental	C/O Theoretical	Difference	C1 [C-C]	C2 [C-O]	C3 [C=O]
PHB	69.99	30.01	2.33	2.00	+0.33	51.89	25.16	22.95
PHB/5PHO	70.19	29.81	2.35	2.10	+0.25	52.81	25.67	21.52
PHB/10PHO	70.90	29.10	2.44	2.20	+0.20	55.46	23.89	20.64
PHB/20PHO	70.82	29.18	2.43	2.40	+0.03	57.96	22.10	19.94
PHB/5TAC	70.49	29.51	2.39	2.03	+0.36	50.91	26.05	23.03
PHB/10TAC	69.89	30.11	2.32	2.05	+0.27	50.22	26.63	23.15
PHB/20TAC	70.39	29.61	2.38	2.10	+0.28	50.77	26.37	22.86

The higher C/O values in comparison to the theoretical ones, measured for all the analyzed PHB films, can be due to hydrocarbon impurities on the surface of the films, as observed in the case of other biopolymers [43]. However, the differences between experimental and theoretical C/O values are lower in the case of modified samples compared to neat PHB, except for PHB/5TAC, where a value close to that of PHB in the limit of experimental error was noticed. Indeed, PHB was used as a powder, which has a high ability to absorb impurities, and the addition of solid or liquid modifiers decreased this tendency. Considering that PHO has double the number of C atoms compared to PHB, the lower difference between experimental and theoretical C/O ratios confirms that PHO did not migrate out of the samples, in good agreement with the TGA and AFM results. Unfortunately, TAC has a C/O ratio close to that of PHB and no conclusion can be drawn from the survey spectra regarding migration. Therefore, high-resolution spectra were analyzed.

Figure 8a–c shows the XPS C1s spectra of neat PHB and PHB films containing the maximum PHO and TAC content.

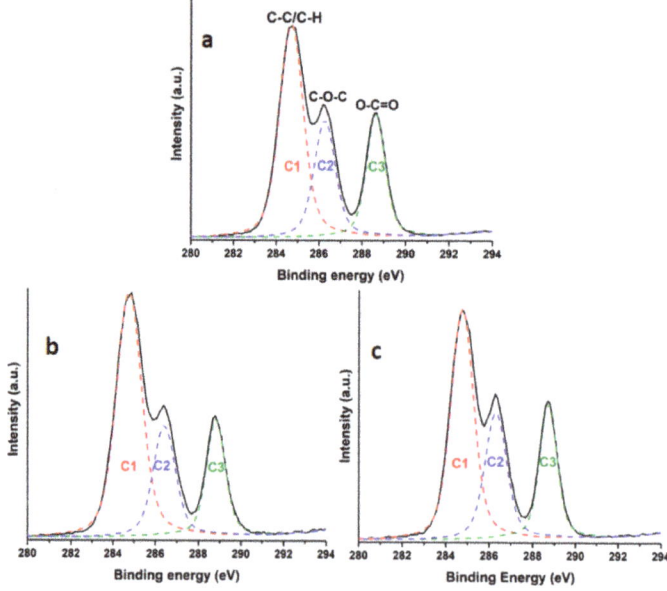

Figure 8. XPS C1s core-level spectrum for neat PHB (**a**), PHB/20PHO (**b**), and PHB/20TAC films (**c**).

Three peaks were detected in all PHB films: the peak at 284.7 eV (C1), attributed to carbon in hydrocarbons (C–C, C–H), the peak at 286.3 eV (C2), associated with the ether bond (C–O), and the peak at higher binding energy, 288.6 eV (C3), corresponding to the carbonyl (C=O) bond. The values of binding energies of these carbon species are in good agreement with the ones found in the literature [44,45]. Based on the proportion of these carbon species, the surface concentration of the different chemical bonds was calculated and the results are given in Table 4. As expected, an increase in C1 proportion was observed with the increase in PHO content in PHB/PHO due to the higher number of C-C bonds in the polymeric modifier and no significant variation in PHB/TAC samples (Table 4). However, the increase in C2 and C3 proportion in PHB/TAC samples shows a higher proportion of TAC on the surface of the films; therefore, the migration of TAC occurred even at room temperature. Contact angle measurements (Figure 9) also show increased hydrophilicity on the surfaces of these films, confirming the XPS results.

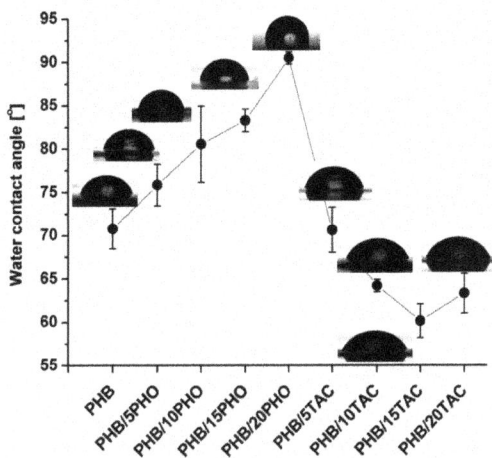

Figure 9. Water contact angle measurements for PHO and TAC modified PHB films.

In summary, the XPS findings support the results of AFM and TGA analysis related to the migration of low molecular weight TAC plasticizer.

3.8. In Vitro Biocompatibility

The attachment and proliferation of L929 cells on the neat and modified PHB films were studied over a period of 1 and 9 days. L929 cells scarcely adhered to the PHB films, regardless of the type and concentration of the toughening agent used, and all tested materials exhibited roughly the same behavior after one day. Images from day 9 (Figure 10) showed a greater number of adhered cells, characterized by fibroblast-like spindle shapes, usually packed in clusters, and tightly anchored on the surfaces. All the tested films allowed the adherence of L929 cells, although some differences were noticed between the modified PHB films. These were related to both the surface roughness determined by AFM (Figure 6) and the surface hydrophobicity characterized by contact angle (Figure 9).

Thus, moderate cell adhesion was observed in the case of plain PHB and PHB-20PHO (Figure 10). The moderate cell adhesion in the case of plain PHB is a well-known behavior [46]. Meanwhile, a high amount of PHO (PHB/20PHO) led to increased hydrophobicity and a contact angle of around 90°, due to the surface-oriented hydrocarbon-containing moieties [47]. This was further reflected in the weaker cell adhesion. Previous studies have shown that better cell adhesion was observed on surfaces with a medium value of the contact angle, of around 70° [46,48]. Therefore, very high or a very low hydrophobicity is not effective in cell adhesion. Indeed, the samples with the best cell adhesion were those with a contact angle between 65° and 75° (PHB/5PHO, PHB/10PHO, PHB/10TAC), therefore around 70°, as generally accepted [46,48]. However, the different behavior of PHB/5TAC supports the idea that there are other factors that influence the biocompatibility. Similarly, PHB/20TAC, with a contact angle close to that of PHB-10TAC, showed slightly poorer cell adhesion. It is worth mentioning that PHB/5TAC and PHB/20TAC showed very low roughness compared to PHB and PHB-PHO films, and it was reported that a lower degree of roughness is not favorable to cell attachment and proliferation [49,50].

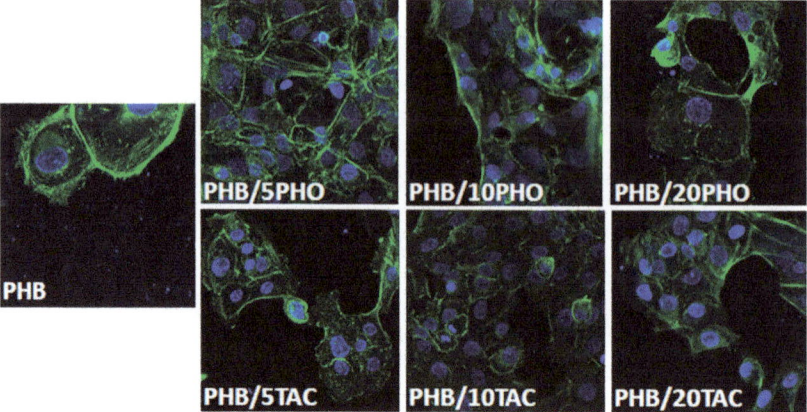

Figure 10. L929 cell adherence and morphology of cells on PHO and TAC modified PHB films.

The chemistry of the surface has also an influence on cell adhesion [49]. The migration of TAC to the surface, highlighted by AFM, contact angle, and TGA results, may contribute to differences in the response of the surface to cell adhesion and proliferation. In summary, the modified PHB films' propensity to support cell proliferation was demonstrated at day 9, when cells tend to cover greater areas, proving good biocompatibility in vitro.

3.9. Pro-Inflammatory Effect Evaluation

To obtain more insight into the biomedical suitability of the modified PHB films, the evaluation of inflammatory susceptibility was assessed by measuring the release of TNF-α (Figure 11).

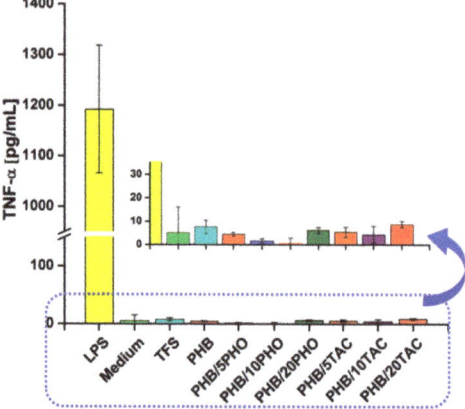

Figure 11. Effect of modified PHB film elution sampleson TNF-α level after an incubation period of 24 h.

As can be noticed from Figure 11, TNF-α concentration was much lower compared to the LPS sample and similar to the control or cell culture medium control, regardless of the composition of the PHB films. Moreover, the presence of TNF-α was undetectable in the case of PHB/10PHO. These preliminary results indicate that PHO and TAC modified PHB films are suitable for biomedical applications since very low levels of inflammatory TNF-α cytokine were detected.

4. Conclusions

PHB films were prepared by solution-casting using two different bio-based toughening agents, a low-molecular one (TAC) and a polymeric one (PHO). The two modifiers induced different effects on the thermal stability of PHB films and especially on $T_{5\%}$. A much higher decrease in $T_{5\%}$, ranging from 39 to 72 °C, was observed in the case of PHB/TAC compared to PHB/PHO films. This different thermal behavior was related to the relatively low boiling point of TAC, which is released from the film starting from 130 °C. Both modifiers led to a slight decrease in the glass transition temperature of PHB along with increased crystallinity and crystal size, favoring the chain motions and the organization of the crystalline phase of PHB. However, PHO led to much better flexibility and elongation of break in PHB/PHO films compared to TAC, being a more efficient toughening agent. In addition, no migration was noticed in the case of PHB/PHO films by AFM and XPS surface analyses. On the contrary, the blurred AFM topographic images of PHB/20TAC suggest the migration of the plasticizer, also confirmed by the increased C2 and C3 proportions in XPS analysis and the decrease in the contact angle value. All the plasticized PHB films supported L929 cell attachment and proliferation, demonstrating good biocompatibility in vitro. The differences in cell adhesion noticed between the plasticized PHB films were related to both surface roughness and surface wettability. In addition, significantly lower levels of TNF-α were detected for PHB films than those of the controls, regardless of the type of the modifier. It can be concluded that PHO is a better modifier for PHB intended for biomedical applications or food packaging due to the lack of migration in the surrounding medium, better thermal stability at the processing temperature of PHB, and better flexibility of the films.

Supplementary Materials: The following are available online at http://www.mdpi.com/2073-4360/12/11/2446/s1, Figure S1: Detailed images of DSC cooling curves of T_g steps for neat PHB and PHB containing PHO (a) and TAC (b); Figure S2: DSC second heating curves for neat PHB and PHB containing PHO (a) and TAC (b).

Author Contributions: A.N.F.: conceptualization, methodology, writing—original draft, review and editing, funding acquisition, project Administration; C.A.N.: methodology, investigation, validation, formal analysis; E.R.: methodology, investigation; M.C.E.: methodology, investigation, validation; M.G.: investigation, validation, I.C.: methodology, investigation; C.M.D.: methodology, investigation, validation; V.T.: methodology, investigation, validation; D.M.P.: conceptualization, writing—review and editing, supervision. All authors have read and agreed to the published version of the manuscript.

Funding: This work was supported by a grant from the Ministry of Research and Innovation, CNCS—UEFISCDI, PN-III-P1-1.1-TE-2016-2164, no. 94/2018, Biocompatible multilayer polymer membranes with tuned mechanical and antiadherent properties (BIOMULTIPOL).

Conflicts of Interest: The authors declare no conflict of interest.

References

1. Nielsen, T.D.; Hasselbalch, J.; Holmberg, K.; Stripple, J. Politics and the plastic crisis: A review throughout the plastic life cycle. *WIREs Energy Environ.* **2020**, *9*, 1–18.
2. Torres-Giner, S.; Hilliou, L.; Melendez-Rodriguez, B.; Figueroa-Lopez, K.J.; Madalena, D.; Cabedo, L.; Vicente, A.A.; Lagaron, J.M. Melt processability, characterization, and antibacterial activity of compression-molded green composite sheets made of poly(3-hydroxybutyrate-co-3-hydroxyvalerate) reinforced with coconut fibers impregnated with oregano essential oil. *Food Packag. Shelf Life* **2018**, *17*, 39–49.
3. Bonartsev, A.P.; Bonartseva, G.A.; Reshetov, I.V.; Kirpichniko, M.P.; Shaitan, K.V. Application of Polyhydroxyalkanoates in Medicine and the Biological Activity of Natural Poly(3-Hydroxybutyrate). *Acta Nat.* **2019**, *11*, 4–16.

4. Mangeon, C.; Michely, L.; Rios de Anda, A.; Thevenieau, F.; Renard, E.; Langlois, V. Natural Terpenes Used as Plasticizers for Poly(3-hydroxybutyrate). *ACS Sustain. Chem. Eng.* **2018**, *6*, 16160–16168.
5. Panaitescu, D.M.; Nicolae, C.A.; Gabor, A.R.; Trusca, R. Thermal and mechanical properties of poly(3-hydroxybutyrate) reinforced with cellulose fibers from wood waste. *Ind. Crops Prod.* **2020**, *145*. [CrossRef]
6. Panaitescu, D.M.; Frone, A.N.; Chiulan, I. Nanostructured biocomposites from aliphatic polyesters and bacterial cellulose. *Ind. Crops Prod.* **2016**, *93*, 251–266.
7. Audic, J.; Lemiègre, L.; Corre, Y. Thermal and mechanical properties of a polyhydroxyalkanoate plasticized with biobased epoxidized broccoli oil. *J. Appl. Polym. Sci.* **2014**, *131*. [CrossRef]
8. Erceg, M.; Kovacic, T.; Klaric, I. Thermal degradation of poly(3-hydroxybutyrate) plasticized with acetyl tributyl citrate. *Polym. Degrad. Stab.* **2005**, *90*, 313–318.
9. Seoane, I.T.; Manfredi, L.B.; Cyras, V.P. Effect of two different plasticizers on the properties of poly(3-hydroxybutyrate) binary and ternary blends. *J. Appl. Polym. Sci.* **2017**, *135*, 46016.
10. Wang, L.; Zhu, W.; Wang, X.; Chen, X.; Chen, G.Q.; Xu, K. Processability modifications of poly(3-hydroxybutyrate) by plasticizing, blending, and stabilizing. *J. Appl. Polym. Sci.* **2008**, *107*, 166–173.
11. Requena, R.; Jiménez, A.; Vargas, M.; Chiralt, A. Effect of plasticizers on thermal and physical properties of compression-moulded poly[(3-hydroxybutyrate)-co-(3-hydroxyvalerate)] films. *Polym. Test.* **2016**, *56*, 45–53. [CrossRef]
12. Garcia-Garcia, D.; Fenollar, O.; Fombuena, V.; Lopez-Martinez, J.; Balart, R. Improvement of mechanical ductile properties of poly(3-hydroxybutyrate) by using vegetable oil derivatives. *Macromol. Mater. Eng.* **2017**, *302*. [CrossRef]
13. Aliotta, L.; Vannozzi, A.; Panariello, L.; Gigante, V.; Coltelli, M.-B.; Lazzeri, A. Sustainable Micro and Nano Additives for Controlling the Migration of a Biobased Plasticizer from PLA-Based Flexible Films. *Polymers* **2020**, *12*, 1366. [CrossRef]
14. Arrieta, M.P.; López, J.; López, D.; Kenny, J.M.; Peponi, L. Development of flexible materials based on plasticized electrospun PLA–PHB blends: Structural, thermal, mechanical and disintegration properties. *Eur. Polym. J.* **2015**, *73*, 433–446. [CrossRef]
15. Kurusu, R.S.; Siliki, C.A.; David, É.; Demarquette, N.R.; Gauthier, C.; Chenal, J.M. Incorporation of plasticizers in sugarcane-based poly(3-hydroxybutyrate)(PHB): Changes in microstructure and properties through ageing and annealing. *Ind. Crops Prod.* **2015**, *72*, 166–174. [CrossRef]
16. Maiza, M.; Benaniba, M.T.; Quintard, G.; Massardier-Nageotte, V. Biobased additive plasticizing Polylactic acid (PLA). *Polímeros* **2015**, *256*, 581–590. [CrossRef]
17. Panaitescu, D.M.; Nicolae, C.A.; Frone, A.N.; Chiulan, I.; Stanescu, P.O.; Draghici, C.; Iorga, M.; Mihailescu, M. Plasticized poly(3-hydroxybutyrate) with improved melt processing and balanced properties. *J. Appl. Polym. Sci.* **2017**, *134*. [CrossRef]
18. Arrieta, M.P.; López, J.; López, D.; Kenny, J.M.; Peponi, L. Biodegradable electrospunbionanocomposite fibers based on plasticized PLA–PHB blends reinforced with cellulose nanocrystals. *Ind. Crops Prod.* **2016**, *93*, 290–301. [CrossRef]
19. Kang, H.; Li, Y.; Gong, M.; Guo, Y.; Guo, Z.; Fang, Q.; Li, X. An environmentally sustainable plasticizer toughened polylactide. *RSC Adv.* **2018**, *8*, 11643–11651. [CrossRef]
20. Corrêa, M.C.S.; Branciforti, M.C.; Pollet, E.; Agnelli, J.A.M.; Nascente, P.A.P.; Avérous, L. Elaboration and Characterization of Nano-Biocomposites Based on Plasticized Poly(Hydroxybutyrate-Co-Hydroxyvalerate) with Organo-Modified Montmorillonite. *J. Polym. Environ.* **2012**, *20*, 283–290. [CrossRef]
21. Dias, A.M.A.; Marceneiro, S.; Braga, M.E.M.; Coelho, J.F.J.; Ferreira, A.G.M.; Simões, P.N.; Veiga, H.I.M.; Tomé, L.C.I.; Marrucho, M.; Esperança, J.M.S.S.; et al. Phosphonium-based ionic liquids as modifiers for biomedical grade poly(vinyl chloride). *Acta Biomater.* **2012**, *8*, 1366–1379. [CrossRef] [PubMed]
22. Nerkar, M.; Ramsay, J.A.; Ramsay, B.A.; Kontopoulou, M. Melt Compounded Blends of Short and Medium Chain-Length Poly-3-hydroxyalkanoates. *J. Polym. Environ.* **2014**, *22*, 236–243. [CrossRef]
23. Panaitescu, D.M.; Lupescu, I.; Frone, A.N.; Chiulan, I.; Nicolae, C.A.; Tofan, V.; Stefaniu, A.; Somoghi, R.; Trusca, R. Medium Chain-Length Polyhydroxyalkanoate Copolymer Modified by Bacterial Cellulose for Medical Devices. *Biomacromolecules* **2017**, *18*, 3222–3232. [CrossRef] [PubMed]

24. Dufresne, A.; Vincendon, M. Poly(3-hydroxybutyrate) and Poly(3-hydroxyoctanoate) Blends: Morphology and Mechanical Behavior. *Macromolecules* **2000**, *33*, 2998–3008. [CrossRef]
25. Chiulan, I.; Panaitescu, D.M.; Frone, A.N.; Teodorescu, M.; Nicolae, C.A.; Casarica, A.; Tofan, V.; Salageanu, A. Biocompatible polyhydroxyalkanoates/bacterial cellulose composites: Preparation, characterization, and in vitro evaluation. *J. Biomed. Mater. Res. A* **2016**, *104*, 2576–2584. [CrossRef]
26. Bragg, W.H.; Bragg, W.L. The reflexion of X-rays by crystals. *Proc. R. Soc. London Ser. A* **1913**, *88*, 428–438.
27. Patterson, A.L. The Scherrer formula for X-ray particle size determination. *Phys. Rev.* **1939**, *56*, 978–982. [CrossRef]
28. Weinmann, S.; Bonten, C. Thermal and rheological properties of modified polyhydroxybutyrate (PHB). *Polym. Eng. Sci.* **2019**, *59*, 1057–1064. [CrossRef]
29. Kim, K.J.; Doi, Y.; Abe, H. Effects of residual metal compounds and chain-end structure on thermal degradation of poly(3-hydroxybutyric acid). *Polym. Degrad. Stab.* **2006**, *91*, 769–777. [CrossRef]
30. Kim, K.J.; Doi, Y.; Abe, H. Effect of metal compounds on thermal degradation behavior of aliphatic poly(hydroxyalkanoic acid)s. *Polym. Degrad. Stab.* **2008**, *93*, 776–785. [CrossRef]
31. Anbukarasu, P.; Sauvageau, D.; Elias, A. Tuning the properties of polyhydroxybutyrate films using acetic acid via solvent casting. *Sci. Rep.* **2016**, *5*, 17884. [CrossRef]
32. Malmir, S.; Montero, B.; Rico, M.; Barral, L.; Bouza, R. Morphology, thermal and barrier properties of biodegradable films of poly (3-hydroxybutyrate-co-3-hydroxyvalerate) containing cellulose nanocrystals. *Comp. Part A Appl. Sci. Manuf.* **2017**, *93*, 41–48. [CrossRef]
33. Siracusa, V.; Ingrao, C.; Karpova, S.G.; Olkhov, A.A.; Iordanskii, A.L. Gas transport and characterization of poly(3 hydroxybutyrate) films. *Eur. Polym. J.* **2017**, *91*, 149–161. [CrossRef]
34. Prakalathan, K.; Mohanty, S.; Nayak, S.K. Reinforcing effect and isothermal crystallization kinetics of poly(3-hydroxybutyrate) nanocomposites blended with organically modified montmorillonite. *Polym. Compos.* **2014**, *35*, 999–1012. [CrossRef]
35. Baltieri, R.C.; Innocentini Mei, L.H.; Bartoli, J. Study of the influence of plasticizers on the thermal and mechanical properties of poly(3-hydroxybutyrate) compounds. *Macromol. Symp.* **2003**, *197*, 33–44. [CrossRef]
36. Iulianelli, G.C.V.; David, G.D.S.; dos Santos, T.N.; Sebastião, P.J.O.; Tavares, M.I.B. Influence of TiO_2 nanoparticle on the thermal, morphological and molecular characteristics of PHB matrix. *Polym. Test.* **2018**, *65*, 156–162. [CrossRef]
37. Chen, J.; Wu, D.; Tam, K.C.; Pan, K.; Zheng, Z. Effect of surface modification of cellulose nanocrystal on nonisothermal crystallization of poly(β-hydroxybutyrate) composites. *Carbohydr. Polym.* **2017**, *157*, 1821–1829. [CrossRef]
38. Hong, S.G.; Gau, T.K.; Huang, S.C. Enhancement of thecrystallization and thermal stability of polyhydroxybutyratebypolymericadditives. *J. Therm. Anal. Calorim.* **2011**, *103*, 967–975. [CrossRef]
39. Garcia-Garcia, D.; Ferri, J.M.; Montanes, N.; Lopez-Martinez, J.; Balart, R. Plasticization effects of epoxidized vegetable oils on mechanical properties of poly(3-hydroxybutyrate). *Polym. Int.* **2016**, *65*, 1157–1164. [CrossRef]
40. Lin, K.W.; Lan, C.H.; Sun, Y.M. Poly[(R)3-hydroxybutyrate] (PHB)/poly(l-lactic acid) (PLLA) blends with poly(PHB/PLLA urethane) as a compatibilizer. *Polym. Degrad. Stab.* **2016**, *134*, 30–40. [CrossRef]
41. Lee, C.W.; Song, B.K.; Jegal, J.; Kimura, Y. Cell adhesion and surface chemistry of biodegradable aliphatic polyesters: Discovery of particularly low cell adhesion behavior on poly(3-[RS]-hydroxybutyrate). *Macromol. Res.* **2013**, *21*, 1305–1313. [CrossRef]
42. Sofińska, K.; Barbasz, J.; Witko, T.; Dryzek, J.; Haraźna, K.; Witko, M.; Kryściak-Czerwenka, J.; Guzik, M. Structural, topographical, and mechanical characteristics of purified polyhydroxyoctanoate polymer. *J. Appl. Polym. Sci.* **2019**, *136*, 47192. [CrossRef]
43. Panaitescu, D.M.; Frone, A.N.; Chiulan, I.; Casarica, A.; Nicolae, C.A.; Ghiurea, M.; Trusca, R.; Damian, C.M. Structural and morphological characterization of bacterial cellulose nano-reinforcements prepared by mechanical route. *Mater. Des.* **2016**, *110*, 790–801. [CrossRef]
44. Wang, C.; Sauvageau, D.; Elias, A. Immobilization of Active Bacteriophages on Polyhydroxyalkanoate Surfaces. *ACS Appl. Mater. Interfaces* **2016**, *8*, 1128–1138. [CrossRef]
45. da Silva, M.G.; Vargas, H.; Poley, L.H.; Rodriguez, R.S.; Baptista, G.B. Structural impact of hydroxyvalerate in polyhydroxyalkanoates (PHAscl) dense film monitored by XPS and photothermal methods. *J. Braz. Chem. Soc.* **2005**, *164*, 790–795. [CrossRef]

46. Lee, C.W.; Horiike, M.; Masutani, K.; Kimura, Y. Characteristic cell adhesion behaviors on various derivatives of poly(3-hydroxybutyrate) (PHB) and a block copolymer of poly(3-[RS]-hydroxybutyrate) and poly(oxyethylene). *Polym. Degrad. Stab.* **2015**, *111*, 194–202. [CrossRef]
47. Qu, X.H.; Wu, Q.; Liang, J.; Zou, B.; Chen, G.Q. Effect of 3-hydroxyhexanoate content in poly(3-hydroxybutyrate-co-3-hydroxyhexanoate) on in vitro growth and differentiation of smooth muscle cells. *Biomaterials* **2006**, *27*, 2944–2950. [CrossRef]
48. Tamada, Y.; Ikada, Y. Fibroblast growth on polymer surfaces and biosynthesis of collagen. *J. Biomed. Mater. Res.* **1994**, *28*, 783–789. [CrossRef]
49. Boyan, B.D.; Hummert, T.W.; Dean, D.D.; Schwartz, Z. Role of material surfaces in regulating bone and cartilage cell response. *Biomaterials* **1996**, *17*, 137–146. [CrossRef]
50. Surmenev, R.A.; Chernozem, R.V.; Syromotina, D.S.; Oehr, C.; Baumbach, T.; Krause, B.; Boyandin, A.N.; Dvoinina, L.M.; Volova, T.G.; Surmeneva, M.A. Low-temperature argon and ammonia plasma treatment of poly-3-hydroxybutyrate films: Surface topography and chemistry changes affect fibroblast cells in vitro. *Eur. Polym. J.* **2019**, *112*, 137–145. [CrossRef]

Publisher's Note: MDPI stays neutral with regard to jurisdictional claims in published maps and institutional affiliations.

© 2020 by the authors. Licensee MDPI, Basel, Switzerland. This article is an open access article distributed under the terms and conditions of the Creative Commons Attribution (CC BY) license (http://creativecommons.org/licenses/by/4.0/).

Article

Study on Cellulose Acetate Butyrate/Plasticizer Systems by Molecular Dynamics Simulation and Experimental Characterization

Weizhe Wang, Lijie Li, Shaohua Jin, Yalun Wang, Guanchao Lan and Yu Chen *

School of Material Science and Engineering, Beijing Institute of Technology, Beijing 100081, China; waxwellwang@163.com (W.W.); lilijie2003@bit.edu.cn (L.L.); jinshaohua@bit.edu.cn (S.J.); 3120181161@bit.edu.cn (Y.W.); lan890805@163.com (G.L.)
* Correspondence: bityuchen@bit.edu.cn; Tel.: +86-10-68912370

Received: 10 April 2020; Accepted: 21 May 2020; Published: 2 June 2020

Abstract: Cellulose acetate butyrate (CAB) is a widely used binder in polymer bonded explosives (PBXs). However, the mechanical properties of PBXs bonded with CAB are usually very poor, which makes the charge edges prone to crack. In the current study, seven plasticizers, including bis (2,2-dinitro propyl) formal/acetal (BDNPF/A or A3, which is 1:1 mixture of the two components), azide-terminated glycidyl azide (GAPA), n-butyl-N-(2-nitroxy-ethyl) nitramine (Bu-NENA), ethylene glycol bis(azidoacetate) (EGBAA), diethylene glycol bis(azidoacetate) (DEGBAA), trimethylol nitromethane tris (azidoacetate) (TMNTA) and pentaerythritol tetrakis (azidoacetate) [PETKAA], were studied for the plasticization of CAB. Molecular dynamics simulation was conducted to distinguish the compatibilities between CAB and plasticizers and to predict the mechanical properties of CAB/plasticizer systems. Considering the solubility parameters, binding energies and intermolecular radical distribution functions of these CAB/plasticizer systems comprehensively, we found A3, Bu-NENA, DEGBAA and GAPA are compatible with CAB. The elastic moduli of CAB/plasticizer systems follow the order of CAB/Bu-NENA>CAB/A3>CAB/DEGBAA>CAB/GAPA, and their processing property is in the order of CAB/Bu-NENA>CAB/GAPA>CAB/A3>CAB/DEGBAA. Afterwards, all the systems were characterized by FT-IR, differential scanning calorimetry (DSC), differential thermogravimetric analysis (DTA) and tensile tests. The results suggest A3, GAPA and Bu-NENA are compatible with CAB. The tensile strengths and Young's moduli of these systems are in the order of CAB/A3>CAB/Bu-NENA>CAB/GAPA, while the strain at break of CAB/Bu-NENA is best, which are consistent with simulation results. Based on these results, it can be concluded that A3, Bu-NENA and GAPA are the most suitable plasticizers for CAB binder in improving mechanical and processing properties. Our work has provided a crucial guidance for the formulation design of PBXs with CAB binder.

Keywords: cellulose acetate butyrate; plasticizer; molecular dynamics simulation; thermal analysis; compatibility

1. Introduction

Cellulose acetate butyrate (CAB), with the molecular formula of $[C_6H_7O_2\text{-}(OCOCH_3)_X\text{-}(OCOC_3H_7)_Y\text{-}(OH)_{3\text{-}X\text{-}Y}]_n$, is the most commonly used binder in polymer bonded explosives (PBXs). In the structure of cellulose ester, hydroxyl groups are co-esterified with acetic acid and butyric acid [1]. Therefore, CAB contains about 12–15% (wt %) of acetyl groups and 26–39% (wt %) of butyryl groups, which endows it with excellent physical and chemical properties including outstanding moisture resistance, UV resistance, temperature resistance, flexibility, transparency, etc. Therefore, many researchers have utilized this excellent cellulose derivative in the field of energetic materials. Li prepared the CL-20 based pressed PBXs with CAB as binder for its excellent compatibility and mechanical properties [2]. Lan ameliorated the HNIW based PBXs with excellent temperature adaptability using CAB and fluorine rubber F2311 as binders [3].

However, the CAB-bonded explosives exhibit some defects and encounter problems in their practical applications. The softening temperature (T_S) and glass transition temperature (T_g) of CAB are relatively high. Therefore, its molecular chain cannot stretch sufficiently and its adhesion is weak in solvents. The PBXs with CAB as binder usually exhibit poor mechanical properties at low temperatures, unsatisfactory insensitivity, and deterioration of cracks [4].

To ameliorate these situations, plasticizers have been added to the CAB binder system to lower its T_S and T_g, increase its plasticity, improve the mechanical properties and insensitivity, and prevent the deterioration of cracks on PBXs [5]. However, the effects of different plasticizers on properties of PBXs with CAB binder remain unclear. Therefore, it is essential to select optimal plasticizers for CAB binder system of PBXs. Experimental tests are time-consuming and inconvenient for repeated comparisons, accompanied by certain risks. In addition, it cannot be used for the in-depth analysis from the microscopic perspective.

Molecular dynamics (MD) simulation has been widely used to predict the compatibility and mechanical properties of blends [6–9]. Compared with the macroscopic experiments, MD simulations in microscopic scales not only can reveal the effects of plasticizer at the atomic and molecular levels, but also is low-cost and time efficient. Therefore, it can be used to study the effects of different plasticizers on CAB binder quickly and efficiently [10]. In the present work, the compatibilities between CAB and different plasticizers were evaluated by MD simulation. The mechanical properties of the plasticized CAB with selected plasticizers were then simulated. Three representative CAB/plasticizer systems were prepared, and their compatibilities, chemical stability and mechanical properties were characterized experimentally, aiming to establish a study method combining MD simulation and experimental characterization for optimizing CAB/plasticizer systems. Our work has provided a guidance and reference for the subsequent modification of CAB binder and the formulation design of PBXs using CAB binder.

2. MD Simulation

Based on their molecular structures (Figure 1), the molecular models of CAB and plasticizers including bis (2,2-dinitro propyl) formal/acetal (BDNPF/A or A3, which is 1:1 mixture of the two components), azide-terminated glycidyl azide (GAPA), n-butyl-N-(2-nitroxy-ethyl) nitramine (Bu-NENA), ethylene glycol bis(azidoacetate) (EGBAA), diethylene glycol bis(azidoacetate) (DEGBAA), trimethylol nitromethane tris (azidoacetate) (TMNTA) and pentaerythritol tetrakis(azidoacetate) (PETKAA) were constructed using the Visualizer module in Material Studio (MS) version 6.0. The butyryl group and acetyl group contents in CAB are approximately set to 37 wt % and 13 wt %, respectively. Anneal and geometric optimizations were then conducted to establish the final molecular models.

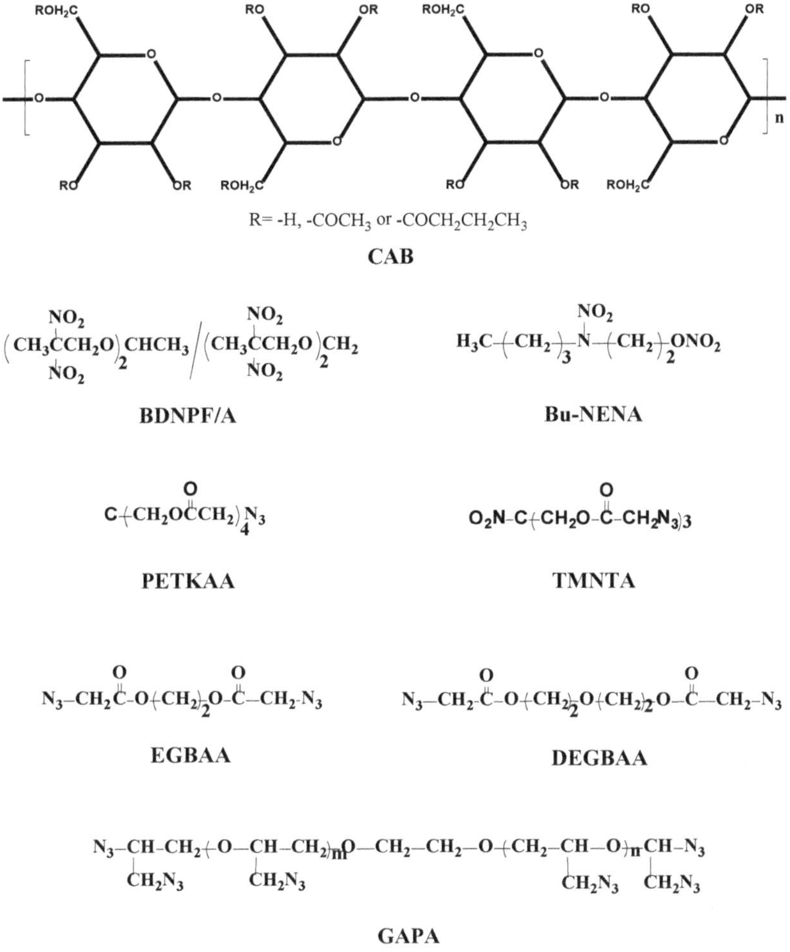
R= -H, -COCH3 or -COCH2CH2CH3
CAB
BDNPF/A
Bu-NENA
PETKAA
TMNTA
EGBAA
DEGBAA
GAPA

(a)

Figure 1. *Cont.*

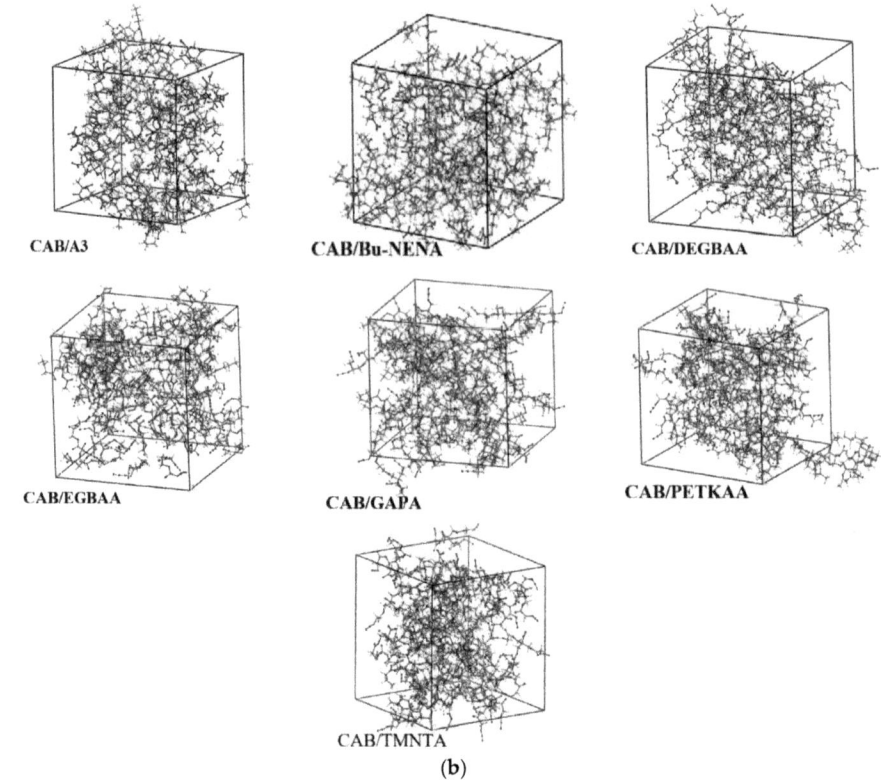

Figure 1. Molecular structures (**a**) and amorphous models (**b**) of CAB/plasticizer systems. CAB = cellulose acetate butyrate.

2.1. Construction of Models

To evaluate the compatibility between CAB and plasticizers and predict the mechanical properties of the plasticized CAB, MD simulation was conducted in the Material Studio with use of COMPASS force field [11].

The next simulation section is the annealing of various molecular models, which aims to relax their configurations, lower the potential energy and eliminate the internal stress of polymer chains. The initial temperature and mid-cycle temperature were set to 298K and 500K, respectively. The temperature range covers different glass transition temperatures (T_g) of CAB and plasticizers (CAB: 403.2K, BDNPF/A: 208.6K, Bu-NENA: 189.7K, EGBAA: 203.7K, DEGBAA: 209.9K, GAPA: 194.4K, PETKAA: 237.8K, TMNTA: 239.1K) [12–18]. The heating ramps per cycle was 25, which was a proper step number. Too rapid cooling often traps the system in a high energy–low density state that cannot represent the glassy state. In the contrast, slow annealing from rubbery state with high T_g can generate more accurate results. After 500 cycles of annealing, the annealed molecular models were obtained for further simulation.

The amorphous models of CAB/plasticizer systems were then established using the Amorphous Cell module as shown in Figure 1. For the construction of these amorphous models, the quality was set to fine, the maximum number of steps was set to 50,000, and the density of the mixed system was calculated based on the mass ratio of CAB/plasticizer. Ewald was set as the electrostatic force, van der Waals was set as the atom based, and cutoff radius was set to 12.50 Å for the simulation. The mass ratio of CAB/plasticizer was set to 2:3, which is a commonly used ratio in PBXs, and the amounts of CAB molecules (3 chains, 734 atoms) in all amorphous models were the same. The plasticizer contents were approximately in the range of 30–60 molecules (1200–1800 atoms) in different systems, depending on the plasticizer.

2.2. Calculations

The geometric structures of CAB/plasticizer amorphous models were optimized by "Geometry Optimization" option in the Forcite module. The MD simulation of 250 ps was performed in NVT ensemble at the temperature of 298K to achieve the optimal amorphous models [19]. COMPASS force field was applied with Andersen thermostat for the simulation. Figure 2 shows the dynamics temperature-time curve and energy-time curves obtained by MD simulation. The simulation is considered to reach the equilibrium state if the fluctuations of dynamic temperature and dynamic energy are less than 5% [20]. The dynamic energy may seem almost constant during the simulation, but it changes slightly with time, which is so-called "aging".

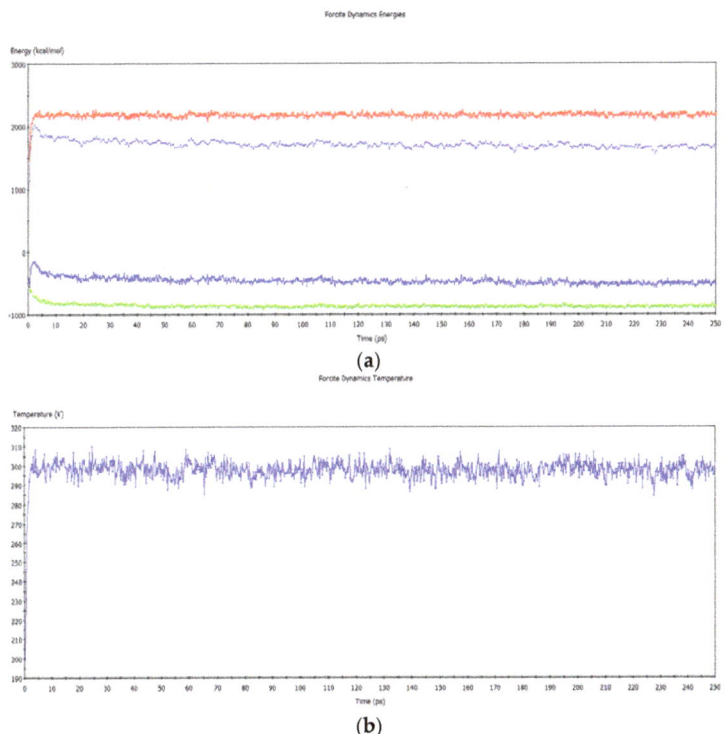

Figure 2. Dynamic energy-time (**a**) and temperature-time (**b**) curves of the molecular dynamics (MD) simulation. The red, light blue, dark blue and green curves in (**a**) represent kinetic energy, total energy, potential energy and non-bonded energy, respectively.

After dynamics simulation reaching the equilibrium state, the calculated dynamic trajectory files were analyzed by the "Cohesive Energy Density" option of the Forcite module to obtain the cohesive energy density and solubility parameters of CAB and plasticizers.

The binding energies of CAB, plasticizers and their mixed systems were produced by the analysis with "Total Kinetic Energy" option in Forcite module. Five frames of each model with lowest energy were selected and their binding energy parameters were obtained with "Energy" option in Forcite, which were then averaged.

Radial distribution functions (RDF) were obtained by the analysis with "Radial distribution function" option of Forcite module. The dynamic trajectory files of the CAB molecules and plasticizers were classified to different sets and analyzed with "Radial distribution function" option to give RDFs.

The mechanical properties of the systems were simulated in a more complex manner. The amorphous models of CAB/plasticizers were firstly subjected to the 250 ps dynamics simulation with NPT ensemble under 1GPa. The last frame model of previous dynamic trajectory files was extracted to subject to another 250 ps dynamics simulation with the NPT ensemble under 1atm. The last frame model was then extracted to undergo the 250 ps dynamics simulation with NVT ensemble and the final dynamic trajectory files were calculated by "Mechanical Properties" option in Forcite module. The simulation tests were performed by the "Constant strain" method. The number of steps for each strain was set to 4 and the maximum strain was 3×10^{-3}. The strain rate was used as the automatic systematical setting.

3. Experiments and Characterization

3.1. Materials

Three representative plasticizers, A3, GAPA and Bu-NENA, were selected for experimental characterization. Analytical grade CAB containing 37 wt % butyryl group and 13 wt % acetyl group was purchased from Eastman Chemical Company (USA). A3 and Bu-NENA (purity > 99.9%) were provided by Liming Chemical Research Institute (Henan, China). GAPA (purity > 99.12%, $\rho = 1.21$ g/cm^3, water content < 0.016%, hydroxyl value ≈ 1.5 mg KOH/g, molecular weight Mn ≈ 800 g/mol) was synthesized in-house.

3.2. Preparation of CAB/Plasticizer Mixed Systems

CAB and plasticizer were mixed at the mass ratio of 2:3. For a typical procedure, 2.00 g CAB was dissolved in 20 mL ethyl acetate in a 50 mL flask by magnetic stirring, and 3.00 g plasticizer was then added into the flask and mixed well with CAB solution. Ethyl acetate was then removed by evaporation under reduced pressure with a rotary evaporator to afford the CAB/plasticizer mixture.

3.3. Characterizations

The structure of CAB/plasticizer was characterized using a Fourier transform infrared spectroscopy (FT-IR) analyzer (NEXUS-470, Nicolet, WI, USA) by the KBr pellet method.

The compatibilities between CAB and plasticizers were respectively evaluated by differential scanning calorimetry (DSC) using a DSC131 Evo instrument (Setaram, France) and differential thermogravimetric analysis (DTA) using a DTA-60 thermal analyzer (Shimadzu, Japan) under 50 mL/min nitrogen flow atmosphere at the heating rate of 10 K/min.

The tensile strengths of the mixed systems were measured using a universal material testing machine (6022, Instron, Norwood, MA, USA) at the loading rate of 100 mm/min until broken. Ten test specimens were prepared and characterized, whose size were 20 mm long and 5 mm wide.

4. Results and Discussion

4.1. MD Simulation

4.1.1. Solubility Parameter

According to the polymer solution theory, the mixture of polymer and plasticizer can be considered as polymer solution system [21]. The necessary thermodynamic condition for a spontaneous dissolution at constant temperature under constant pressure is:

$$\Delta G_M = \Delta H_M - T\Delta S_M < 0 \tag{1}$$

where ΔG is the free energy of mixing, ΔS is the entropy of mixing, ΔH is the heat of mixing, and T is the temperature of dissolution. As a polymer is mixed with a plasticizer, the polymer itself is in a chaotic state. Therefore, the magnitude of ΔG mainly depends on ΔH [22]. Hildebrand introduced the concept of solubility parameter (δ) [23], which was defined as the square root of the cohesive energy density (CED) (Equation (2)).

$$\delta = (\Delta E/V)^{1/2} = [(\Delta H - RT)/V]^{1/2} \qquad (2)$$

The heat of mixing ΔH_M of the polymer dissolution can be expressed using the Hildebrand formula of small molecules:

$$\Delta H_M = V_M \Phi_1 \Phi_2 \left[(\Delta E_1/V_1)^{1/2} - (\Delta E_2/V_2)^{1/2} \right]^2 \qquad (3)$$

where Φ_1 and Φ_2 are the volume fractions of the two components, respectively, and V is the molar volume of the mixture. If the square root of CED is replaced with δ, Equation (3) can be re-written as:

$$\Delta H_M = V_M \Phi_1 \Phi_2 (\delta_1 - \delta_2)^2 \qquad (4)$$

Therefore, the value of ΔH_M is determined by δ_1 and δ_2. The closer δ_1 and δ_2 are, the smaller ΔH_M is and the better the compatibility of the two components will be. For an energetic material, its polymer and plasticizer are generally considered compatible if $|\Delta \delta| < 3.68 \sim 4.06$ ($J^{1/2} \cdot cm^{-3/2}$) [24].

Table 1 lists the solubility parameters (δ_{MD}) of CAB and each plasticizer, and the respective differences between the solubility parameters of CAB and those of different plasticizers ($|\Delta \delta_{MD}|$) obtained by the MD simulation. The difference between the simulation δ of CAB (17.29) and reported δ (18.87) is very small [25,26], indicating that although MD simulation produces statistical uncertainties from the ideal simulation conditions and imperfection algorithm, the simulation results are mostly consistent with reality.

Table 1. Solubility parameters (δ_{MD}) of CAB and each plasticizer and the respective differences ($|\Delta \delta_{MD}|$) between the solubility parameters of CAB and different plasticizers.

Component	$\delta_{MD}/(J^{1/2} \cdot cm^{-3/2})$	$|\Delta \delta_{MD}|/(J^{1/2} \cdot cm^{-3/2})$
CAB	17.29	0
A3	20.01	2.72
GAPA	20.56	3.27
EGBAA	23.59	6.20
DEGBAA	18.72	1.43
TMNTA	24.55	7.26
Bu-NENA	20.84	3.55
PETKAA	23.30	6.01

The $|\Delta \delta_{MD}|$ values between CAB and A3, GAPA, DEGBAA and Bu-NENA plasticizer are less than 3.55, suggesting that these plasticizers are compatible with CAB. However, the values of $|\Delta \delta_{MD}|$ between CAB and EGBAA, TMNTA and PETKAA are greater than 6.00, indicating that these plasticizers are unsuitable for the plasticization of CAB.

4.1.2. Binding Energy

The good compatibility between a polymer and a different compound implies strong intermolecular interaction(s) between them. The strength of such interaction can be quantitatively described with binding energy (E_{bind}'). The average interaction energy between CAB and a plasticizer (E_{inter}) can be defined as:

$$E_{inter} = E_{total} - \left(E_{CAB} + E_{plasticizer} \right) \qquad (5)$$

where E_{total} is the total energy of the mixed system, and E_{CAB} and $E_{plasticizer}$ are the average energies of CAB and plasticizer, respectively. The binding energy E_{bind} is the negative value of the average interaction energy E_{inter}, e.g.:

$$E_{bind} = -E_{inter} \qquad (6)$$

Because the average molecular weights of the mixed systems are different, the binding energy is converted into per unit mass E_{bind}' for comparison purpose as shown in Equation (7):

$$E_{bind}' = E_{bind}/M_n \qquad (7)$$

where M_n is the average molecular weight calculated based on the mass ratio of CAB/plasticizer.

Table 2 lists the E_{bind} and E_{bind}' of each CAB/plasticizer system. No chemical bonds contribute to the binding energy. The binding energies are numerically equal to the corresponding non-bonding energies and follow the order of CAB/Bu-NENA≈CAB/EGBAA> CAB/DEGBAA> CAB/TMNTA> CAB/PETKAA> CAB/GAPA> CAB/A3. It is known that the system with greater E_{bind}' has better compatibility [27]. Therefore, CAB/A3 is the least compatible system and all other CAB/plasticizer systems are more compatible. The discrepancy with Solubility Parameter simulation results is mainly generated from the ideal simulation conditions and imperfection algorithms. Although the solubility parameter is the most reliable criteria for compatibility, binding energy can also provide supplementary information.

Table 2. The binding energies (E_{bind}) and per unit mass binding energies (E_{bind}') between CAB and different plasticizers.

System	$E_{valence}$/kcal·mol^{-1}	E_{vdw}/kcal·mol^{-1}	E_{elect}/kcal·mol^{-1}	E_{bind}/kcal·mol^{-1}	E_{bind}'/kcal·g^{-1}
CAB/A3	0	−448.44	−149.96	598.40	0.71
CAB/GAPA	0	−565.11	−116.36	681.47	0.91
CAB/PETKAA	0	−596.22	−157.32	753.55	1.08
CAB/EGBAA	0	−578.03	−144.21	722.25	2.03
CAB/Bu-NENA	0	−550.03	−115.85	665.88	2.03
CAB/TMNTA	0	−600.74	−153.22	753.96	1.25
CAB/DEGBAA	0	−562.56	−138.11	700.67	1.65

4.1.3. Radial Distribution Function (RDF)

In statistical mechanics, the RDF of a system of different particles (atoms, ions, molecules, etc.) is applicable to the structural investigations of both solid and liquid packing (local structure) for studying specific interactions, such as hydrogen bonding. It measures the probability of finding a target particle at the distance of r around the given reference particle [28]. In other words, it describes how the density varies with distance from a reference particle [29]. In a blend of various molecules, two different components tend to be compatible with each other if the RDF curve of mixed system is higher than each of their own [30]. Figure 3 shows the radial distribution functions (RDFs) of different CAB/plasticizer systems and those of their individual components.

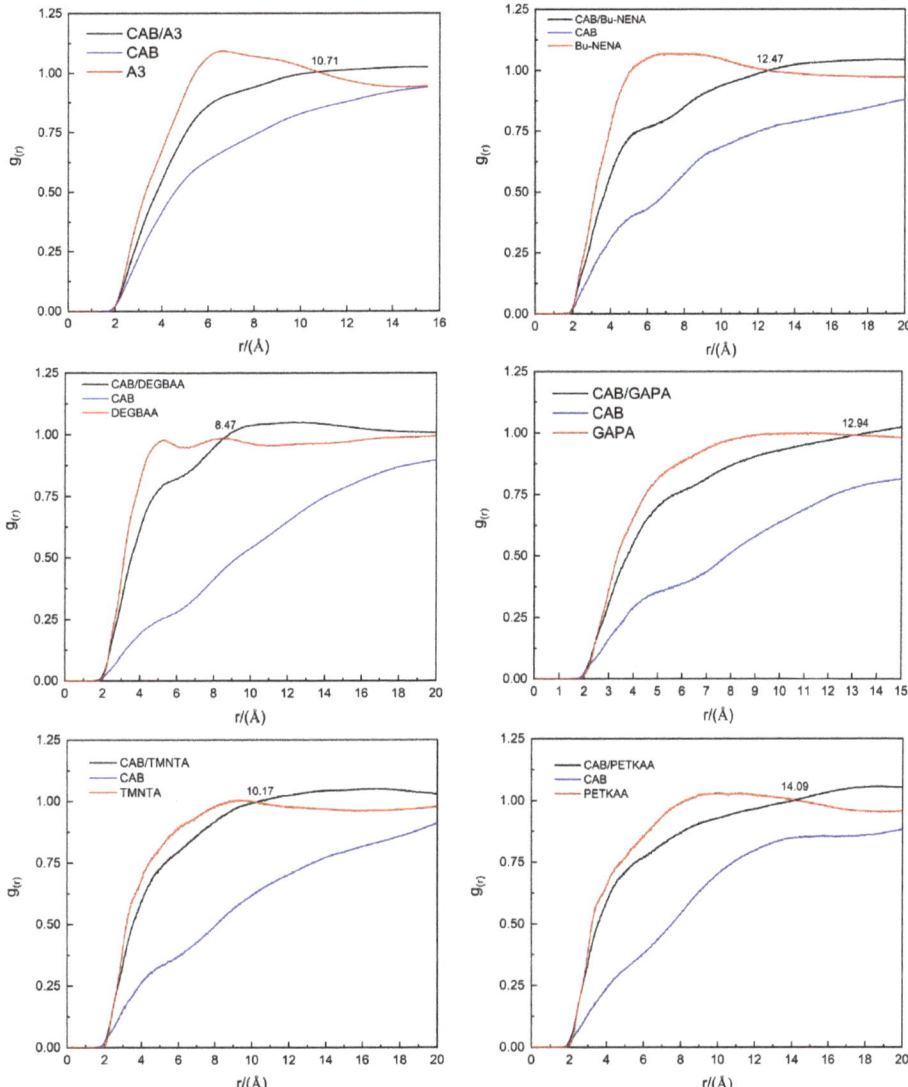

Figure 3. Radial distribution functions (RDF) curves of different CAB/plasticizer systems.

The RDFs of CAB and all plasticizers reach the peak values at the distances slightly shorter than 5.00–6.00 Å. It is well known that the distance ranges of hydrogen bond and van der Waals force are 0–3.10 Å and 3.10–5.00 Å, respectively [31–34]. Therefore, van der Waals force is the major intermolecular forces of CAB-CAB and plasticizer-plasticizer pairs. The RDF of CAB/plasticizer is lower than that of plasticizer, but higher than CAB itself in the range of 0–6.00 Å for all systems, suggesting that the compatibility between CAB and plasticizers seems not so good.

The RDF curves of CAB/Bu-NENA, CAB/GAPA and CAB/PETKAA are higher than themselves at the distances longer than 12.47 Å, 12.94 Å and 14.09 Å, which are far beyond the range of intermolecular forces, indicating these plasticizers are not compatible with CAB. The nitro and azide groups in their structure may be the main reason of the weak physical interactions between these groups and cellulose chains of CAB.

The RDF curves of CAB/A3, CAB/DEGBAA, and CAB/TMNTA are higher than those of the corresponding plasticizers at distances longer than 10.71 Å, 8.41 Å and 10.17 Å, indicating that these plasticizers are compatible with CAB. These results suggest that compatibilities of CAB/A3, CAB/DEGBAA, and CAB/TMNTA are better than those of CAB/GAPA, CAB/Bu-NENA and CAB/PETKAA.

Considering three compatibility criteria including solubility parameter, binding energy and radial distribution function (RDF) comprehensively, it can be concluded that GAPA, Bu-NENA, A3 and DEGBAA are compatible with CAB, and PETKAA, TMNTA and EGBAA are incompatible with CAB.

4.1.4. Simulation of Mechanical Properties

The mechanical properties of an energetic material greatly affect the safety and storage performance of its explosive products, which thus are of significant importance [35]. Bulk modulus (K), shear modulus (G), Poisson's ratio (μ), Young's modulus (E), etc. are usually used to describe the mechanical properties of energetic materials.

The volume of a material decreases under uniform pressure P. Therefore, bulk modulus is defined as:

$$K = \frac{P}{-\Delta V/V} = -\frac{PV}{\Delta V} \tag{8}$$

where V is the volume of material and ΔV is the volume change under pressure P. Bulk modulus is a measure of compressibility and breaking strength of a material. The larger the bulk modulus, the higher the breaking strength [36].

Shear modulus (G) is the ratio of shear stress (σ) to shear strain (γ):

$$G = \frac{\sigma}{\gamma} = \frac{P/A}{tg\theta} \tag{9}$$

It is a measure of the stiffness of a material. The greater the value of shear modulus, the higher the material hardness and the smaller the deformation.

The ratio of bulk modulus to shear modulus (K/G) reflects the extent of plastic change (elongation in tension) of the material. The greater the value of K/G is, the better the ductility of the material is [37]. Young's modulus (E) is defined as the ratio of tensile stress (σ) to tensile strain (ε_1):

$$E = \frac{\sigma}{\varepsilon_1} \tag{10}$$

It can be used to evaluate the capability of a material to resist deformation and volume change caused by external stresses.

Poisson's ratio (μ) is defined as the ratio of transverse shrinkage deformation (ε_2) to longitudinal stretch deformation (ε_1):

$$\mu = -\frac{\varepsilon_2}{\varepsilon_1} \tag{11}$$

It is an elastic constant reflecting the transverse deformation of a material. In general, the μ less than 0.5 under tensile stress results in volume increases. The materials with the Poisson's ratios in the range between 0.2 and 0.4 are generally considered to have good plasticity [38–41].

The moduli of a typical isotropic material satisfy the following relationship [42]:

$$E = 2G(1+\mu) = 3K(1-2\mu) \tag{12}$$

Table 3 lists the bulk moduli (K), shear moduli (G) and other calculated mechanical parameters of the CAB/plasticizer systems obtained by MD simulation. CAB/Bu-NENA exhibits the largest bulk modulus (K), indicates that its breaking strength is the highest. The shear moduli (G) of the

CAB/plasticizer systems are similar, within the range of 0.8–0.9 GPa. The Poisson's ratio (μ) and the K/G value of CAB/Bu-NENA are also the greatest, indicating its ductility and formation property are the best. The Young's modulus (E) of the CAB/Bu-NENA system is greatest, suggesting its resistance to deformation is the highest. The discrepancy between simulation results and actual mechanical parameters is mainly generated from the ideal simulation conditions and imperfection algorithms.

Table 3. The mechanical parameters of CAB/plasticizers.

System	K/GPa	G/GPa	K/G	μ	E/GPa
CAB/A3	2.32	0.86	2.70	0.34	2.29
CAB/GAPA	2.50	0.78	3.20	0.36	2.13
CAB/DEGBAA	1.54	0.91	1.70	0.25	2.27
CAB/Bu-NENA	5.23	0.89	5.87	0.42	2.53
CAB/TMNTA	2.53	1.02	2.49	0.32	2.69
CAB/PETKAA	2.26	1.34	1.68	0.25	3.36
CAB/EGBAA	2.24	0.83	2.69	0.33	2.22

Based on the simulated mechanical properties obtained above, it can be concluded that the mechanical properties of CAB/Bu-NENA are the best, followed by those of CAB/A3 and CAB/GAPA. The mechanical properties of CAB/DEGBAA are poorest.

4.2. Experimental Characterization

Based on the simulation results, CAB/A3, CAB/GAPA and CAB/Bu-NENA mixed systems were prepared, and their compatibilities and mechanical properties were characterized by FT-IR, DSC, DTA and tensile strength tests for comparison purpose. The feasibility to optimize the plasticity of binder and predict the mechanical properties of the plasticized binder by combining theoretical simulation and experimental characterization was further demonstrated.

4.2.1. Chemical Stability

The chemical stability of an energetic material can be determined by FT-IR. If the characteristic peaks of mixture are the same as those of raw materials, it can be considered no chemical change occurs during the mixing [43]. Figure 4 shows the FT-IR spectra of CAB, plasticizers and their mixed products.

The CAB/GAPA system is analyzed as an example. CAB exhibits a broad and strong absorption peak at 3445.86 cm^{-1} due to the stretching vibration of -OH. The doublet peak at 2964.33 cm^{-1} and 2877.88 cm^{-1} are attributed to the antisymmetric and symmetric stretching vibrations of methylene group. The strong absorption peak at 1742.35 cm^{-1} can be assigned to the stretching vibration of -C=O. The peaks at 1165.11 cm^{-1} and 1065.15 cm^{-1} are ascribed to the stretching vibration of the unique -COCOC-polyether structure in CAB.

GAPA shows a doublet peak at 2929.89 cm^{-1} and 2877.42 cm^{-1} that can be assigned to the antisymmetric and symmetric stretching vibration of methylene. The strong peak at 2100.68 cm^{-1} and the peak at 1281.22 cm^{-1} are the characteristic absorption peaks of GAPA caused by the stretching vibration and bending vibration of -N$_3$, respectively. The peak at 1128.16 cm^{-1} can be assigned to the antisymmetric stretching vibration of ether bond.

The stretching vibration of -OH remains at 3462.72 cm^{-1} in the CAB/GAPA mixture. The peaks of the mixture at 2934.93 cm^{-1} and 2879.95 cm^{-1} are due to the superposition of the antisymmetric and symmetric stretching vibration absorption peaks of the methylene groups in CAB and GAPA. The stretching vibration peak of -N$_3$ of GAPA shifts to 2097.35 cm^{-1}. The stretching vibration peak of -C=O of CAB is found at 1730.42 cm^{-1}. The peak at 1275.13 cm^{-1} is attributed to the bending vibration of the -N$_3$ of GAPA. The stretching vibration of -COCOC- polyether structure of CAB results in the absorption peak at 1202.30 cm^{-1}. Based on these results, it can be concluded that no chemical reaction occurs during the mixing, and thus the chemical properties of the mixture systems are relatively stable.

Similar results are obtained for the CAB/Bu-NENA and CAB/A3 systems. Therefore, mixing CAB with plasticizers does not change the chemical properties of the individual components, and the CAB/plasticizer systems are chemically stable.

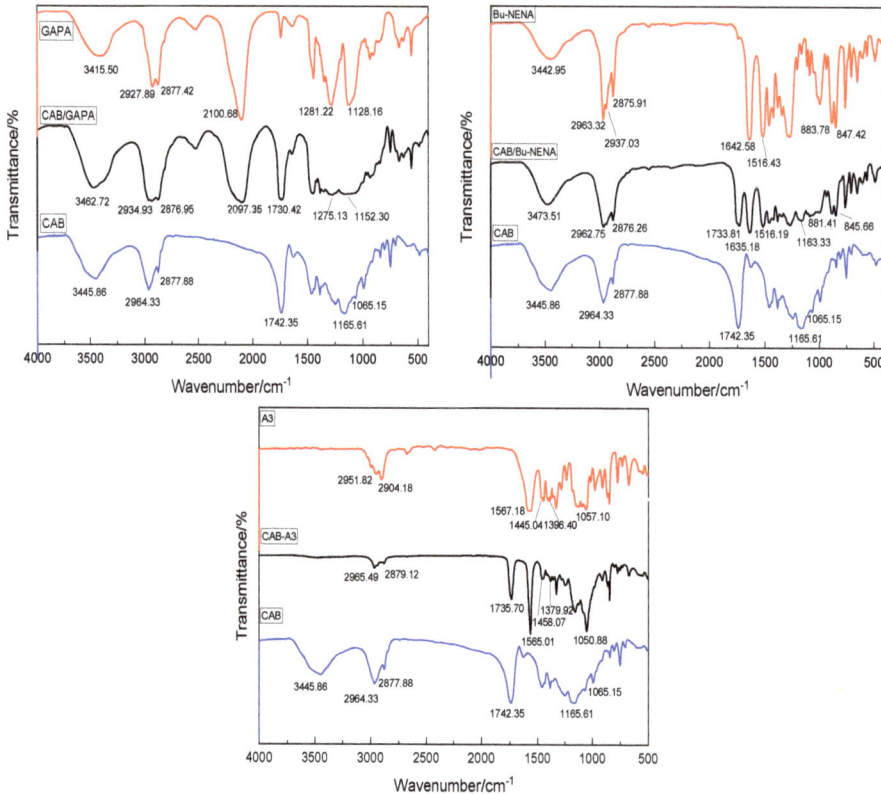

Figure 4. FT-IR spectra of CAB, plasticizer and CAB/plasticizer systems.

4.2.2. Compatibility

The compatibilities between an explosive and its contacting materials can be evaluated by DCS and DTA. According to the National Military Standard of China GJB 772A-97 502.1, the compatibility can be classified into four levels: level A with $\Delta T_p \leq 2.0/°C$ and $\Delta E/E_a \leq 20\%$, the system is compatible or highly compatible; level B with $\Delta T_p \leq 2.0/°C$ and $\Delta E/E_a > 20\%$, the system is slightly sensitized or fairly compatible; level C with $\Delta T_p > 2.0/°C$ and $\Delta E/E_a \leq 20\%$, the system is sensitized or poorly compatible; level D with $\Delta T_p > 2.0/°C$ and $\Delta E/E_a > 20\%$, the system is hazardous. In the standard, ΔT_p is the change of the decomposition exothermal temperature, and $\Delta E/E_a$ is the changing rate of the apparent activation energy.

Figure 5 shows the DSC and DTA curves of CAB, the plasticizers and the CAB/plasticizer systems at the heating rate of 10 K/min. Table 4 lists the compatibilities of CAB with three energetic plasticizers obtained by DSC and DTA.

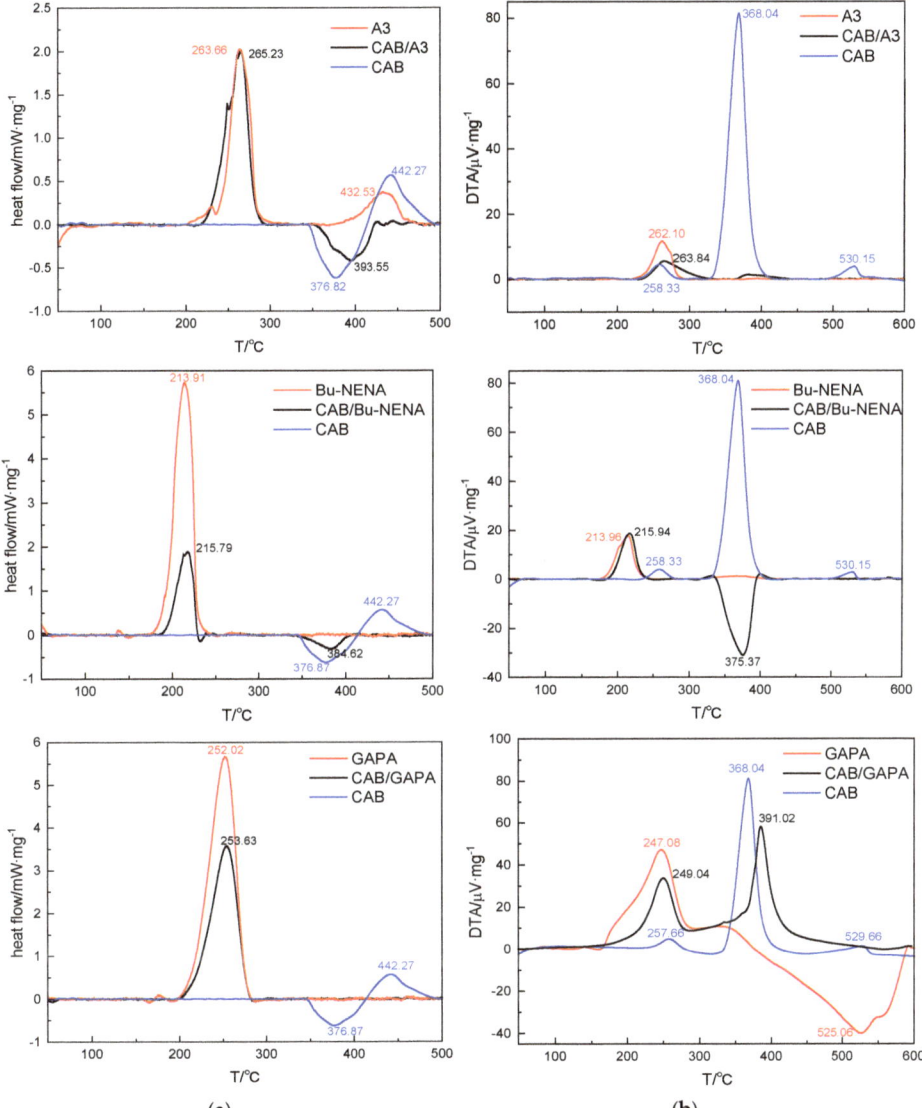

Figure 5. Differential scanning calorimetry (DSC) (**a**) and differential thermogravimetric analysis (DTA) (**b**) curves of CAB, three plasticizers and the CAB/plasticizer systems at the heating rate of 10 K/min.

Table 4. ΔT_p and $\Delta E/E_a$ of different CAB/plasticizer systems and the corresponding compatibility obtained from DSC/DTA.

System	ΔT_p/°C		$(\Delta E/E_a)$/%		Compatibility	
	DSC	DTA	DSC	DTA	DSC	DTA
CAB/A3	1.57	1.74	17.25	19.48	Very good	Very good
CAB/Bu-NENA	1.88	1.98	16.62	11.82	Very good	Very good
CAB/GAPA	1.61	1.96	18.05	18.80	Very good	Very good

The first exothermic peak of A3 appears at 263.66 °C, which is caused by the main decomposition. The secondary decomposition of the partial decomposition products appeared at 432.53 °C. The main exothermic peak of CAB/A3 is found at 265.23 °C. The peak temperature shift ΔT_p is within 2 °C and the $\Delta E/E_a$ is calculated to be 17.25%, indicating the compatibility of CAB/A3 system reaches level A. The exothermic peak temperature of Bu-NENA is 213.91 °C, and the main exothermic peak of CAB/Bu-NENA is at 215.79 °C. The peak temperature shift ΔT_p is within 2 °C and the calculated $\Delta E/E_a$ is 16.62%, suggesting the compatibility of CAB/Bu-NENA system is also level A. Similarly, the compatibility of CAB/GAPA system is found to be level A with the exothermic peak temperatures of GAPA and CAB/Bu-NENA respectively at 252.02 and 253.63 °C, and the $\Delta E/E_a$ of 18.05%.

Similar results are also obtained by DTA (Table 4). The peak temperature shifts of all three CAB/plasticizer systems are within 2 °C and the corresponding $\Delta E/E_a$ values are less than 20.00%. These results suggest that A3, Bu-NENA and GAPA are compatible with CAB at level A, and thus can be safely used in explosive design. In addition, the experimental compatibility results obtained by DSC/DTA are consistent with MD simulation results, indicating MD simulation is applicable to the characterization of the plasticized binders for the design of energetic materials.

4.2.3. Mechanical Properties

Figure 6 shows the tensile strengths (σ), Young's moduli (E) and strains at break (ε) of the three CAB/plasticizer systems measured experimentally. The stress-strain curves of CAB/plasticizer systems were measured firstly and the values of σ, E and ε were calculated from the first linear part of the stress-strain curves (elastic domain), respectively.

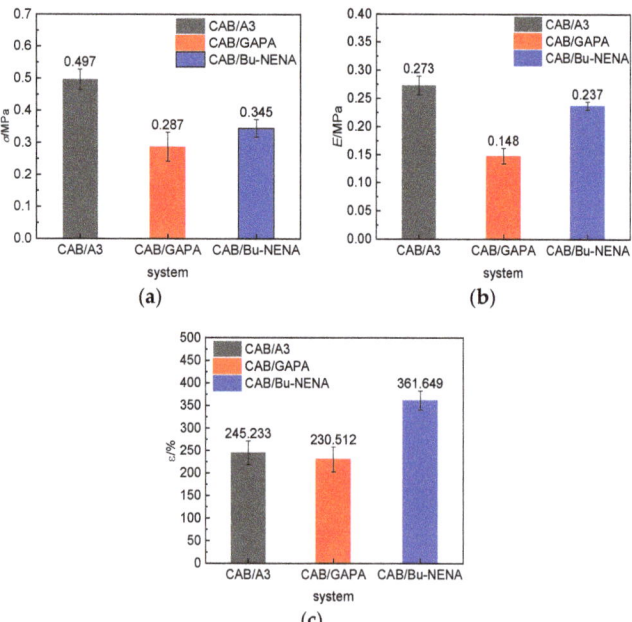

Figure 6. Mechanical properties of CAB/plasticizer systems measured experimentally: (**a**) Tensile strengths (σ), (**b**) Young's moduli (E), and (**c**) strains at break ε.

The tensile strengths (σ) and Young's moduli (E) of CAB/plasticizer systems are in the order of CAB/A3>CAB/Bu-NENA>CAB/GAPA, indicating that the mechanical strength and resistance capacity to deformation of CAB/A3 are best, significantly better than CAB/Bu-NENA and CAB/GAPA. However, the strain at break (ε) of CAB/Bu-NENA are higher than CAB/A3 and CAB/GAPA, which means better deformability, but, worst stiffness. Tensile strength test results are consistent with MD simulation results, which further confirm the applicability of MD simulation to the design of energetic materials.

5. Conclusions

In the present work, the compatibilities between CAB and seven plasticizers were firstly evaluated numerously by MD simulation and compatible plasticizers are approximately selected. Afterwards, the mechanical properties of CAB with selected plasticizers were calculated by MD simulation. The mixed systems with suitable plasticizers were further characterized experimentally. Simulation results suggest that GAPA, Bu-NENA, A3 and DEGBAA are compatible with CAB well and other plasticizers including PETKAA, TMNTA and EGBAA are incompatible with CAB. The shear moduli (G) of optimal CAB/plasticizer systems are similar, but their bulk moduli (K), Poisson's ratios (μ), K/G ratios and Young's moduli (E) are all in the order of CAB/Bu-NENA >CAB/A3 >CAB/GAPA, suggesting the mechanical properties of CAB/Bu-NENA are the best.

FT-IR characterization suggests that no chemical reaction occurs during the mixing procedure of CAB and plasticizers, and these CAB/plasticizer systems are chemically stable. DSC/DTA analysis further demonstrates that A3, GAPA and Bu-NENA plasticizers are compatible with CAB at level A, and thus these energetic plasticizers are safe for the explosive designs with CAB as binder. It is found by tensile strength measurements that tensile strengths (σ) and Young's moduli (E) of CAB/plasticizer systems are in the order of CAB/A3>CAB/Bu-NENA> CAB/GAPA, indicating that the mechanical strength and resistance capacity to deformation of CAB/A3 are best, significantly better than CAB/Bu-NENA and CAB/GAPA. However, the strain at break (ε) of CAB/Bu-NENA are higher than CAB/A3 and CAB/GAPA, which means better deformability, but, worst stiffness.

All experimental results are consistent with MD simulation results well, indicating that MD simulation is a suitable study method for energetic material designs. Based on both simulation and experimental results, it can be concluded that A3, Bu-NENA and GAPA are the suitable plasticizers of CAB binder to improve the poor mechanical properties and processing properties of PBXs. Our work has provided a crucial guidance and reference for the formulation of PBXs using CAB as the binder.

Author Contributions: The work presented here was carried out in collaboration with all authors. W.W., L.L. and Y.C. designed the works and wrote the paper; W.W., S.J. and Y.W. performed the molecular dynamics simulation and data analysis; W.W. and G.L. compared the simulation results with the experimental results. All authors have read and agreed to the published version of the manuscript.

Funding: This work was supported by "the Fundamental Research Funds for the Central Universities".

Acknowledgments: This work was supported by "the Fundamental Research Funds for the Central Universities".

Conflicts of Interest: The authors declare no conflict of interest. The sponsors had no role in the design, execution, interpretation, or writing of the study.

References

1. Podshivalov, A.; Besson, F.; Budtova, T.; Bronnikov, S. Morphology and improved impact strength of cellulose acetate butyrate blended with polyethylene copolymer. *eXPRESS Polym. Lett.* **2018**, *12*, 856–866. [CrossRef]
2. Li, X.; Chen, S.; Wang, X.; Shang, F.; Dong, W.; Yu, Z.; Yu, Y.; Zou, H.; Jin, S.; Chen, Y. Effect of polymer binders on safety and detonation properties of ε-CL-20-based pressed-polymer-bonded explosives. *Mater. Express* **2017**, *7*, 209–215. [CrossRef]
3. Lan, G.; Jin, S.; Jing, B.; Chen, Y.; Wang, D.; Li, J.; Wang, N.; Chen, M. Investigation into the Temperature Adaptability of HNIW-based PBXs. *Propellants Explos. Pyrotech.* **2019**, *44*, 327–336. [CrossRef]
4. Dandekar, A.; Roberts, Z.A.; Paulson, S.; Chen, W.; Son, S.F.; Koslowski, M. The effect of the particle surface and binder properties on the response of polymer bonded explosives at low impact velocities. *Comput. Mater. Sci.* **2019**, *166*, 170–178. [CrossRef]
5. Badgujar, D.; Talawar, M.; Zarko, V.; Mahulikar, P. New directions in the area of modern energetic polymers: An overview. *Cpmbust. Explo. Shock+* **2017**, *53*, 371–387. [CrossRef]

6. Singh, A.; Radhakrishnan, S.; Vijayalakshmi, R.; Talawar, M.; Kumar, A.; Kumar, D. Screening of polymer-plasticizer systems for propellant binder applications: An experimental and simulation approach. *J. Energy Mater.* **2019**, *37*, 365–377. [CrossRef]
7. Bergh, M.; Caleman, C. A validation study of the general amber force field applied to energetic molecular crystals. *J. Energy Mater.* **2016**, *34*, 62–75. [CrossRef]
8. Jangid, S.K.; Radhakrishnan, S.; Solanki, V.J.; Singh, M.K.; Pandit, G.; Vijayalakshmi, R.; Sinha, R.K. Evaluation studies on partial replacement of RDX by spherical NTO in HTPB-based insensitive sheet explosive formulation. *J. Energy Mater.* **2019**, *37*, 320–330. [CrossRef]
9. Cruz, J.N.; Moraes, E.S.; Pantoja, R.P.; Pereira, T.S.; Mota, G.V.; Neto, A.M.J.C. Sensors Using the Molecular Dynamics of Explosives in Carbon Nanotubes Under External Uniform Electric Fields. *J. Nanosci. Nanotechnol.* **2019**, *19*, 5687–5691. [CrossRef]
10. Yılmaz, G.A.; Şen, D.; Kaya, Z.T.; Tinçer, T. Effect of inert plasticizers on mechanical, thermal, and sensitivity properties of polyurethane-based plastic bonded explosives. *J. Appl. Polym. Sci.* **2014**, *131*. [CrossRef]
11. Sun, H. COMPASS: An ab initio force-field optimized for condensed-phase applications overview with details on alkane and benzene compounds. *J. Phys. Chem. B* **1998**, *102*, 7338–7364. [CrossRef]
12. El-Sakhawy, M.; Kamel, S.; Salama, A.; Sarhan, H.A. Carboxymethyl cellulose acetate butyrate: A review of the preparations, properties, and applications. *J. Drug Deliv.* **2014**, *2014*, 575969. [CrossRef]
13. Wingborg, N.; Eldsater, C. 2,2-Dinitro-1,3-Bis-Nitrooxy-Propane (NPN): A New Energetic Plasticizer. *Propellants Explos. Pyrotech.* **2002**, *27*, 314–319. [CrossRef]
14. Drees, D.; Loffel, D.; Messmer, A.; Schmid, K. Synthesis and Characterization of Azido Plasticizer. *Propellants Explos. Pyrotech.* **1999**, *24*, 159–162. [CrossRef]
15. Kumari, D.; Singh, H.; Patil, M.; Thiel, W.; Pant, C.S.; Banerjee, S. Synthesis, characterization, thermal and computational studies of novel tetra-azido esters as energetic plasticizers. *Thermochim. Acta* **2013**, *562*, 96–104. [CrossRef]
16. Zhao, B.; Gao, F.; Wang, Y.; Liu, Y.; Chen, B.; Pan, Y. Azido Energetic Plasticizers for Gun and Rocket Propellants. *Prog. Chem.* **2019**, *31*, 475–490. [CrossRef]
17. Nair, U.R.; Asthana, S.N.; Rao, A.S.; Gandhe, B.R. Advances in High Energy Materials. *Def. Sci. J.* **2010**, *60*, 137–151. [CrossRef]
18. Amim, J., Jr.; Blachechen, L.S.; Petri, D.F.S. Effect of sorbitan-based surfactants on glass transition temperature of cellulose esters. *J. Therm. Anal. Calorim.* **2012**, *107*, 1259–1265. [CrossRef]
19. Khanniche, S.; Mathieu, D.; Barthet, C.; Pereira, F.; Hairault, L. Molecular dynamics simulation of gaseous nitroaromatic compounds interacting with silica surfaces under various humidity conditions. *Appl. Surf. Sci.* **2018**, *455*, 533–542. [CrossRef]
20. Mendonça, F.B.; Gonçalves, R.F.; Urgessa, G.S.; Iha, K.; Domingues, M.; Rocco, J.A. Computational Chemistry Employment in Verification and Validation of Detonation Pressure of Plastic Explosive-Pbx. *Quim. Nova* **2018**, *41*, 310–314. [CrossRef]
21. Sadeghi, A.; Nazem, H.; Rezakazemi, M.; Shirazian, S. Predictive construction of phase diagram of ternary solutions containing polymer/solvent/nonsolvent using modified Flory-Huggins model. *J. Mol. Liq.* **2018**, *263*, 282–287. [CrossRef]
22. Kundu, B.K.; Mobin, S.M.; Mukhopadhyay, S. Mechanistic and thermodynamic aspects of a pyrene-based fluorescent probe to detect picric acid. *New J. Chem.* **2019**, *43*, 11483–11492. [CrossRef]
23. Dahiwale, S.; Bhongale, C.; Roy, S.; Navle, P.; Asthana, S. Studies on ballistic parameters of di-butyl phthalate-coated triple base propellant used in large caliber artillery gun ammunition. *J. Energy Mater.* **2019**, *37*, 98–109. [CrossRef]
24. Singh, A.; Kumar, R.; Soni, P.K.; Singh, V. Compatibility and thermokinetics studies of octahydro-1, 3, 5, 7-tetranitro-1, 3, 5, 7-tetrazocine with polyether-based polyurethane containing different curatives. *J. Energy Mater.* **2019**, *37*, 141–153. [CrossRef]
25. Ramanaiah, S.; Rani, P.R.; Sreekanth, T.; Reddy, K.S. Determination of Hansen solubility parameters for the solid surface of cellulose acetate butyrate by inverse gas chromatography. *J. Macromol. Sci. Part B Phys.* **2011**, *50*, 551–562. [CrossRef]
26. Lindblad, M.S.; Keyes, B.M.; Gedvilas, L.M.; Rials, T.G.; Kelley, S.S. FTIR imaging coupled with multivariate analysis for study of initial diffusion of different solvents in cellulose acetate butyrate films. *Cellulose* **2008**, *15*, 23–33. [CrossRef]

27. Cappello, M.; Lamia, P.; Mura, C.; Polacco, G.; Filippi, S. Azidated Ether-Butadiene-Ether Block Copolymers as Binders for Solid Propellants. *J. Energy Mater.* **2016**, *34*, 318–341. [CrossRef]
28. Bafghi, S.M.A.T.; Kamalvand, M.; Morsali, A.; Bozorgmehr, M.R. Radial distribution function within the framework of the Tsallis statistical mechanics. *Physical A* **2018**, *506*, 857–867. [CrossRef]
29. Larsen, M.L.; Shaw, R. A method for computing the three-dimensional radial distribution function of cloud particles from holographic images. *Atmos. Meas. Tech.* **2018**, *11*, 4261. [CrossRef]
30. García-Negrón, V.; Oyedele, A.D.; Ponce, E.; Rios, O.; Harper, D.P.; Keffer, D.J. Evaluation of nano-and mesoscale structural features in composite materials through hierarchical decomposition of the radial distribution function. *J. Appl. Crystallogr.* **2018**, *51*, 76–86. [CrossRef]
31. Liu, N.; Li, Y.N.; Zeman, S.; Shu, Y.J.; Wang, B.Z.; Zhou, Y.S.; Zhao, Q.L.; Wang, W.L. Crystal morphology of 3, 4-bis (3-nitrofurazan-4-yl) furoxan (DNTF) in a solvent system: Molecular dynamics simulation and sensitivity study. *CrystEngComm* **2016**, *18*, 2843–2851. [CrossRef]
32. Chen, F.; Zhou, T.; Li, J.; Wang, X.L.; Cao, D.L.; Wang, J.L.; Yang, Z.J. Crystal morphology of dihydroxylammonium 5, 5'-bistetrazole-1, 1'-diolate (TKX-50) under solvents system with different polarity using molecular dynamics. *Comput. Mater. Sci.* **2019**, *168*, 48–57. [CrossRef]
33. Shu, Y.; Zhang, S.W.; Shu, Y.J.; Liu, N.; Yi, Y.; Huo, J.C.; Ding, X.Y. Interactions and physical properties of energetic poly-(phthalazinone ether sulfone ketones)(PPESKs) and ε-hexanitrohexaazaisowurtzitane (ε-CL-20) based polymer bonded explosives: A molecular dynamics simulations. *Struct. Chem.* **2019**, *30*, 1041–1055. [CrossRef]
34. Li, J.; Jin, S.S.; Lan, G.C.; Ma, X.; Ruan, J.; Zhang, B.; Chen, S.S.; Li, L.J. Morphology control of 3-nitro-1, 2, 4-triazole-5-one (NTO) by molecular dynamics simulation. *CrystEngComm* **2018**, *20*, 6252–6260. [CrossRef]
35. Sekkar, V.; Alex, A.S.; Kumar, V.; Bandyopadhyay, G.G. Theoretical evaluation of crosslink density of chain extended polyurethane networks based on hydroxyl terminated polybutadiene and butanediol and comparison with experimental data. *J. Energy Mater.* **2018**, *36*, 38–47. [CrossRef]
36. Lin, H.; Chen, J.F.; Cui, Y.M.; Zhang, Z.J.; Yang, D.D.; Zhu, S.G.; Li, H.Z. A DFT-D study on structural, electronic, thermodynamic, and mechanical properties of HMX/MPNO cocrystal under high pressure. *J. Energy Mater.* **2017**, *35*, 157–171. [CrossRef]
37. Ahmadi, S.H.; Keshavarz, M.H.; Hafizi Atabak, H.R. Introducing Laser Induced Breakdown Spectroscopy (LIBS) as a Novel, Cheap and Non-destructive Method to Study the Changes of Mechanical Properties of Plastic Bonded Explosives (PBX). *Z. Anorg. Allg. Chem.* **2018**, *644*, 1667–1673. [CrossRef]
38. Lan, G.C.; Jin, S.S.; Li, J.; Wang, J.Y.; Lu, Z.Y.; Wu, N.N.; Li, L.J.; Wang, D.X. Miscibility, glass transition temperature and mechanical properties of NC/DBP binary systems by molecular dynamics. *Propellants Explos. Pyrotech.* **2018**, *43*, 559–567. [CrossRef]
39. Li, J.; Jin, S.S.; Lan, G.C.; Chen, S.S.; Li, L.J. Modelling. Molecular dynamics simulations on miscibility, glass transition temperature and mechanical properties of PMMA/DBP binary system. *J. Mol. Graph. Model.* **2018**, *84*, 182–188. [CrossRef]
40. Shu, Y.; Yi, Y.; Huo, J.C.; Liu, N.; Wang, K.; Lu, Y.Y.; Wang, X.C.; Wu, Z.K.; Shu, Y.J.; Zhang, S.W. Interactions between poly-(phthalazinone ether sulfone ketone)(PPESK) and TNT or TATB in polymer bonded explosives: A molecular dynamic simulation study. *J. Mol. Model.* **2017**, *23*, 334. [CrossRef]
41. Lu, Y.Y.; Shu, Y.J.; Liu, N.; Shu, Y.; Wang, K.; Wu, Z.K.; Wang, X.C.; Ding, X.Y. Theoretical simulations on the glass transition temperatures and mechanical properties of modified glycidyl azide polymer. *Comput. Mater. Sci.* **2017**, *139*, 132–139. [CrossRef]
42. Hang, G.Y.; Yu, W.L.; Wang, T.; Wang, J.T.; Li, Z. Theoretical investigations on stabilities, sensitivity, energetic performance and mechanical properties of CL-20/NTO cocrystal explosives by molecular dynamics simulation. *Theor. Chem. Acc.* **2018**, *137*, 114. [CrossRef]
43. Gumieniczek, A.; Galeza, J.; Berecka, A.; Mroczek, T.; Wojtanowski, K.; Lipska, K.; Skarbek, J. Chemical stability and interactions in a new antihypertensive mixture containing indapamide and dihydralazine using FT-IR, HPLC and LC-MS methods. *RSC Adv.* **2018**, *8*, 36076–36089. [CrossRef]

© 2020 by the authors. Licensee MDPI, Basel, Switzerland. This article is an open access article distributed under the terms and conditions of the Creative Commons Attribution (CC BY) license (http://creativecommons.org/licenses/by/4.0/).

www.ingramcontent.com/pod-product-compliance
Lightning Source LLC
LaVergne TN
LVHW070438100526
838202LV00014B/1617

MDPI
St. Alban-Anlage 66
4052 Basel
Switzerland
Tel. +41 61 683 77 34
Fax +41 61 302 89 18
www.mdpi.com

Polymers Editorial Office
E-mail: polymers@mdpi.com
www.mdpi.com/journal/polymers